9-3-69

Holonyak

D0848582

HANDBOOK
OF
RADAR
MEASUREMENT

PRENTICE-HALL INTERNATIONAL, INC., *London*
PRENTICE-HALL OF AUSTRALIA, PTY. LTD., *Sydney*
PRENTICE-HALL OF CANADA, LTD., *Toronto*
PRENTICE-HALL OF INDIA PRIVATE LTD., *New Delhi*
PRENTICE-HALL OF JAPAN, INC., *Tokyo*

HANDBOOK
OF
RADAR
MEASUREMENT

DAVID K. BARTON
and
HAROLD R. WARD
Raytheon Company

PRENTICE-HALL, INC.

Englewood Cliffs, New Jersey

13–380683–9
Library of Congress Catalog Card Number 76–77553

Printed in the United States of America

This book is dedicated to
Irvin L. McNally
who created an environment
in which the work could be done

Preface

A radar set, by definition, is intended to perform the tasks of RAdio Detection And Ranging on target objects. While the detection process is of prime importance, and is a prerequisite for other functions, the measurement process has also been inherent in radar since its inception. Initially extended to include angular coordinate data as well as range, the measurement process is now applied also to frequency (or Doppler shift), to target amplitude, and to other parameters.

The development and application of radar measurement theory has lagged almost a decade behind comparable work in target detection. Two early milestones in the field were the 1950 paper by Woodward and Davies, of the British Telecommunications Research Establishment (see References, Appendix G), and Woodward's subsequent book (1953), both of which applied information theory to the processes of radar measurement. In the United States, the reports by Peter Swerling, of RAND Corporation (1956, 1959), covered search-radar accuracy and other phases of parameter estimation. These were followed by an increasing number of studies in this field, notable among which was the work, at the Massachusetts Institute of Technology's Lincoln Laboratory, of Roger Manasse and M. I. Skolnik (1955–1960). A more design-oriented approach to the subject was represented by an earlier study by G. M. Kirkpatrick, at General Electric Company, which established the ideal measurement sensitivity of a monopulse antenna, and the extent to which practical illumination functions could approach this ideal. His results, originally available only as classified reports or in a brief summary in the *Transactions of the Institute of Radio Engineers* (1953), have recently been declassified but are still not widely available.

The purpose of this handbook is to present, in unified and consistent form, results of these and other theoretical studies of radar measurement, and to relate these results to design parameters of practical equipment. Although we identify the theory and applications with the field of radar, many of the results are equally applicable to the closely related problems of satellite communications and telemetry. Emphasis is placed on equations, curves, and tabular data which permit the system designer to estimate the measurement accuracy of a given system, to identify the principal sources of error, and to determine the degree to which the theoretical limits of accuracy are approached. With this knowledge, he can arrive more rapidly at an optimum compromise between performance, cost, and complexity, and can anticipate and solve critical problems before committing the equipment to construction and test.

The first chapter introduces the basic problems and theories of target measurement in an environment of noise, clutter, and undesired echoes. The significance of signal-to-noise energy ratio, ambiguity, and resolution is discussed with respect to the measurement problem. The detailed application of these theories to measurement of angle, range, and Doppler parameters of the target in noise is presented in the three subsequent chapters. Chapter 2 relates angular measurement capability to aperture illumination functions, as well as to beamwidth, difference slope, and sidelobe ratio of the antenna. Relationships are established which unite Manasse's theory of ideal angular accuracy, the results of Swerling and Skolnik for scanning antennas, and those of Hannan, Barton, and Develet for practical monopulse trackers. These are presented in forms which are adapted both to conventional tracking antennas and to array antenna systems. Chapters 3 and 4 present the corresponding expressions for range (time delay) and Doppler velocity (frequency), respectively. As with angular measurement, the ideal performance is used as a reference against which practical circuits, waveforms, and signal spectra are compared.

In Chapter 5, we describe in some detail the errors resulting from our inability to resolve completely the desired target signal from other signals: those reflected from adjacent targets, from large targets with considerable separation, from target signals reaching the antenna indirectly via ground reflection paths, and from clutter echoes. This material is followed, in Chapter 6, by a summary of effects produced by the target itself: glint in range, angle, and Doppler; amplitude scintillation, and measuring system errors induced by scintillation. Chapter 7 contains a description of the effects of discrete processes in the measurement system. These include sampling of the signal in time and space domains, and quantization of the signal samples into discrete amplitude levels. These effects are introduced in array radar systems, and in digital signal processors.

In Chapter 8, we review the procedures by which the many components of error are combined in a system error analysis. Description of the error in terms of rms magnitude, amplitude distribution, frequency spectrum, and correlation time are discussed. A general procedure for error analysis is developed and applied to angle, range, and Doppler measurement problems. Examples are given which illustrate the application of material from the previous chapters and from the appendices.

The final section of the book consists of seven appendices. Appendix A presents plotted and tabular data on antennas, with common illumination functions and the corresponding far-field patterns plotted on several different scales. Linear, rectangular, and circular aperture shapes are covered, and significant parameters such as half-power beamwidth, gain, and rms aperture width are tabulated. Both even (sum-channel) and odd (difference-channel) functions are given. Appendix B applies these same data to the case of waveforms and frequency spectra. Appendix C covers the application of waveform-spectrum pairs to data filtering (smoothing and differentiation).

In Appendix D, the effects of the atmosphere on range, angle, and Doppler measurements are analyzed. Much of this information has been available in reports prepared by the National Bureau of Standards group at Boulder, Colorado, but the various effects of the troposphere are now described here in a more unified (though perhaps overly simplified) manner.

Appendix E contains tables of a few functions which are of particular value in radar system analysis. A complete list of the symbols used in the handbook is given in Appendix F, with reference to the section in which the symbol is first used or defined. Finally, in Appendix G, we list the references on which we have drawn the basic material for the entire handbook.

Much information on radar design and performance has been omitted in the interests of conserving space. The subject of radar range calculation has been treated very briefly in Chapter 1, and target detection theory has been excluded entirely, along with information on the physical characteristics and detailed design of radar circuits, components, and subsystems. Many of the equations are presented without derivation and proof, although an attempt has been made to justify their general reasonableness in each application. For more rigorous proof, the reader is directed to the original sources.

Through the compiling and presentation of this collection of theory and data, in concise and consistent form, the authors have hoped to simplify the task of radar system design and analysis, while encouraging more accurate prediction and description of radar performance. During preparation of the material, it has been possible to establish a few new relationships which serve to correlate and unify the results of previous workers in the field. Except for this connective material, the authors claim no credit for the originality of the ideas presented here.

D. K. B.
H. R. W

Contents

HANDBOOK
OF
RADAR
MEASUREMENT

The basic radar functions are detection and measurement. Evaluation of radar performance in these two functions is the fundamental problem of radar system design and analysis. Both functions require that the radar resolve the desired target signal from its surroundings, so we may regard resolution as a third essential process in radar. Before proceeding with a detailed discussion of measurement problems, we will review here the factors which limit detection, resolution, and measurement performance, and will show how these three processes, which might be thought independent, are actually related very closely.

1.1 Detection

Detection is the first step in target location. Radar cost, size, and weight are largely determined by the detection range which is required, and thus it is important to detect efficiently. Theories developed by North (1943) and others point the way to optimum detection, using "matched-filter" receivers. We will discuss detection theory briefly, because the approach to optimum detection is fundamental to the measurement process and to measurement accuracy. The remainder of this section states a very simple principle, and shows how it forms the basis for defining the ideal radar.

The simple principle shows how we must combine signal samples to achieve the best performance in detection. Each complex signal sample, A_i, has associated with it an independent, complex noise sample, N_i. The samples are weighted by complex factors, H_i, and added linearly to obtain an output voltage ψ:

$$\psi = \sum_i H_i(A_i + N_i). \qquad (1.1)$$

The resulting signal-to-noise power ratio, S/N, at the output of the adder is given by

1

Introduction

to

Measurement

Theory

$$\frac{S}{N} = \frac{|\sum_i H_i A_i|^2}{\sum_i |H_i N_i|^2}. \tag{1.2}$$

To maximize this ratio, the weighting of the ith signal sample must be

$$H_i = \frac{A_i^* C}{|N_i|^2}, \tag{1.3}$$

where A_i^* indicates the complex conjugate of the signal sample, and C is a constant with dimension of A_i, inserted to make the weighting factors H_i dimensionless.

This result is deceptively simple, and seems too simple to serve as the basis of detection and measurement theory. The secret of its success lies in the application, which requires that we choose signal samples in such a way that the accompanying noise samples are independent. Two examples will illustrate this.

Application of the principle to waveforms leads to the matched filter. Here, the signal is assumed to be a radar echo accompanied by broadband noise. Representing the signal by its voltage spectrum, $A(f)$, is equivalent to taking a sample of the signal at each frequency. There is also a noise sample at each frequency, which is independent of the noise samples at all other frequencies. The processor which weights these samples by $H(f)$ is the signal filter in the radar receiver. The filter transfer function, $H(f)$, implies the addition of signal components, following the weighting of the signal component at frequency f by the factor H. The output voltage reaches a maximum at a reference time $t = 0$, when the amplitude is

$$\psi(0) = \int_{-\infty}^{\infty} H(f)[A(f) + N(f)] \, df. \tag{1.4}$$

We see that this voltage is the sum of the weighted samples of signal plus noise. Hence the filter is performing the operation described in Eq. (1.1).

To apply our principle, we must let

$$H(f) = \frac{A^*(f)C}{|N^2(f)|} = \frac{A^*(f)C}{N_o}. \tag{1.5}$$

Generally, the noise power spectrum $N^2(f)$ is constant with frequency, and can be replaced with N_o, the noise spectral density in the band of interest. We are left with the classical description of the matched filter. This is the linear filter which will cause the signal peaks at its output to rise higher above the rms (root-mean-square) noise level than any other filter we could select.

An important property of this matched filter is that the maximum output signal-to-noise power ratio is equal to E/N_o, where E is the total signal energy. It is not a function of such other signal parameters as bandwidth, spectral distribution, or time duration. Hence, the potential detectability in noise is a function of signal energy and no other signal parameter.

An antenna can also be thought of as a filter. It collects, weights, and sums the signals impinging on the aperture. In an array antenna, each element receives a signal sample and an independent sample of noise from the environment. Noise originating in the receiver can be equated with background noise of a given intensity or temperature. Since the element signal components are equal in amplitude, the antenna illumination function, which weights the signal samples, should be uniform. The phase of the element signals depends on the angle of arrival. These phase shifts must be removed before the signals are summed, so that the signal components will add in phase. Phase removal is equivalent to steering the beam to the angle of arrival of the signal.

For maximum signal-to-noise ratio, then, the antenna aperture must be uniformly illuminated, the beam must be pointed directly toward the target, and a matched filter must be used in the receiver. The *response* of the matched radar determines its ability to resolve and locate targets.

1.2 Resolution

A target is said to be resolved if its signal is separated by the radar from those of other targets, in at least one of the coordinates used to describe it. For example, a tracking radar may describe a target by two angles, time delay, and frequency. A second target signal from the same angle and at the same frequency, but with different time delay, may be resolved if the separation is greater than the delay resolution of the radar.

Resolution, then, is determined by the relative response of the radar to targets separated from the target to which the radar is matched. The antenna and receiver are configured to match a target signal at a particular angle, delay, and frequency. The radar will respond with reduced gain to targets at other angles, delays, and frequencies. This "response function" can be plotted as a surface in a four-dimensional coordinate system. Because four-dimensional surfaces are impossible to plot, and because angle response is almost always independent of the delay-frequency response, these pairs of coordinates are usually separated.

In angle, the response function $\psi(u, v)$ is simply the antenna pattern. It is found by measuring the system response as a function of the angle from the beam center. The angular response, in general, is a two-dimensional surface of the type shown in Fig. 1.1. It has a main lobe in the direction to which it is matched, and sidelobes extending over all visible space. Angular resolution, or main-lobe width in the u- and v-coordinates, is generally taken to be the distance between the 3-db points of the pattern. The width, amplitude, and location of the lobes are determined by the aperture illumination (weighting) functions in the two coordinates across the aperture.

Because the matched antenna is uniformly illuminated, its response has relatively high sidelobes, which are objectionable in most radar applications. To avoid these, the antenna illumination may be mismatched slightly, with resulting loss in gain and broadening of the main lobe.

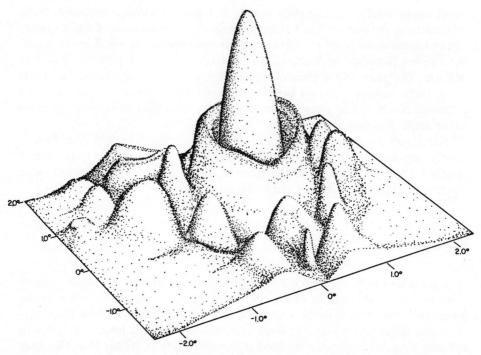

Fig. 1.1   Typical angular response function for a directive antenna (from Howard, 1967).

Time delay and frequency can be viewed in the same way as the two angular coordinates. There is a two-dimensional surface which describes the filter response to a given signal as a function of the time delay, t_d, and the frequency shift, f_d, of the signal relative to some reference point. Points on the surface are found by recording the receiver output voltage, while varying these two target coordinates. The response function $\psi(t_d, f_d)$ is given, for any filter and signal, by

$$\psi(t_d, f_d) = \int_{-\infty}^{\infty} H(f) A(f - f_d) \exp\left(j2\pi f t_d\right) df, \tag{1.6}$$

or

$$\psi(t_d, f_d) = \int_{-\infty}^{\infty} h(t_d - t) a(t) \exp\left(j2\pi f_d t\right) dt, \tag{1.7}$$

where the functions $A(f)$ and $a(t)$, $H(f)$ and $h(t)$, are Fourier transform pairs describing the signal and filter, respectively.

To obtain the matched filter, with noise of uniform spectral density, we set

$$H(f) = \frac{A^*(f)}{C} \qquad \text{(dimensionless voltage gain)}, \tag{1.8}$$

$$h(-t) = \frac{a^*(t)}{C} \qquad (\text{sec}^{-1}), \tag{1.9}$$

where C is an arbitrary gain constant with the dimension of volt-seconds (V-sec). The corresponding matched-filter response function is

$$\psi_o(t_d, f_d) = \frac{1}{C} \int_{-\infty}^{\infty} A^*(f) A(f - f_d) \exp (j2\pi f t_d) \, df, \tag{1.10}$$

or

$$\psi_o(t_d, f_d) = \frac{1}{C} \int_{-\infty}^{\infty} a^*(t - t_d) a(t) \exp (j2\pi f_d t) \, dt. \tag{1.11}$$

This function has the same shape as Woodward's *ambiguity function*, but differs in dimensions and in the absence of normalization. To obtain Woodward's normalization to unit energy,† we set $C = 1.0$ V-sec, and let

$$\int_{-\infty}^{\infty} |A(f)|^2 \, df = \int_{-\infty}^{\infty} |a(t)|^2 \, dt = 1.0 \text{ W-sec.} \tag{1.12}$$

For this condition, $\psi_o(0, 0) = 1$ V, and Woodward's work shows that total volume under the surface $|\psi_o(t_d, f_d)|^2$ is also unity (one watt, in our dimensions), regardless of the waveform. In the presence of randomly-phased signals, distributed uniformly in delay and in frequency with power density S_o W/Hz, the power output in watts will then be equal to S_o. Similarly, for noise of power density N_o W/Hz, the output noise power in watts will equal N_o.

The response function $\psi(t_d, f_d)$ for a mismatched filter has the same shape as the *cross-ambiguity function* used by Rihaczek (1965), again differing in dimensions and lacking normalization. The normalization constraints for unit volume under the surface $|\psi(t_d, f_d)|^2$ are simple, and easily interpreted in terms of signal energy and filter parameters:

$$\int_{-\infty}^{\infty} |A(f)|^2 \, df \int_{-\infty}^{\infty} |H(f)|^2 \, df = \int_{-\infty}^{\infty} |a(t)|^2 \, dt \int_{-\infty}^{\infty} |h(t)|^2 \, dt = 1 \text{ W.} \tag{1.13}$$

Thus, if the energy under the signal envelope is one watt-second, as given by Eq. (1.12), the product of center-frequency power gain and filter noise bandwidth must be one hertz. The noise output in watts is again equal to N_o in W/Hz. The peak signal response at the point $t_d = 0$, $f_d = 0$ is no longer unity, however, but is reduced by an efficiency factor to be defined in Chap. 3. As in the antenna case, the broadening of the response region, relative to the matched case, has been accompanied by reduced peak gain, when the response is normalized for equal noise outputs.

---

† If $a(t)$ represents a modulating waveform, the integrals of Eq. (1.12) equal twice the signal energy.

Resolution in delay, frequency, or angle may be described in several ways: by the half-power width of the central response; by Woodward's time and frequency resolution constants (extended, by analogy, to angle); by the target spacing needed to ensure a given level of accuracy in measurement; or by other measurement and detection criteria. For the purposes of this handbook, the half-power (3-db) width of the response functions will be used to measure resolution. Considerable control may be exercised over this 3-db resolution width, by application of coding to the basic pulse waveforms, for example. However, when all regions of response are considered, the transform-pair relationships in time and frequency will be found to impose strict limits on combined resolution in these two coordinates. This is unlike the case of angular resolution, where the transform relationships are between the angular response function and the aperture illuminations in the two planes. The angular resolution is thus limited only by aperture area, measured in square wavelengths.

To compare the 3-db resolution measure with Woodward's values, we can refer to the detailed discussion by Burdic (1968, pp. 181–205). For a rectangular pulse of width τ, Woodward's time resolution constant is 0.67τ. For a Gaussian pulse, the resolution constant is 1.5 times the 3-db width, while for a $(\sin x)/x$ pulse it is 1.13 times the 3-db width. Thus the ratios of the two measures lie between $2:3$ and $3:2$ for the extreme cases, and should be essentially unity for most practical waveforms. The 3-db width is easily determined from either theoretical or experimental data. The various measures of resolution are compared further in App. A, Table A.22 (p. 337).

We have shown in this discussion that resolution in all radar coordinates is controlled by the weighting functions, which are chosen to obtain a peak response for a target at a particular point in four-dimensional space. The relative response for targets at other locations then describes the radar's ability to resolve targets. Examining these ideas further, we will see that the measurement accuracy is also determined by the response function ψ.

1.3 Measurement

Measurement begins when a target is detected within a given resolution cell. This cell is described by the 3-db widths of the radar response function in all four radar coordinates. The measurement problem normally requires that we find, with the least error, the location of the target within this cell. In this section we discuss the ideal parameter estimator, how it is realized, and what limits its accuracy.

Since the target is assumed to have been detected by a system matched to a particular set of target coordinates, its position will be known to within half the resolution width in each coordinate. To obtain an improved estimate of target location in any coordinate, we must add to the radar system an estimator which produces another output, measuring the target position within the resolution cell.

The ideal estimator is the device which does this job most accurately. Noise accompanying the signal imposes a limit on this accuracy, and hence the ideal estimator is contrived to do the best job of minimizing noise effects.

It has been shown that the ideal estimator is the device which senses the derivative of the response function ψ, in the coordinate of interest. In the waveform example used above, the matched filter was defined by the signal waveform, and its response was given in the delay coordinate by

$$\psi_o(t_d, 0) = \frac{1}{C} \int_{-\infty}^{\infty} a^*(t - t_d)a(t)\, dt. \tag{1.14}$$

The optimum estimator should find the peak of this response function. Mathematically, we would find the peak by taking the derivative of the function and locating the point where this derivative is zero. For the delay coordinate,

$$\frac{\partial \psi_o(t_d, 0)}{\partial t_d} = \frac{1}{C} \int_{-\infty}^{\infty} a(t) \frac{\partial}{\partial t_d}[a^*(t - t_d)]\, dt. \tag{1.15}$$

Notice that this is the same as the response function of a filter matched to the derivative of the signal in the time coordinate.

In general, the derivative will follow an S-shaped curve near the matched point, with the zero crossing at that point. For a target at any other point, the amplitude of the derivative may be used to estimate the separation of the target from the matched point. This is of practical concern because, in most cases, the estimator's zero point is centered at the predicted value of the variable to be measured. The actual radar signal then supplies an estimate of the error in prediction.

Some examples of how the ideal estimator is realized may clarify these ideas. There are two common approaches to building the differentiator. The first approximates a match to the derivative of the response function directly. This is illustrated by a multimode monopulse feed in an antenna, where the angle error output is derived from an antenna pattern with odd symmetry. In the waveform case, the estimator would be a filter matched to the derivative of the signal in time or frequency.

The second and perhaps more common means of approximating the ideal estimator is to build two or more filters, matched to different points in the measured coordinate. These points are separated, typically, by one resolution interval. The ratio of difference to sum of two adjacent outputs will approximate a derivative. A simple amplitude-monopulse feed, where two beams are formed and squinted apart by about one beamwidth, is an example of this approach. The outputs from these two beams are compared, and an estimate is obtained from the ratio of the difference to the sum of the two voltages. A Doppler filter bank is another example of this technique applied to frequency estimation. This is a bank of filters matched to adjacent points in frequency. The filter with the largest response locates the target frequency to an accuracy of half the resolution element. A better estimate is obtained by interpolating between filters.

We said earlier that the ideal estimator was that which gave the most accurate answer in the presence of noise. We will now answer the question of how accurate the estimate can be. Woodward (1953) has shown that the minimum rms error for estimating time delay is given by

$$\sigma_t = \frac{1}{\beta\sqrt{\mathscr{R}}}. \tag{1.16}$$

From the earlier discussion, we would expect that a formula of the same type can be applied to any of the radar coordinates. There are two factors used in this formula. $\mathscr{R} = 2E/N_o$ is the ratio of peak signal to noise power at the output of a filter matched to the target location. The parameter β is defined by

$$\beta^2 = \frac{\int_{-\infty}^{\infty} \left| \frac{\partial}{\partial t} a(t) \right|^2 dt}{\int_{-\infty}^{\infty} |a(t)|^2 dt} = \frac{\int_{-\infty}^{\infty} (2\pi f)^2 |A(f)|^2 df}{\int_{-\infty}^{\infty} |A(f)|^2 df}. \tag{1.17}$$

Notice that this is the normalized second moment of the signal energy spectrum, and is a measure of signal bandwidth.

Measurement accuracy for ideal estimators in the other coordinates will be given by equations similar in form to Eq. (1.16). In each case, we will use a parameter which measures the second moment, or "rms spread" of the signal weighting function which determines resolution in the measured coordinate. The three basic relationships, with definitions of the symbols used, are given in Table 1.1.

Table 1.1

BASIC MEASUREMENT RELATIONSHIPS

Measured Coordinate	Equation for Rms Error	Resolution Element	Radar Parameter in Transform Coordinate	Rms Spread in Transform Coordinate
Angle, θ (rad)	$\sigma_\theta = \dfrac{\lambda}{\mathscr{L}\sqrt{\mathscr{R}}}$ (1.18)	Beamwidth, θ_3	Aperture width, w/λ, (wavelengths)	Rms aperture width, $\mathscr{L}/\lambda$
Delay time, t_d (sec)	$\sigma_t = \dfrac{1}{\beta\sqrt{\mathscr{R}}}$ (1.19)	Effective pulse duration, τ_3	Bandwidth of signal, B_{3a}	Rms bandwidth of signal, β
Frequency, f (Hz)				
Coherent case	$\sigma_f = \dfrac{1}{\alpha\sqrt{\mathscr{R}}}$ (1.20)	Spectral line width, B_{3a}	Observation time, t_o	Rms signal duration, α
Noncoherent case	$\sigma_f = \dfrac{1}{\alpha_1\sqrt{\mathscr{R}}}$ (1.21)	Bandwidth of signal, B_{3a1}	Pulse duration, τ_3	Rms pulse duration, α_1

1.4 Application to Practical Systems

Received Energy Ratio

The preceding discussion has emphasized the potential performance of a radar system which is ideally matched to the target signal. For such a system, the energy ratio $\mathscr{R}_o$ is found from the radar equation in a form which considers the total signal energy E incident on the antenna aperture area, A, during the observation:

$$E = \int_{t_0} P_r \, dt,$$

$$P_r = \frac{P_t G_t A \sigma}{(4\pi)^2 R^4},$$

$$N_o = kT_i = kT_o \overline{NF_o},$$

$$\mathscr{R}_o = \frac{2E}{N_o} = \frac{2P_{av} t_o G_t A \sigma}{(4\pi)^2 R^4 kT_i}. \tag{1.22}$$

(The symbols are as defined in App. F.)

The observation interval, t_o, is the period over which the signal is integrated in a tracking servo system of noise bandwidth β_n, whose output the error is to be evaluated:

$$t_o = \frac{1}{2\beta_n}. \tag{1.23}$$

For a search radar system where the single-scan error is to be determined, we take the interval as the conventional "time-on-target" between the one-way half-power points of the antenna pattern, θ_3, at a scan rate ω:

$$t_o = \frac{\theta_3}{\omega}. \tag{1.24}$$

In pulse radar, this interval is often converted into a "number of pulses integrated," $n = f_r t_o$:

$$n = \frac{f_r}{2\beta_n} \quad \text{(tracking system)}, \tag{1.25}$$

$$n = \frac{f_r \theta_3}{\omega} \quad \text{(search system)}. \tag{1.26}$$

Practical Loss Factors

Actual radar systems use antennas which are imperfectly matched to the target signal, either for reasons of sidelobe reduction or for economy and simplicity. The energy which is extracted by the aperture from the incident wave, for use by the receiver, is reduced by an aperture efficiency η_a.

$$\mathcal{R} = \mathcal{R}_o \eta_a = \frac{2P_{av}t_o G_t A_r \sigma}{(4\pi)^2 R^4 k T_i L_1} = \frac{2P_{av}t_o G_t G_r \lambda^2 \sigma}{(4\pi)^3 R^4 k T_i L_1}, \qquad (1.27)$$

$$P_{av}t_o = P_t \tau n, \qquad (1.28)$$

$$A_r \equiv A\eta_a = \frac{G_r \lambda^2}{4\pi} \qquad \text{[see Eq. (2.3)]}.$$

The loss factor L_1 has been included in Eq. (1.27) to allow for received energy losses other than aperture efficiency (e.g., transmission line loss or beamshape loss). Relationships between η_a and antenna illumination will be described in detail in Chap. 2.

The single-pulse energy ratio $\mathcal{R}_1$ is found from Eq. (1.27) by substituting $n = 1$ in Eq. (1.28).

$$\mathcal{R}_1 = \frac{2P_t \tau G_t A_r \sigma}{(4\pi)^2 R^4 k T_i L_1} = \frac{2P_t \tau G_t G_r \lambda^2 \sigma}{(4\pi)^3 R^4 k T_i L_1}. \qquad (1.29)$$

This value and the corresponding S/N ratio are most useful in performance calculations of noncoherent radars which integrate many pulses after envelope detection. In such radars, the IF filter is matched approximately to the spectrum of each individual pulse, and a post-detection filter is used to integrate data over the observation interval t_o.

Signal-to-Noise Ratio

The practical receiver cannot be matched precisely to the signal spectrum and waveform, so the maximum output S/N power ratio will fall below the matched-filter value $E/N_o = \mathcal{R}/2$. The following special cases are encountered most frequently.

Single-Pulse Ratio

$$\frac{S}{N} = \frac{\mathcal{R}}{2nL_m} = \frac{\mathcal{R}_1}{2L_m} \qquad (1.30)$$

Single Pulse, Wideband IF $(B_n \tau \gg 1)$

$$\frac{S}{N} \cong \frac{\mathcal{R}}{2nB_n \tau} \cong \frac{\mathcal{R}_1}{2B_n \tau} \qquad (1.31)$$

Coherent Pulsed Radar, Fine-Line Filter

$$\left(\frac{S}{N}\right)_f = \frac{\mathcal{R}}{2L_m} = \frac{\mathcal{R}}{2L_{m1}L_{mf}} = \frac{nS/N}{L_{mf}} \qquad (1.32)$$

CW Radar, Wideband Filter $(B_n t_o \gg 1)$

$$\left(\frac{S}{N}\right)_{av} = \frac{\mathscr{R}}{2L_m} \cong \frac{\mathscr{R}}{2B_n t_o} \qquad (1.33)$$

In the above equations, L_m represents the IF filter matching loss (see Chap. 3). In the coherent radar case, this is divided into L_{m1} for mismatch of the single-pulse spectrum and L_{mf} for mismatch to the fine-line structure (Sec. 4.6).

The form of the radar equation which is most commonly used for computation of the S/N ratio, in the noncoherent echo case, combines Eqs. (1.27)–(1.31):

$$\frac{S}{N} = \frac{P_t \tau G_t G_r \lambda^2 \sigma}{(4\pi)^3 R^4 k T_i L_1 L_m} \cong \frac{P_t G_t G_r \lambda^2 \sigma}{(4\pi)^3 R^4 k T_i B_n L_1}. \qquad (1.34)$$

For signals emitted by beacons, the equivalent expressions are

$$\frac{S}{N} = \frac{P_b \tau G_b G_r \lambda^2}{(4\pi)^2 R^2 k T_i L_r L_m} \cong \frac{P_b G_b G_r \lambda^2}{(4\pi)^2 R^2 k T_i B_n L_r}. \qquad (1.35)$$

Here, L_r includes those losses which lie between the beacon transmitter and the radar receiver.

There are many expressions which may be used to obtain the signal-to-noise ratio for either echo or beacon case, but we have chosen those which separate the receiving gain (or aperture) from transmitting gain. This separation will be important in our later discussion of angle measurements, where the slope of a monopulse difference pattern is referred to the output voltage of an "ideal" receiving antenna using the same aperture to receive the same incident wave. The corresponding energy ratio $\mathscr{R}_o$ is that which could be extracted from the incident wave by the ideal antenna and receiver, given the target illumination or beacon power. Thus, the aperture efficiency η_a is applied only to the receiving system, and the same equations may be used to describe echo and beacon measurements.

Losses in Search and Scanning

For search radar, the radar equation gives signal-to-noise ratios $\mathscr{R}_m$ and $(S/N)_m$ for a target on the axis of the beam. A beam pattern factor, or beamshape loss, is then applied to find the energy actually received as the beam scans across the target:

$$\mathscr{R} = \frac{\mathscr{R}_m}{L_p} = \frac{2n(S/N)_m L_m}{L_p}. \qquad (1.36)$$

Similarly, for a tracker which uses an offset beam, the S/N ratio and energy ratio for a target on the tracking axis are reduced by a crossover loss factor, L_k, relative to the values calculated for the nose of the beam:

$$\mathscr{R} = \frac{\mathscr{R}_m}{L_k} = \frac{2n(S/N)_m L_m}{L_k}. \qquad (1.37)$$

Values of L_p and L_k depend upon whether we consider two-way (echo) or one-way (beacon) operation, and on details of the scan patterns (see Sec. 2.5).

Measurement Sensitivity

In addition to the loss in the S/N ratio encountered in practical systems, there will be departures from the measurement sensitivity which would have been obtained by matching to the derivative of the signal. Where the "rms width" of the aperture, spectrum, or waveform duration was used in Table 1.1, we must use instead a different (and generally smaller) value to express the error in a practical (mismatched) system. The calculation of this value is the major subject of the next three chapters.

Measurement Ambiguity

In practical systems, we must also evaluate the effects of undesired signal echoes and reflections from the ground or from clutter. These problems require us to consider the amplitude of the response function in regions outside the 3-db contour of the conventional resolution element, and to guard against "ambiguity" as well as normal error in measurement. We use the term *ambiguity* in cases where the apparent location of the target falls outside the immediate vicinity of the actual target, in a region not connected with the normal spread of measurement. There are four general types of ambiguity:

1. Repetitive Response Lobes. The use of repetitive waveforms introduces a series of equal response lobes in time and frequency coordinates. Examples are "second-time-around" echoes from objects whose delay exceeds the pulse repetition period, and the Doppler ambiguity at multiples of the repetition rate. Similar effects occur in angular response of interferometers and arrays with grating lobes, both the result of omission of significant signal samples between the antenna elements.

2. Sidelobes. The target signal may be strong enough to be detected in a region of sidelobe response, quite disconnected from the main lobe in at least one coordinate. In our later discussions of practical measurement systems, we will describe the magnitude of sidelobe response, so that the system designer can estimate the probability of such an ambiguity in his particular application.

3. "Ghosts." The system may be so designed as to combine two measurements on two or more targets, producing spurious indications of targets at locations removed from either true target. Common examples are ambiguities in range and Doppler measurements, when using triangular-wave FM transmissions, and the ghosts produced by scanning with two fan beams which are oriented at right angles to each other. In both cases, the ambiguities are the result of detection of two or more targets by the main response lobe, but the spurious locations may fall far beyond the normal spread of error for either target.

4. If the energy ratio $\mathscr{R}$ is too low, random noise may be detected as a false target in any portion of the observed space. In normal error analysis, we assume that such erroneous readings are ruled out by requiring $\mathscr{R}$ to be sufficient to exceed a detection threshold which corresponds to a very low false-alarm probability. In a typical case, $\mathscr{R}$ will be greater than 20, and often 100 or more. However, if false detections do occur, their effect on accuracy must be evaluated for the particular type of processor which is used. Some systems can exclude the bad points from the averaging process, while others may be subject to large errors from a single false alarm.

1.5 Summary

This chapter was intended to review and relate the fundamental ideas of detection, resolution, and measurement. The important points may be summarized as follows:

1. All the target coordinates can be viewed in a similar way by the radar designer.

2. Matching the radar response to a point in these coordinates optimizes the detectability at that point.

3. The response of the matched radar to targets at other locations describes radar resolution in all coordinates.

4. The ideal measurement device differentiates the response function in order to estimate distance from the center of the resolution cell.

5. The limiting noise error is a function of signal-to-noise ratio and of the parameter which measures resolution in that coordinate.

6. Practical systems may be described with reference to the ideal radar by applying loss factors to reduce the effective signal-to-noise ratio, and by reducing the value of the resolution parameter.

7. A major limiting factor in pactical systems, in addition to noise, is the ambiguity produced by response at undesired regions of space. To guard against this ambiguity, we must generally degrade the detection efficiency and the measurement accuracy for targets in the main lobe of the response.

The factors which determine the ability of a radar system to measure the angle of arrival of a particular signal are:

1. The dimensions of the antenna aperture relative to the wavelength of the signal;

2. The degree to which different portions of the aperture are coupled to (illuminated by) the radar receiver and transmitter;

3. The ratio of total signal energy received during the period of measurement to receiver noise spectral density; and

4. The method used to scan the beam, or to process the outputs of received beam positions observed simultaneously or sequentially.

In this chapter, we summarize the relationships between antenna size, illumination function, beam width, signal-to-noise ratio, and angular error, for commonly used measurement systems and procedures.

In the course of the discussion, we will attempt to clarify and relate several performance measures which have been used in past literature to describe angular measurement capability of different types of radar:

(a) The rms aperture width $\mathscr{L}$, introduced by Manasse (1960) and Skolnik (1960) to relate angular measurement theory to Woodward's previous results for range and Doppler;

(b) The difference slope K, introduced by Kirkpatrick (1953) and redefined by Hannan (1961) to describe the measurement sensitivity of an aperture which uses an odd (difference-channel) illumination function;

(c) The normalized monopulse slope k_m, used by Barton (1964) to describe a measured monopulse difference pattern in terms of its associated sum-pattern gain and beamwidth, rather than in terms of aperture size and illumination;

(d) The conical-scan error slope k_s, expressing the fractional scan modulation of the signal per beamwidth of tracking error;

2

Angular Measurement in Noise

(e) Several constants which relate beamwidth, squint angle of offset beams, crossover loss, and number of hits per beamwidth (in search radar) to angular sensitivity.

The past literature is not consistent in definitions, usage, and symbols, but we will attempt to relate the several systems of describing performance to a consistent theory of measurement, and to show that there is no substantial disagreement in previous results.

2.1 Antenna Parameters

The terms, symbols, and basic relationships which describe the performance of an antenna in its measurement function are summarized here. Additional information is contained in App. A.

Coordinate System

An aperture plane is established, which may be the actual plane of the antenna elements or an imaginary surface located immediately in front of a curved antenna. The antenna axis is normal to this plane (Fig. 2.1). The following symbols are used:

x = horizontal coordinate in aperture plane,
y = coordinate normal to x in aperture plane (often the vertical coordinate),
z = coordinate normal to aperture plane (along the antenna axis),
w = total width of antenna along x-axis,
h = total dimension of antenna along y-axis (often height),
A = projected area of antenna on xy-plane,
θ = angular coordinate measured from z-axis in a plane rotated by an angle ϕ from the xz-plane,
ϕ = angular coordinate measured around z-axis from the x-axis.

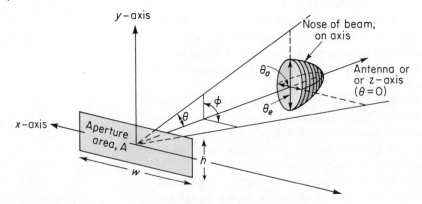

Fig. 2.1   Antenna coordinate system.

Description of Reference Beam

The single beam of a scanning antenna, and the reference (or sum) beam of a monopulse antenna, are characterized by even functions which describe the gain patterns in θ and the illumination of the aperture in x and y. We will assume that the antenna is properly phased to form a beam along the z-axis, so that the absolute magnitude of the illumination function is the important factor. The effect of phase errors on measurement performance will be considered in Chap. 7. The usual narrowband assumption for signals will be made, and we will consider patterns in θ which depend upon illumination in the x-coordinate ($B\lambda \ll c$, $\phi = 0$, $\theta \ll 1$ rad, $w/\lambda \gg 1$). The following symbols are used:

$F(\theta) =$ voltage gain (pattern) function,

$G(\theta) = F^2(\theta) =$ power gain (pattern) function,[†]

$g(x, y) =$ aperture illumination function (voltage),

$\theta_3 =$ half-power, one-way beamwidth,

$\theta_a =$ half-power beamwidth in plane containing x-axis ($\phi = 0$),

$\theta_e =$ half-power beamwidth in plane containing y-axis ($\phi = 90°$),

$\theta_o =$ half-power beamwidth for uniform illumination $g(x, y) = 1$,

$\theta_k =$ offset angle of beam axis from tracking axis,

$G_m = G(0) =$ maximum gain (defines beam axis),

$G_o =$ on-axis gain for uniform illumination,

$G_{sr} =$ ratio of G_m to sidelobe gain (sidelobe ratio),

$\eta_a \equiv G_m/G_o =$ aperture efficiency, or sum-channel gain ratio (Hannan, 1961),

$A_r \equiv \eta_a A =$ effective receiving aperture area,

$K_\theta \equiv \theta_3/\theta_o =$ sum-channel beamwidth ratio (Hannan, 1961).

Basic relationships between these functions and parameters, used in measurement theory, are as follows:

$$\boxed{\begin{array}{c}
\textit{Reference-Beam Parameters} \\[2mm]
G_m = \dfrac{4\pi A_r}{\lambda^2} = \dfrac{4\pi}{\lambda^2} \dfrac{\left| \int_A g(x, y)\, dA \right|^2}{\int_A |g(x, y)|^2\, dA} \qquad (2.1) \\[6mm]
G_o = \dfrac{4\pi A}{\lambda^2} \qquad (2.2) \\[6mm]
\eta_a \equiv \dfrac{G_m}{G_o} = \dfrac{A_r}{A} = \dfrac{\left| \int_A g(x, y)\, dA \right|^2}{A \int_A |g(x, y)|^2\, dA} \qquad (2.3)
\end{array}}$$

[†] The term *power* is used in its general sense as the square of voltage, not to distinguish power gain from directivity.

In many cases, the illumination function may be separated into x- and y-components: $g(x, y) = g(x)g(y)$. Using such "separable" illumination functions, the relationships for rectangular apertures are simplified.

Reference-Beam Parameters
for Separable Illumination

$$G_m = \frac{4\pi w_r h_r}{\lambda^2} \tag{2.4}$$

$$w_r = \eta_x w = \frac{\left| \int_{-w/2}^{w/2} g(x)\, dx \right|^2}{\int_{-w/2}^{w/2} |g(x)|^2\, dx} \tag{2.5}$$

$$h_r = \eta_y h = \frac{\left| \int_{-h/2}^{h/2} g(y)\, dy \right|^2}{\int_{-h/2}^{h/2} |g(y)|^2\, dy} \tag{2.6}$$

$$\eta_a = \eta_x \eta_y \tag{2.7}$$

Error Sensitivity

When a target has been detected in the reference (sum) beam, its angular position can be estimated by finding the beam position which gives the peak response. This can be done by scanning across or around the target, or by placing the target in the null of a second (difference-channel) pattern which approximates the derivative of the sum pattern. We will consider first the sensitivity of monopulse difference-channel measurements, and then show that the scanning systems form the same type of derivative patterns as functions of time, by taking signal samples sequentially in the region near the peak of the sum beam.

For simplicity, we again consider only the patterns in the θ coordinate, with $\phi = 0$. The following difference-beam functions are defined:

$F_d(\theta)$ = voltage gain of difference pattern,

$G_d(\theta) = F_d^2(\theta)$ = power gain of difference pattern,

$g_d(x, y)$ = aperture illumination producing difference pattern,

G_{se} = ratio of G_m (sum-channel) to sidelobe level in difference channel.

Error sensitivity is described, in various applications, using the following parameters.

Fundamental Definitions of Difference-Beam Parameters

$$K \equiv \frac{1}{\sqrt{G_o}} \frac{\partial F_d}{\partial \theta}\bigg|_{\theta=0} = \text{Relative difference slope (Hannan, 1961)} \tag{2.8}$$

$K_o \equiv$ Maximum possible value of K for given aperture

$$K_r \equiv \frac{K}{K_o} = \text{Difference slope ratio (Hannan, 1961)}$$

$$k_m \equiv \frac{\theta_3}{\sqrt{G_m}} \frac{\partial F_d}{\partial \theta}\bigg|_{\theta=0} = \theta_3 K \sqrt{G_o/G_m}$$

$$= \text{Normalized monopulse slope (Barton, 1964)} \tag{2.9}$$

$$k_s \equiv \frac{\theta_3}{\sqrt{G(\theta_k)}} \frac{\partial F}{\partial \theta}\bigg|_{\theta=\theta_k} = \text{Conical-scan error slope (Barton, 1964)} \tag{2.10}$$

In monopulse applications, the error sensitivity is proportional to the slope of the difference pattern near its null point. This is approximately the second derivative of the sum pattern at its peak, since the difference pattern resembles the derivative of the sum pattern. For the present, we will consider that this relationship is exact, giving the following expressions for difference pattern and illumination:

$$F_d(\theta) = \frac{\lambda}{\mathscr{L}_s} \frac{\partial F}{\partial \theta}, \tag{2.11}$$

$$g_d(x, y) = \frac{2\pi x}{\mathscr{L}_s} g(x, y). \tag{2.12}$$

The difference-channel illumination is expressed as the product of the sum-channel illumination and a "linear-odd" function in x. The normalizing constant $\mathscr{L}_s$ is the rms width of the sum-channel illumination power in the x-coordinate, or the "effective aperture width" (denoted by γ in Skolnik, 1962, p. 477):

$$\mathscr{L}_s \equiv \left[\frac{\int_A (2\pi x)^2 \, |g(x, y)|^2 \, dA}{\int_A |g(x, y)|^2 \, dA} \right]^{1/2}. \tag{2.13}$$

Its inclusion in Eqs. (2.11) and (2.12) ensures that the total illumination power and integrated power gain under the pattern will be equal for the sum and difference channels. As we will see later, the effective (power) width of the illumination function is also significant in establishing the sensitivity of a scanning antenna, which was the case analyzed by Skolnik.

The sensitivity of a monopulse antenna is controlled not by $\mathscr{L}_s$ but by the slope parameter K, which can be expressed as follows.

Relative Difference Slope
(for derivative difference pattern)

$$K = \frac{1}{\sqrt{G_o}} \frac{\partial^2 F}{\partial \theta^2}\bigg|_{\theta=0} = \frac{2\pi}{\lambda} \frac{\left| \int_A x^2 \, g(x, y) \, dA \right|}{\left[A \int_A x^2 \, |g(x, y)|^2 \, dA \right]^{1/2}}$$

$$= \frac{\mathscr{L}_\theta^2 \sqrt{\eta_a}}{\mathscr{L}_s \lambda} \tag{2.14}$$

$$\mathscr{L}_\theta \equiv \left[\frac{\left|\int_A (2\pi x)^2 g(x, y)\, dA\right|}{\left|\int_A g(x, y)\, dA\right|}\right]^{1/2} = \text{Effective (voltage) aperture width.} \quad (2.15)$$

The effective (voltage) aperture width, $\mathscr{L}_\theta$, is greater than $\mathscr{L}_s$ for weighted (tapered) illumination functions. In addition to its definition in terms of the illumination function, Eq. (2.15), $\mathscr{L}_\theta$ may be interpreted physically as the curvature of the sum-pattern function near the peak of the beam:

$$\left(\frac{\mathscr{L}_\theta}{\lambda}\right)^2 = \frac{-1}{\sqrt{G_m}} \frac{\partial^2 F}{\partial \theta^2}\bigg|_{\theta=0} = \text{Normalized second derivative of sum pattern.} \quad (2.16)$$

As noted earlier, Eq. (2.14) is based on the approximation that the difference pattern follows the derivative of the sum pattern.

Maximum sensitivity is obtained when no weighting is applied to the sum illumination.

Maximum Difference Slope

$$g(x, y) = 1, \qquad g_d(x, y) = \frac{2\pi x}{\mathscr{L}_o}$$

$$K_o = \frac{2\pi}{\lambda}\left[\frac{1}{A}\int_A x^2\, dA\right]^{1/2} = \frac{\mathscr{L}_o}{\lambda} \qquad\qquad (2.17)$$

$$\mathscr{L}_o \equiv \left[\frac{(2\pi)^2}{A}\int_A x^2\, dA\right]^{1/2} \qquad\qquad (2.18)$$

The rms width of the unweighted aperture, $\mathscr{L}_o$, is the parameter used by Manasse (1960) in his discussion of ideal accuracy of angular measurements. The quantity $(\mathscr{L}_o/\lambda)^2$ is the normalized second derivative of the sum pattern resulting from uniform illumination of the aperture, by Eq. (2.16).

Having expressed the maximum difference slope, we can now give the "difference slope ratio," K_r, in terms of effective aperture widths or of integrals of the illumination function:†

† In his original analysis of angular accuracy capabilities of antennas, Kirkpatrick used a normalized angle equivalent to u' of our App. A:

$$u' = \frac{2\pi w \theta}{\lambda}.$$

Denoting his difference slope parameter by K', we can write

$$K' = \frac{\partial F_d}{\partial u'} = \frac{\lambda}{2\pi w}\frac{\partial F_d}{\partial \theta}\bigg|_{\theta=0} = \frac{\lambda\sqrt{G_o}}{2\pi w} K.$$

The ratio K/K_o, however, is the same whether slopes are expressed in u' or in θ.

$$K_r = \frac{K}{K_o} = \frac{\mathscr{L}_\theta^2}{\mathscr{L}_o \mathscr{L}_s} \sqrt{\eta_a}$$

$$= \frac{\left| \int_A x^2 g(x,y)\, dA \right|}{\left[\int_A x^2 |g(x,y)|^2\, dA \int_A x^2\, dA \right]^{1/2}}. \tag{2.19}$$

We have seen that the error slope parameters for an antenna may be expressed in three ways:

(a) As integrals of the illumination function, weighted by x^2;

(b) As ratios of effective or rms aperture widths; or

(c) As curvatures (second derivatives) of the sum patterns which can be generated by the aperture with and without taper.

For completeness, we will repeat the expressions for K, K_o, and K_r as they apply to the special case of a rectangular aperture using separable illumination functions in the two coordinates, and for the general case where the sum and difference illuminations are independent, or not governed by Eq. (2.12).

Rectangular Aperture, Separable Illuminations

$$K = \frac{2\pi}{\lambda} \frac{\left| \int_{-w/2}^{w/2} x^2 g(x)\, dx \right|}{\left[\dfrac{w}{\eta_y} \int_{-w/2}^{w/2} x^2 |g(x)|^2\, dx \right]^{1/2}} = \frac{\mathscr{L}_\theta^2 \sqrt{\eta_a}}{\mathscr{L}_s \lambda} \tag{2.20}$$

$$K_o = \frac{2\pi}{\lambda} \left[\frac{1}{w} \int_{-w/2}^{w/2} x^2\, dx \right]^{1/2} = \frac{\pi w}{\sqrt{3}\,\lambda} \tag{2.21}$$

$$K_r = \frac{\left| \int_{-w/2}^{w/2} x^2 g(x)\, dx \right|}{\left[\dfrac{w^3}{12\eta_y} \int_{-w/2}^{w/2} x^2 |g(x)|^2\, dx \right]^{1/2}} = \frac{\mathscr{L}_\theta^2 \sqrt{\eta_a}}{\mathscr{L}_o \mathscr{L}_s} \tag{2.22}$$

General Case, Difference Illumination $g_d(x,y)$

$$K = \frac{2\pi}{\lambda} \frac{\left| \int\!\int_A x g_d(x,y)\, dA \right|}{\left[A \int_A |g_d(x,y)|^2\, dA \right]^{1/2}} \tag{2.23}$$

$$K_o = \frac{2\pi}{\lambda} \left[\frac{1}{A} \int_A x^2\, dA \right]^{1/2} = \frac{\mathscr{L}_o}{\lambda} \tag{2.24}$$

$$K_r \equiv \frac{K}{K_o} = \frac{\left| \int\!\int_A x g_d(x,y)\, dA \right|}{\left[\int_A |g_d(x,y)|^2\, dA \int_A x^2\, dA \right]^{1/2}} \tag{2.25}$$

Because no specific sum-channel illumination is assumed in the general case, the effective widths of the illumination voltage and power cannot be used to describe the slope factors. It would be perfectly possible to select a hypothetical pattern function for which the actual difference pattern was the derivative, and to use the corresponding illumination function in Eqs. (2.13)–(2.19) to find values of K and K_r. To the extent that the actual sum-channel pattern differs from the hypothetical, some of the error expressions to be derived below would require adjustment.

2.2 Expressions for Noise Error

Ideal Case

The minimum error obtainable with a given aperture and incident field intensity has been stated by Manasse.

$$
\boxed{
\begin{array}{c}
\textit{Ideal Minimum Error} \\[2mm]
(\sigma_\theta)_{min} = \dfrac{\lambda}{\mathscr{L}_o\sqrt{\mathscr{R}_o}} = \dfrac{1}{K_o\sqrt{\mathscr{R}_o}}
\end{array}
}
\tag{2.26}
$$

Here, $\mathscr{R}_o$ is the energy ratio which would be obtained with uniform aperture illumination (antenna gain G_o). To achieve this accuracy, we must use a sum channel with uniform illumination and a difference channel with linear-odd illumination in the axis which controls the pattern in θ.

Monopulse System with Taper

When the actual monopulse antenna has tapered illumination functions, rather than uniform and linear-odd, the sidelobe levels in both patterns will be reduced and the thermal noise error will be increased.

$$
\boxed{
\begin{array}{c}
\textit{Monopulse Error} \\[2mm]
\sigma_\theta = \dfrac{1}{K\sqrt{\mathscr{R}_o}} = \dfrac{\sqrt{\eta_a}}{K\sqrt{\mathscr{R}_m}} = \dfrac{\lambda}{K_r\mathscr{L}_o\sqrt{\mathscr{R}_o}}
\end{array}
}
\tag{2.27}
$$

The on-axis energy ratio $\mathscr{R}_m$ in the sum channel is the value obtained when the gain G_m is inserted for G_r in the radar equation, Eq. (1.27).

If we wish to express the measurement sensitivity in terms of an rms aperture width $\mathscr{L}$ in Eq. (1.18), the value for a tapered illumination will be

$$
\mathscr{L} = \frac{\lambda K}{\sqrt{\eta_a}} = \frac{\mathscr{L}_\theta^2}{\mathscr{L}_s}.
\tag{2.28}
$$

Although less than $\mathscr{L}_o$, this quantity is greater than $\mathscr{L}_s$, and hence the monopulse error for tapered illumination will be smaller than the error derived by Skolnik (1962, p. 477) for scanning antennas, even when the actual received energy is the same. The monopulse error can be written in terms of the effective widths of illumination voltage and power:

$$\sigma_\theta = \frac{\lambda \mathscr{L}_s}{\mathscr{L}_\theta^2 \sqrt{\mathscr{R}_m}}. \tag{2.29}$$

This can be compared with the expression for scanning beams, given in Sec. 2.5 below.

2.3 Error Normalized to Beamwidth

The expression of rms (effective) aperture width in terms of the curvature of the beam pattern leads to particularly convenient relationships between error slope and 3-db beamwidths. Because all practical patterns have approximately the same shape within the 3-db contours, we may express the effective (voltage) width of the aperture illumination for all aperture shapes and for all even illumination functions in terms of the 3-db beamwidth:

$$\frac{\mathscr{L}_\theta \theta_3}{\lambda} \cong 1.63 \qquad (\pm 2\%). \tag{2.30}$$

This invariant property of antennas was pointed out by Spencer (1946). Actual values are given in Tables 2.1 and 2.2, below, for several aperture shapes and illuminations, and data on other functions appear in App. A. Even for the extreme case of the two-element interferometer, the constant given in Eq. (2.30) is only four percent high, the true value being $\pi/2$.

By combining Eqs. (2.20) and (2.30) we can express the relative difference slope of a derivative difference pattern in a form normalized to beamwidth:

$$K\theta_3 \cong 1.63 \frac{\mathscr{L}_\theta}{\mathscr{L}_s} \sqrt{\eta_a}. \tag{2.31}$$

The normalized error for a uniformly illuminated aperture of any shape, with the linear-odd difference illumination, can now be written

$$\frac{(\sigma_\theta)_{min}}{\theta_o} = \frac{\lambda}{\mathscr{L}_o \theta_o \sqrt{\mathscr{R}_o}} \cong \frac{1}{1.63 \sqrt{\mathscr{R}_o}}. \tag{2.32}$$

For arbitrary tapered illumination with a derivative difference pattern, we have

$$\frac{\sigma_\theta}{\theta_3} = \frac{\lambda(\mathscr{L}_s/\mathscr{L}_\theta)}{\mathscr{L}_\theta \theta_3 \sqrt{\mathscr{R}_m}} \cong \frac{(\mathscr{L}_s/\mathscr{L}_\theta)}{1.63 \sqrt{\mathscr{R}_m}}. \tag{2.33}$$

This leads to a simple expression for error in terms of the normalized monopulse slope defined in Eq. (2.9).

$$\boxed{\begin{array}{c} \textit{Normalized Monopulse Error} \\[6pt] \dfrac{\sigma_\theta}{\theta_3} = \dfrac{1}{k_m \sqrt{\mathscr{R}_m}} \qquad \text{(for all apertures)} \end{array}} \qquad (2.34)$$

Relationships between the different slope constants and ratios are as follows:

$$k_m = \frac{K\theta_3}{\sqrt{\eta_a}} = \frac{K_r K_o \theta_3}{\sqrt{\eta_a}} = \frac{K_r K_o K_\theta \theta_o}{\sqrt{\eta_a}} \cong 1.63 \frac{K_r K_\theta}{\sqrt{\eta_a}}. \qquad (2.35)$$

Also, for the derivative difference pattern,

$$k_m \cong 1.63 \frac{\mathscr{L}_\theta}{\mathscr{L}_s}. \qquad (2.36)$$

Values for this condition are shown in Tables 2.1 and 2.2.

Of the various slope parameters defined above, k_m is the only one which can be applied to the Gaussian pattern and illumination functions. Although $\mathscr{L}_\theta$ and $\mathscr{L}_s$ are also defined for this case, values of G_o, $\mathscr{L}_o$, K_o and K_θ are infinite, while K, K_r, θ_o and η_a are zero. We will thus be able to use k_m, $\mathscr{L}_\theta$, and $\mathscr{L}_s$ to relate the Gaussian-beam results of Swerling and Develet to those of Kirkpatrick, Manasse, and Hannan. Note, also, that k_m increases with the beamwidth θ_3 when the difference illumination is held constant and increasing taper is applied to the sum illumination. For this reason, k_m is not a true measure of angle sensitivity for a given aperture, but only for a given pair of sum and difference patterns.

2.4 Monopulse Antenna Performance

Ideal Illumination: $g(x, y) = 1$, $g_d(x, y) = 2\pi x/\mathscr{L}_o$

Although this case is far from ideal in terms of sidelobe levels and actual measurement accuracy, it may be taken as a reference for antenna gain and angle sensitivity. The significant parameters for four aperture shapes are given in Table 2.1.

Table 2.1

IDEAL MONOPULSE PARAMETERS

Aperture Shape	$\theta_o w/\lambda$	$\mathscr{L}_o/w$	$k_m = \mathscr{L}_o\theta_o/\lambda = K_o\theta_o$
Interferometer	0.500	π	$\pi/2$
Rectangular	0.886	$\pi/\sqrt{3}$	1.607
Circular	1.028	$\pi/2$	1.617
Triangular	1.276	$\pi/\sqrt{6}$	1.636

Optimum Illumination: $g_d(x, y) = (2\pi x/\mathscr{L}_s)g(x, y)$

The parameters for this case, corresponding to a difference pattern which is the derivative of the sum pattern, are given in Table 2.2. In general, the derivative-pattern relationship is possible only in phased arrays which permit completely independent control of sum and difference illumination functions. Where values of $\theta_3 w/\lambda$ in Table 2.2 differ from those given in the radar literature (Silver, 1949; Skolnik, 1962), careful checking of accuracy has verified the correctness of the figures to the last digit shown here.

Table 2.2

OPTIMUM MONOPULSE PARAMETERS

Sum Illumination Function	$\dfrac{\theta_3 w}{\lambda}$	η_x	$\dfrac{\mathscr{L}_\theta}{w}$	$\dfrac{\mathscr{L}_s}{w}$	$\dfrac{\mathscr{L}_\theta \theta_3}{\lambda}$	$\dfrac{\mathscr{L}_s \theta_3}{\lambda}$	$\dfrac{K}{K_o}$	k_m	$\dfrac{G_{sr}}{\text{(db)}}$
Circular, $g(x) = \sqrt{1 - (2x/w)^2}$	1.028	0.924	1.571	1.406	1.617	1.446	0.932	1.81	17.6
Parabolic,* $g(x) = 1 - 2x^2/w^2$	0.972	0.969	1.66	1.53	1.613	1.481	0.983	1.75	17.1
Parabolic,* $g(x) = 1 - 4x^2/w^2$	1.155	0.833	1.407	1.188	1.624	1.372	0.836	1.92	20.6
Triangular, $g(x) = 1 - \|2x/w\|$	1.276	0.750	1.28	0.994	1.636	1.268	0.79	2.08	26.4
Cosine, $g(x) = \cos(\pi x/w)$	1.189	0.812	1.37	1.136	1.629	1.350	0.812	1.96	23.0
Cosine^2, $g(x) = \cos^2(\pi x/w)$	1.441	0.667	1.134	0.89	1.636	1.283	0.653	2.08	32.0
Cosine^4, $g(x) = \cos^4(\pi x/w)$	1.853	0.515	0.886	0.669	1.645	1.240	0.466	2.18	48.0
Gaussian, $g(x) = \exp(-x^2/2\sigma_x^2)$	**	0.0	**	**	1.662	1.177	0.0	2.35	∞
Gaussian, assume $w = 6\sigma_x$	1.59	0.591	1.045	0.74	1.662	1.177	0.628	2.35	∞

* The first parabolic distribution shown has a pedestal level $\Delta = 0.5$ at the edge of the aperture, while the second has $\Delta = 0$ (no pedestal).

** For Gaussian illumination, $\theta_3 \sigma_x/\lambda = 0.265$, $\mathscr{L}_\theta/\sigma_x = 2\pi$, $\mathscr{L}_s/\sigma_x = \sqrt{2}\,\pi$.

In Figs. 2.2 and 2.3 we have plotted the sum-channel efficiency and the difference slope ratio as functions of the sum-channel sidelobe ratio, G_{sr}, for various illuminations. The optimum difference illumination, Eq. (2.12), is assumed in calculation of slopes. This normally gives difference-channel sidelobe levels 6 db above the corresponding sidelobes in the sum channel.† It may be noted that the greatest slope factor and efficiency, for a given sidelobe ratio, are obtained with the Taylor functions, and that the truncated Gaussian functions are almost as efficient. The (cos)^n functions are less efficient, but their sidelobe levels fall off much more rapidly after the first lobe.

Data from these figures and from Table 2.2 may be used to find efficiency and slope when separable two-dimensional illuminations are applied in x- and y-coordinates. Values of η_x and η_y are taken from the functions in these coordinates. Their product gives the overall aperture efficiency η_a in accordance with Eq. (2.7). The difference slope ratio in the plane containing the x-axis ($\phi = 0$) is

$$K'_{rx} = K_{rx}\sqrt{\eta_y},$$

† An exception occurs in the Taylor illumination cases, where some of the sidelobes are narrowed, producing larger derivative values.

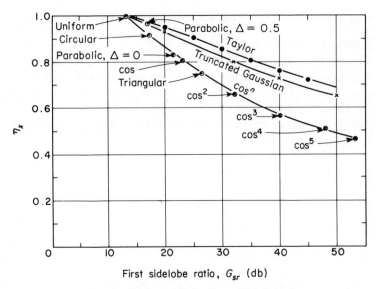

Fig. 2.2 Aperture efficiency vs. sidelobe ratio.

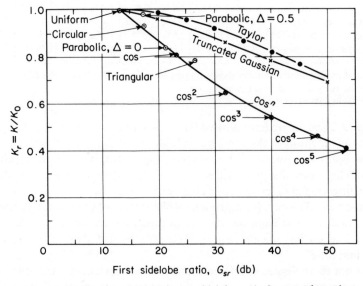

Fig. 2.3 Difference slope ratio vs. sidelobe ratio in sum channel.

where K_{rx} is the value read from the table or figure for the x-coordinate with uniform y illumination. The slope ratio in the plane containing the y-axis is found in a similar way by interchanging x and y subscripts and functions.

Figure 2.4 shows three related slope parameters as functions of G_{sr}, again with optimum difference illuminations. As noted earlier, k_m increases as beamwidth is

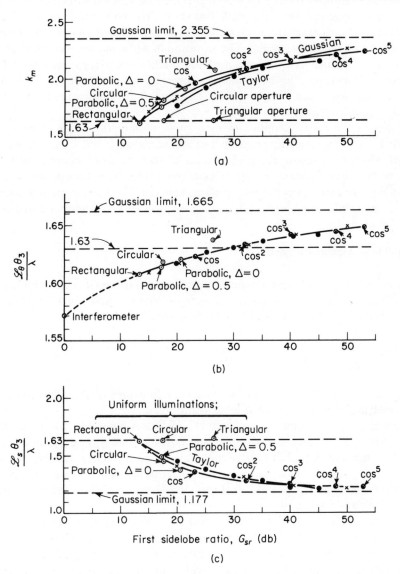

Fig. 2.4 Antenna slope parameters vs. sidelobe ratio in sum channel.

broadened by tapering. The relatively invariant parameter $\mathscr{L}_\theta \theta_3/\lambda$ also increases, but only by three percent for G_{sr} between 13 db and infinity. The parameter $\mathscr{L}_s \theta_3/\lambda$, applicable to scanning beams, decreases steadily with increasing taper and sidelobe ratio. It is interesting that all three of these parameters are uniquely determined for rectangular apertures by the sidelobe ratio and not by the shape of the illumination function or the edge value used to obtain this ratio. This is significant in system analysis, in that the performance parameters may be selected without knowing the exact illumination function. The data also show that the uniform rectangular and Gaussian functions are extreme cases, with characteristics of all practical antennas falling between the two extremes. The variation in performance between the Gaussian function (which is often assumed for mathematical convenience) and any practical illumination may be estimated from Fig. 2.4. Parameters of this figure are not dependent upon the presence of any particular illumination in the second coordinate, as long as it is separable.

Although the data on optimum monopulse illuminations cannot be applied directly to radar systems using reflectors or lens antennas, it will serve as a reference against which to compare the performance of such horn-fed systems, and other antennas whose characteristics approximate those listed here. Except for the effects of spillover, we can expect the relationships between efficiency, slope, and sidelobe ratio to apply with equal force to illuminations produced by monopulse feed horns.

Multihorn Monopulse Feeds

Practical design of horn-fed reflectors and lenses has been discussed by Hannan (1961). Considering a rectangular aperture of width w and focal length Z, the null-to-null width of the received signal lobe in the focal plane will be

$$2\Delta = \frac{2\lambda Z}{w}.$$

Figures 2.5 and 2.6 give antenna parameters as functions of the total feed width, a, normalized to this focal-plane lobe width.

To find the performance of an antenna which uses separable, two-dimensional illuminations in x- and y-coordinates, we take values of η_x and η_y, and proceed in accordance with Eq. (2.7) to obtain overall aperture efficiency as their product. The error slopes in the two planes are then determined as described in the preceding section. Hannan's paper also includes details as to spillover ratio and amplitude of the second sidelobe.

The curves show a basic conflict between the optimum feed dimensions for sum illumination ($a \cong 1.4\lambda Z/w$) and difference illumination ($a \cong 2.4\lambda Z/w$). The necessity to compromise between these values can be avoided by using additional horns or multimode horns, interconnected with appropriate microwave networks. Hannan has tabulated results for several such configurations, comparing them to a conventional four-horn feed whose dimensions are set to optimize sum-channel gain. Table 2.3 gives his results, plus derived values of k_m and sidelobe levels. The performance of the several feeds is compared in Fig. 2.7 to that of the optimum monopulse illuminations for given sidelobe levels. For G_{sr} near 20 db, we see that the complex feeds (twelve-horn, or four-horn multimode) can approach closely the optimum slope without the loss in efficiency of the sum channel which prevailed with the simple feed, and with values of G_{se} which are actually better than those

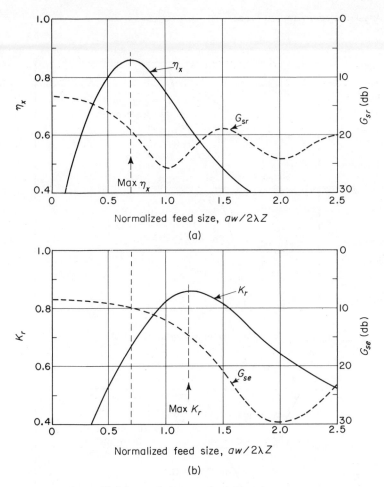

Fig. 2.5 Efficiency and slope vs. feed size (after Hannan).

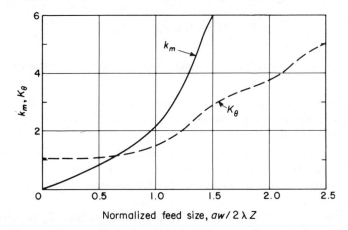

Fig. 2.6 Normalized slope and beamwidth ratio vs. feed size (after Hannan).

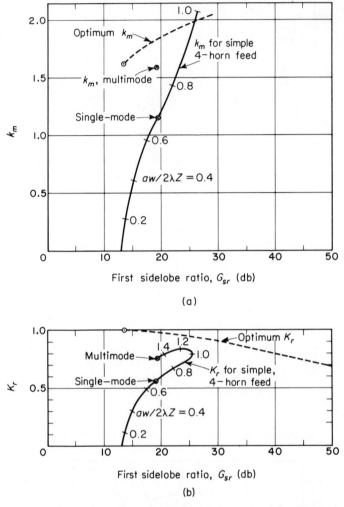

Fig. 2.7 Comparison of monopulse feed performance (after Hannan).

applicable to the optimum illuminations. The loss in efficiency in the simple four-horn case, adjusted for equal difference slope ($a \cong 2\lambda Z/w$), is about 2 db, and the difference sidelobe ratio is about 12 db.

A different analysis of monopulse parameters is given by Develet (1963), who describes the performance of an amplitude-comparison monopulse system in terms of two separate, squinted, Gaussian beams, characterized by their individual beamwidths α_3, crossover level $L_k = 10 \log_{10} [G(0)/G(\theta_k)]$, and sum-channel beamwidth ψ (corresponding to our θ_3). For the Gaussian beam, L_k is related to the squint angle θ_k and the beamwidth α_3 by

Table 2.3

MONOPULSE FEED HORN PERFORMANCE

Type of Horn	η_a	H-plane $K_r\sqrt{\eta_y}$	k_m	E-plane $K_r\sqrt{\eta_x}$	k_m	G_{sr} (db)	G_{se} (db)	*Feed Shape*
Simple four-horn	0.58	0.52	1.2	0.48	1.2	19	10	
Two-horn dual-mode	0.75	0.68	1.6	0.55	1.2	19	10	
Two-horn triple-mode	0.75	0.81	1.6	0.55	1.2	19	10	
Twelve-horn	0.56	0.71	1.7	0.67	1.6	19	19	
Four-horn triple-mode	0.75	0.81	1.6	0.75	1.6	19	19	

$$L_k = \frac{40 \ln 2}{\ln 10} \left(\frac{\theta_k}{\alpha_3}\right)^2. \tag{2.37}$$

For the strong-signal case, Develet's results may be written

$$\frac{\sigma_\theta}{\theta_3} = \left[\frac{5}{(2L_k \ln 2 \ln 10)\mathscr{R}_k}\right]^{1/2} \left(\frac{\alpha_3}{\theta_3}\right) = \frac{1}{k_m\sqrt{\mathscr{R}_k}}, \tag{2.38}$$

$$k_m = \left[\frac{2L_k \ln 2 \ln 10}{5}\right]^{1/2} \left(\frac{\theta_3}{\alpha_3}\right) = (4 \ln 2)\frac{\theta_k \theta_3}{\alpha_3^2}. \tag{2.39}$$

Here, $\mathscr{R}_k$ is the energy ratio produced by the sum of both beams at the crossover point. Figure 2.8 shows the relationship between k_m, θ_k, θ_3, and α_3 for various crossover levels. The curves show that relatively large values of k_m are possible for large squint angles, but the fall-off in energy ratio restricts the practical squint angles to about half the individual beamwidth. The feeds required for generation of slightly squinted beams must overlap in the focal plane, and Hannan's approach to analysis is then more accurate. However, the results of Hannan and Develet are in good agreement for feed dimensions near $2\lambda Z/w$, with $L_k \cong 3$ db.

Another common approach to monopulse analysis requires that θ_3 and k_m be determined from actual pattern measurements of sum and difference channels. These values may be compared to the previously tabulated results to find whether full use is being made of the aperture, within limitations set by sidelobe levels. In one such case (George and Zamanakos, 1959) the patterns were found to follow closely the equations

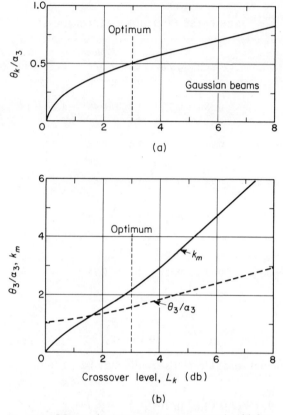

Fig. 2.8   Monopulse parameters vs. crossover level (after Develet).

$$F(\theta) \cong \sqrt{G_m} \, \cos^2 \left(\frac{1.18\theta}{\theta_3} \right), \tag{2.40}$$

$$F_d(\theta) \cong \sqrt{\frac{G_m}{2}} \, \sin \left(\frac{2.36\theta}{\theta_3} \right). \tag{2.41}$$

Hence,

$$k_m = \frac{\theta_3}{\sqrt{G_m}} \left. \frac{\partial F_d}{\partial \theta} \right|_{\theta = 0} \cong \frac{2.36}{\sqrt{2}} = 1.67.$$

In this case, the form of the difference pattern follows that of the optimum (derivative of the sum pattern), with an amplitude predicted by Eq. (2.11). This result was obtained in an early four-horn feed design with multimode structure. Expressions of this form can be applied to analysis of errors based on F_m (or G_m) and θ_3, without further knowledge of antenna dimensions or illumination function.

2.5 Performance of Scanning Beams

Search Radar Case

The optimum accuracy for a single beam, scanning at a constant rate across a target, is found by analogy to time-delay measurement on waveforms (Skolnik, 1962). The appropriate rms aperture width for this case is the power-weighted value, $\mathcal{L}_s$, defined in Eq. (2.13).

$$\sigma_\theta = \frac{\lambda}{\mathcal{L}_s\sqrt{\mathcal{R}}} \qquad \text{(one-way pattern)}, \qquad (2.42)$$

$$\mathcal{R} = \frac{\mathcal{R}_m}{L_p} = \frac{2n(S/N)_m}{L_p} \qquad \text{[matched filter, see Eq. (1.36)]}.$$

Normalized Error in Search, One-Way Case
$$\frac{\sigma_\theta}{\theta_3} = \frac{\lambda}{\mathcal{L}_s\theta_3\sqrt{\mathcal{R}}} = \frac{\sqrt{L_{p1}}}{k_{p1}\sqrt{2n(S/N)_m}}$$

$$(2.43)$$

In the above, $k_{p1} = \mathcal{L}_s\theta_3/\lambda$ is the slope factor for a scanning antenna receiving its signal from a radiating source of constant power. Table 2.2 contained values for typical illumination functions.

If the same antenna is used to illuminate a reflecting target, we use a larger effective rms width $\mathcal{L}_{s2}$ to express the aperture illumination which would have generated a pattern equal to the observed two-way pattern.

Normalized Error in Search, Two-Way Case
$$\frac{\sigma_\theta}{\theta_3} = \frac{\lambda}{\mathcal{L}_{s2}\theta_3\sqrt{\mathcal{R}}} = \frac{\sqrt{L_{p2}}}{k_{p2}\sqrt{2n(S/N)_m}}$$

$$(2.44)$$

For the two-way case, $k_{p2} = \mathcal{L}_{s2}\theta_3/\lambda$, where the value of $\mathcal{L}_{s2}$ is calculated from the antenna illumination convolved with itself. The numerical value of the term $\sqrt{L_{p2}}/(k_{p2}\sqrt{2})$ is in excellent agreement with Swerling's results for the scanning Gaussian beam, after conversion of his units to those used here (see Barton, 1964, p. 51). The two extreme cases given in Table 2.4 fall within fifteen percent of the following approximation for error, both in one-way and two-way operation:

$$\frac{\sigma_\theta}{\theta_3} \simeq \frac{0.5}{\sqrt{n(S/N)_m}}. \qquad (2.45)$$

Table 2.4

MEASUREMENT CONSTANTS FOR SEARCH RADAR

Illumination Type	Gaussian	Rectangular
Illumination function, $g(x)$	$\exp\left[-\dfrac{x^2}{2\sigma_x^2}\right]$	Uniform over w
Beam pattern, $F(\theta)$	$\exp\left[-\dfrac{1.386\theta^2}{\theta_3^2}\right]$	$\dfrac{\sin(2.78\theta/\theta_3)}{2.78\theta/\theta_3}$
One-way operation:		
Rms width, $\mathscr{L}_s$	$\sqrt{2}\,\pi\sigma_x$	$1.815w$
Beamshape loss, L_{p1}	0.94	0.886
Slope, k_{p1}	1.18	1.607
Constant, $\sqrt{L_{p1}}/(k_{p1}\sqrt{2})$	0.58	0.414
Two-way operation:		
Beam pattern, $F^2(\theta)$	$\exp\left[-\dfrac{2.77\theta^2}{\theta_3^2}\right]$	$\left[\dfrac{\sin(2.78\theta/\theta_3)}{2.78\theta/\theta_3}\right]^2$
Equivalent illumination	$\exp\left[-\dfrac{x^2}{\sigma_x^2}\right]$	Triangular over $2w$
Rms width, $\mathscr{L}_{s2}$	$2\pi\sigma_x$	$1.988w$
Beamshape loss, L_{p2}	1.328	1.329
Slope, k_{p2}	1.665	1.76
Constant, $\sqrt{L_{p2}}/(k_{p2}\sqrt{2})$	0.490	0.463

Results with other illumination functions can be represented by other constants within the limits 0.414 and 0.58 set by the uniform and the Gaussian function for the one-way case.

Since the quantity $n(S/N)_m$ must exceed 25 for reasonable target detectability, the thermal noise error should not exceed $\theta_3/10$ when an efficient signal processor is used with the search radar.

Conical-Scan Tracker

The expression for thermal noise error in a conical-scan tracker is similar in form to Eq. (2.34) for monopulse and Eq. (2.43) for search radar.

$$
\boxed{
\begin{array}{c}
\textit{Normalized Conical-Scan Error} \\[6pt]
\dfrac{\sigma_\theta}{\theta_3} = \dfrac{\sqrt{2}}{k_s\sqrt{\mathscr{R}}} = \dfrac{\sqrt{2L_k}}{k_s\sqrt{\mathscr{R}_m}} = \dfrac{\sqrt{L_k}}{k_s\sqrt{n(S/N)_m}}
\end{array}
}
\qquad (2.46)
$$

The crossover loss L_k is determined by the offset angle of the beam axis from the tracking axis:

$$L_k = \frac{G_m}{G(\theta_k)} = \left[\frac{F_m}{F(\theta_k)}\right]^2. \tag{2.47}$$

The additional factor of $\sqrt{2}$ in the numerator of the error equation is equivalent to another 3-db loss in measurement performance. This can be attributed to the division of signal energy into pairs of sidebands, produced by the amplitude modulation of the signal at the scan frequency (Barton, 1964, p. 276), or to the fact that only half the energy during each scan cycle is available for measurement of each angular coordinate.

The normalized sensitivity of the conical-scan system may be expressed as the ratio $k_s/\sqrt{L_k}$, which is determined directly from the defining equation (2.10):

$$\frac{k_{s1}}{\sqrt{L_{k1}}} = \frac{\theta_3}{F_m}\frac{\partial F}{\partial \theta}\Big|_{\theta=\theta_k} \qquad \text{(one way)} \tag{2.48}$$

$$\frac{k_{s2}}{\sqrt{L_{k2}}} = \frac{\theta_3}{G_m}\frac{\partial G}{\partial \theta}\Big|_{\theta=\theta_k} = \frac{2k_{s1}}{L_{k1}} \qquad \text{(two-way).} \tag{2.49}$$

Figures 2.9 and 2.10 show conical-scan error sensitivities and crossover losses for both one-way and two-way operation, with rectangular and Gaussian illumination functions. Because the normalized slope increases very slowly as it approaches its maximum, the optimum point for radar system operation is usually on the lower-loss side of the maximum, as shown.

Relationships between k_s and optimum monopulse k_m may be written in terms of offset angle and rms aperture widths, for small offset angles ($\theta_k < 0.3\theta_3$):

$$\frac{k_{s1}}{\sqrt{L_{k1}}} \cong \frac{\theta_3\theta_k}{F_m}\frac{\partial^2 F}{\partial \theta^2}\Big|_{\theta=0} = \left[\frac{\mathscr{L}_\theta\theta_3}{\lambda}\right]^2\left(\frac{\theta_k}{\theta_3}\right) \cong 2.65\frac{\theta_k}{\theta_3}, \tag{2.50}$$

$$\frac{k_{s1}}{\sqrt{L_{k1}}} \cong \frac{\mathscr{L}_s\theta_k}{\lambda}k_m \leq 0.6\,k_m, \tag{2.51}$$

$$\frac{k_{s2}}{\sqrt{L_{k2}}} \cong 5.3\frac{\theta_k}{\theta_3} = \frac{2\mathscr{L}_s\theta_k}{\lambda\sqrt{L_{k1}}}k_m \leq 0.8k_m. \tag{2.52}$$

A comparison of actual values in Fig. 2.9 with those of Fig. 2.4 shows that the conical-scan sensitivities for one-way and two-way operation lie forty to eighty percent below optimum k_m, not including the 3-db factor caused by the modulation process.

2.6 Effects of Signal Processing

Expressions for angular error have been given above for cases in which the receiving system extracts all the angular information furnished by the antenna, without degradation except for the receiver's contribution to input noise. Ideally, this

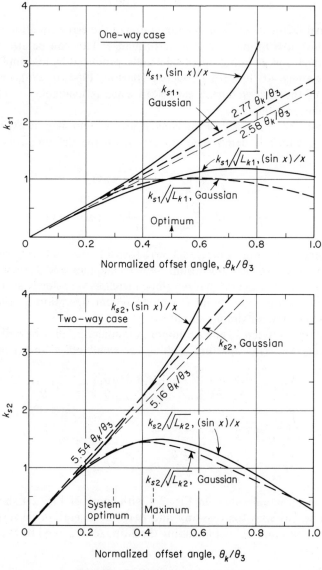

Fig. 2.9 Conical-scan error slopes.

would require coherent integration of the signal over the entire period of observation or measurement, t_o, prior to detection or other nonlinear processing. Such a process may also be described in terms of a filter matched to the signal spectrum, followed by detection and sampling at the instant of maximum output. Since target detection dictates a total energy ratio $\mathscr{R}$ well above unity (typically 20 or greater),

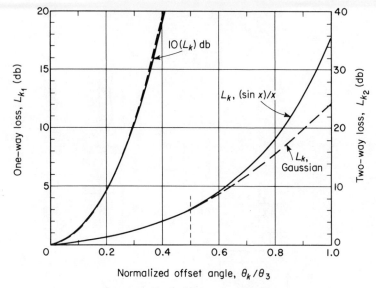

Fig. 2.10   Conical-scan crossover loss.

there will be only negligible detector loss, or "small-signal suppression," in this coherent, matched-filter case.

In other cases, it may be necessary to introduce the detector prior to final integration, as shown in the typical monopulse configuration of Fig. 2.11. Here we show the error detector preceded by two steps: normalization of the error signal with respect to sum-channel signal amplitude, and IF bandpass filtering. After detection, the video error signal is filtered, sampled by a range gate, and integrated in a low-pass filter, which may be a tracking servo loop. The order of processing steps may be changed (e.g., range gating at IF rather than at video) and some steps may be omitted in certain systems, but the general procedure will be much the same in any measurement device.

Noncoherent Processing

The first factor of importance in estimating processing loss is the reduction in attainable IF S/N ratio by the factor n, equal to the number of pulses received during the observation time, in a noncoherent system. When successive pulses are not predictable in phase (coherent), the ratio S/N is determined not by the energy ratio $\mathscr{R}$, but by the single-pulse energy ratio $\mathscr{R}_1$:

$$2\frac{S}{N} \le \mathscr{R}_1 = \frac{\mathscr{R}}{n} \qquad \text{[see Eq. (1.30)]}.$$

The factor of n may be regained in post-detection processing, provided that S/N remains high enough to avoid detector loss, as described below.

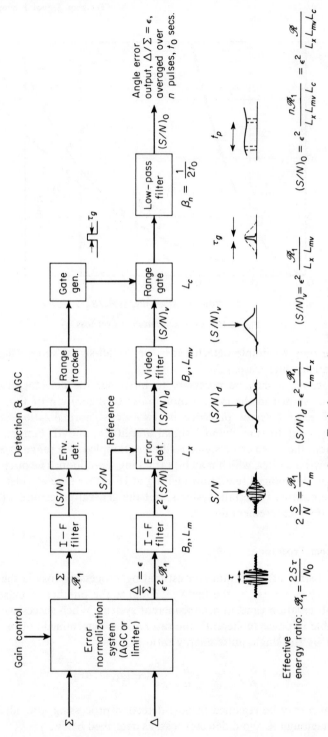

Fig. 2.11 Typical monopulse error processor.

IF Filter Loss, L_m

This loss factor indicates the extent to which the IF S/N ratio at the detector, at the instant of maximum signal envelope, falls below the optimum single-pulse value:

$$L_m \equiv \frac{\mathscr{R}_1}{2(S/N)} \qquad \text{[see Eq. (1.30)]}.$$

Plots of L_m vs. $B_n\tau$ for several filter and pulse shapes are given in Chap. 3. It is important to note that this loss serves as an intermediate term in computation of detector loss, and is superseded, in final computation, by the system matching loss L_{mv}, which includes video filter effects.

Detector Loss, L_x

There is an irreversible loss of information when the S/N ratio (of the sum channel in a monopulse system) is not very large. This loss is given by

$$L_x = C_a = \frac{(S/N) + 1}{(S/N)} = \frac{S + N}{S} \qquad \text{(monopulse)}, \tag{2.53}$$

$$L_x = C_d = \frac{2(S/N) + 1}{2(S/N)} = \frac{2S + N}{2S} \qquad \text{(conical scan)}. \tag{2.54}$$

This loss must be calculated for that value of S/N which exists at the detector at the time the error sample is taken. The loss is generally negligible for $(S/N) > 10$, provided that the post-detection signal processing is reasonably efficient. The detector factors C_a and C_d are discussed by Barton (1964, pp. 278–281), based on the early work of Lawson and Uhlenbeck (1950) and of Davenport and Root (1958).

When we apply L_x to reduce the effective value of $\mathscr{R}$ in Eq. (2.27), etc., the results are in agreement with the full expressions of Manasse (1960), Develet (1961), and Barton (1964) for both monopulse and conical-scan radar.

Monopulse Radar, $B_n\tau > 2$

$$\frac{\sigma_\theta}{\theta_3} = \frac{\sqrt{(S/N) + 1}}{k_m(S/N)\sqrt{2nB_n\tau}} \tag{2.55}$$

Conical-Scan Radar, $B_n\tau > 2$

$$\frac{\sigma_\theta}{\theta_3} = \frac{\sqrt{2(S/N) + 1}}{k_s(S/N)\sqrt{2nB_n\tau}} \tag{2.56}$$

For search applications, a calculation of L_x from Eq. (2.53), based on the average ratio $(S/N)_m/L_p$, will yield results in excellent agreement with those of Swerling. For the case where the IF filter is matched to the individual pulse, we have the following.

Search Radar, $L_m = 1$

$$\frac{\sigma_\theta}{\theta_3} = \frac{\sqrt{L_p L_x}}{k_p \sqrt{2n(S/N)_m}} = \frac{\sqrt{(S/N)_m + L_p}}{k_p (S/N)_m} \times \frac{\sqrt{L_p}}{\sqrt{2n}} \tag{2.57}$$

Search Radar, One-Way Case,
Gaussian Beam, $(S/N)_m \ll 1$

$$\frac{\sigma_\theta}{\theta_3} \simeq \frac{L_{p1}}{k_{p1}(S/N)_m \sqrt{2n}} = \frac{0.563}{(S/N)_m \sqrt{n}} \tag{2.58}$$

Search Radar, Two-Way Case,
Gaussian Beam, $(S/N)_m \ll 1$

$$\frac{\sigma_\theta}{\theta_3} \simeq \frac{L_{p2}}{k_{p2}(S/N)_m \sqrt{2n}} = \frac{0.565}{(S/N)_m \sqrt{n}} \tag{2.59}$$

The constant 0.565 in this last equation compares with the value 0.580 obtained when Swerling's symbols are translated to agree with ours (see Barton, 1964, p. 51). The small discrepancy in value of this constant is attributable to the fact that angle information is obtained from the sloping sides of the beam envelope, where the loss averages higher than L_p.

System Matching Loss, L_{mv}

In cases where the range gate takes a short sample of error signal at the time of maximum amplitude, the system matching loss will be determined by the band-

Table 2.5

LOSS FOR NARROW RANGE GATE, $\tau_g/\tau \ll 1/2B_v\tau$

Bandwidth Relationships	Matching Loss
$B_n\tau \gg 2B_v\tau > 2$	$L_{mv} = 2B_v\tau$
$2B_v\tau \gg B_n\tau > 2$	$L_{mv} = B_n\tau$
$B_n\tau \gg 2B_v\tau$	Use $2B_v\tau$ in Fig. 3.17.
$2B_v\tau \gg B_n\tau$	Use $B_n\tau$ in Fig. 3.17.
$B_n\tau \cong 2B_v\tau$	Use $B_n\tau/\sqrt{2} \cong \sqrt{2}\,B_v\tau$ in Fig. 3.17.
Filter matched to spectral envelope	$L_{mv} \cong 1.0$

width preceding the range gate. The loss for this case is calculated as shown in Table 2.5, where B_v is the video bandwidth.

Filtering After Range Gate

Where the range gate is made broad enough to pass more than one noise sample, its width will establish an effective video bandwidth equal to $1/2\tau_g$. In such cases, where $2B_v\tau \gg \tau/\tau_g \ge 1$, the term τ/τ_g will replace $2B_v\tau$ in calculation of L_{mv}. Resulting loss calculations are shown in Table 2.6.

Table 2.6

LOSS FOR MEDIUM RANGE GATE, $1/2B_v\tau \ll \tau_g/\tau \le 1$

Bandwidth and Gate-Width Relationships	Matching Loss
$B_n\tau \gg \dfrac{\tau}{\tau_g} \ge 1$	$L_{mv} = \dfrac{\tau}{\tau_g}$
$\dfrac{\tau}{\tau_g} \gg B_n\tau > 2$	$L_{mv} = B_n\tau$
$\dfrac{\tau}{\tau_g} \gg B_n\tau$	Use $B_n\tau$ in Fig. 3.17.
$B_n\tau \simeq \dfrac{\tau}{\tau_g} \ge 1$	Use $\dfrac{B_n\tau}{\sqrt{2}} = \dfrac{\tau}{\sqrt{2}\,\tau_g}$ in Fig. 3.17.
$\dfrac{\tau}{\tau_g} < 1$	See below.

Range-Gate Collapsing Loss L_c

Where the range gate is broad enough to permit integration of noise samples adjacent to the signal pulse, the loss will be dependent upon the characteristics of the detector, the normalization circuit time-constant, location of the gate (in IF or video portion of the processor), and S/N ratio. Several significant cases are listed in Table 2.7, from which the performance of most practical systems can be estimated.

Normalization Error for Off-Axis Targets

The normalization circuits are intended to adjust system gain in accordance with target signal amplitude. When thermal noise is present, it will introduce error in the gain setting, thereby changing the apparent position of targets which lie off the antenna axis. The case of rapid normalization, in which the target signal amplitude is measured over the same period as the target position, has been analyzed by Sharenson (1962). His analysis shows the presence of an additional noise

Table 2.7

LOSS FOR WIDE RANGE GATE, $\tau_g/\tau > 1$

Processing System	Collapsing Loss
Phase-sensitive error detector, IF gain held constant during gate, IF or video range gate, $B_n\tau_g \gg 1$	$L_c = \dfrac{\tau_g}{\tau}$
Gating at IF, followed by IF filter, $B_n\tau_g < 1$, all detectors	$L_c = \dfrac{\tau_g}{\tau}$ (use also in calculating L_x)
Linear envelope detector, IF gain constant during gate, IF or video gate, $B_n\tau_g \gg 1$, $S/N \gg 1$	$L_c = 1 + \dfrac{\tau_g - \tau}{\tau}\left(2 - \dfrac{\pi}{2}\right)$
Same, but $S/N \leq 1$	$L_c = \dfrac{\tau_g}{\tau}$
Envelope detector followed by circuit to hold peak value, $S/N > 10$	$L_c \cong 1$
Phase-sensitive error detector following IAGC or hard-limiting normalizer, IF gate, $B_n\tau_g \gg 1$ (also video gate, $B_v\tau \gg 1$)	$L_c \cong 1 + \dfrac{\tau_g - \tau}{\tau}\left(\dfrac{S}{N}\right)$

(*Note:* Systems with IAGC or hard limiting must use $\tau_g/\tau \leq 1$ to avoid serious degradation.)

term at the output of the measurement channel, proportional to the off-axis angle θ, for targets within the linear region of error slope ($\theta < \theta_3/2$).

Monopulse Error for Off-Axis Targets

$$\frac{\sigma_\theta}{\theta_3} = \frac{\sqrt{\sigma_1^2 + \sigma_2^2}}{\theta_3} = \frac{\sqrt{L_\theta[1 + (k_m\theta/\theta_3)^2]}}{k_m\sqrt{\mathscr{R}_m}}$$

$$(2.60)$$

The first term, σ_1, is similar to the error of Eq. (2.34) but is increased by the presence of the loss L_θ, which describes the reduced signal received from the off-axis target:

$$\frac{\sigma_1}{\theta_3} = \frac{\sqrt{L_\theta}}{k_m\sqrt{\mathscr{R}_m}} = \frac{1}{k_m\sqrt{\mathscr{R}}}. \tag{2.61}$$

The second term, σ_2, represents modulation of the error signal by noise in the normalization process:

$$\frac{\sigma_2}{\theta_3} = \frac{(\theta/\theta_3)\sqrt{L_\theta}}{\sqrt{\mathscr{R}_m}} = \frac{(\theta/\theta_3)}{\sqrt{\mathscr{R}}}. \tag{2.62}$$

The loss L_θ for the one-way case may be expressed as $G_m/G(\theta)$, while for two-way

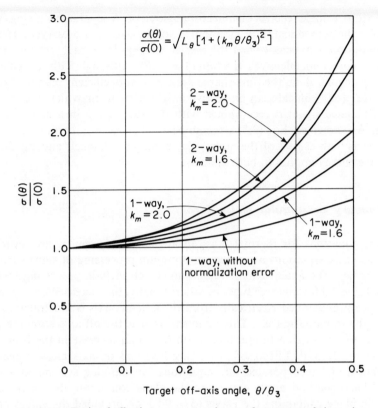

Fig. 2.12   Ratio of off-axis error to on-axis error for a target of given size.

(radar echo) operation it is the square of this value. Figure 2.12 shows the ratio of error for an off-axis target to the error which would be observed for the same target on the axis (equal $\mathscr{R}_m$).

When the gain-normalizing circuits use a time-constant shorter than that used for the measurement itself, the amplitude and rate of change of modulation will both increase, but the output standard deviation, after low-pass filtering, will remain essentially the same. When a longer time-constant is used in the normalization circuit, such that the noise cannot induce a change in gain during the period of measurement, there will be no additional noise component σ_2, and the thermal noise at the output will follow the lowest curve of Fig. 2.12. However, there may be other, larger errors in the output, such as those caused by target scintillation, or bias error resulting from improper calibration of the error curve at the selected gain setting.

Time-Shared Monopulse Systems

Where the effect of thermal noise on angular measurement is not critical, the two difference channels of the antenna may be combined by time-sharing or com-

mutation into a single receiver channel for angle error. A separate receiving channel
is used for the sum signal, and monopulse normalization is preserved. The effect
of commutation is to remove half the received energy from each coordinate measurement channel (an effective reduction of n or $\mathcal{R}$). Although 3 db less efficient in
recovering angular data, the time-shared system is more efficient than the "pseudo-
monopulse" (or "conical-scan-on-receive-only") system, in which the two-coordinate difference signal is recombined with the sum signal in a single receiving
channel. As shown in Eq. (2.51), the slope of the one-way conical-scan system
cannot be as high as that of the monopulse difference channel, and the 3-db loss
of effective energy applies to both systems.

2.7 Summary of Angular Noise Error

The basic equations for thermal noise in angular measurement are repeated in
Table 2.8. These equations are based on optimum processing of signals from targets on or near the antenna axis. Equations which include processing losses are
given in Table 2.9 for all three types of radar. Finally, the increased error produced
when a beam is scanned electrically from the antenna axis is illustrated in Table
2.10, for the monopulse case. This table shows that the off-axis scanning case is
equivalent to a reduction in aperture width (or to an increase in the beamwidth)
by the factor $1/\cos\theta$. Effects of processing loss and off-axis measurement may
be combined and incorporated into equations for scanning systems in a similar
fashion. The equations apply equally in the plane containing the x-axis ($\phi = 0$)
and in the plane containing the y-axis ($\phi = 90$ deg), provided the values of slope
factors and rms widths are chosen in the proper axis.

Summary of Procedure: Monopulse Systems

(a) If the antenna area and other radar-target parameters are known, find the
energy ratio $\mathcal{R}_o$ for a perfectly matched system from Eq. (1.22), or from a corresponding expression for beacon signal reception. If the on-axis signal-to-noise
ratio $(S/N)_m$ is the starting point, calculate $\mathcal{R}_m$ from Eqs. (1.30)–(1.33).

(b) Then, if the difference-channel illumination, $g_d(x, y)$, is known, find the
relative difference slope K from Eq. (2.23) and calculate the error for optimum
signal processing from Eq. (2.27).

(c) If the type of monopulse horn feed is known instead of the illumination
function, estimate K from Hannan's results (Figs. 2.5 to 2.7).

(d) If only the sum-channel beamwidth θ_3 and the sidelobe ratio G_{sr} are known,
and the system has been designed for minimum tracking error, assume an optimum
(derivative) difference pattern and read the slope k_m from Fig. 2.4. Calculate error
from Eq. (2.34).

(e) If measured sum and difference patterns are available, find k_m graphically,
using Eq. (2.9). In the absence of specific data, assume $k_m \cong 1.5$.

Table 2.8

BASIC ERROR EQUATIONS

Ideal

$$(\sigma_\theta)_{min} = \frac{\lambda}{\mathscr{L}_o\sqrt{\mathscr{R}_o}} = \frac{1}{K_o\sqrt{\mathscr{R}_o}} \tag{2.26}$$

Monopulse

$$\sigma_\theta = \frac{1}{K\sqrt{\mathscr{R}_o}} = \frac{\sqrt{\eta_a}}{K\sqrt{\mathscr{R}_m}} \tag{2.27}$$

Monopulse (normalized to beamwidth)

$$\frac{\sigma_\theta}{\theta_3} = \frac{1}{k_m\sqrt{\mathscr{R}_m}} \tag{2.34}$$

Search (one-way)

$$\frac{\sigma_\theta}{\theta_3} = \frac{\lambda}{\mathscr{L}_s\theta_3\sqrt{\mathscr{R}}} = \frac{\sqrt{L_{p1}}}{k_{p1}\sqrt{\mathscr{R}_m}} \cong \frac{0.5}{\sqrt{n(S/N)_m}} \tag{2.43}, (2.45)$$

Search (two-way)

$$\frac{\sigma_\theta}{\theta_3} = \frac{\lambda}{\mathscr{L}_{s2}\theta_3\sqrt{\mathscr{R}}} = \frac{\sqrt{L_{p2}}}{k_{p2}\sqrt{\mathscr{R}_m}} \cong \frac{0.5}{\sqrt{n(S/N)_m}} \tag{2.44}, (2.45)$$

Conical Scan

$$\frac{\sigma_\theta}{\theta_3} = \frac{\sqrt{2L_k}}{k_s\sqrt{\mathscr{R}_m}} \tag{2.46}$$

Assumptions

Beam axis is normal to aperture: $\theta \ll 1$ rad
Optimum processing: $S/N > 10$, $L_{mv} \cong 1.0$, $L_c \cong 1.0$
No normalization error
Energy ratio and rms error evaluated over the same measurement period, t_o:
   $\mathscr{R} = 2B_n t_o(S/N)_{av} = 2n(S/N)$

Brief Definitions of Symbols

   σ_θ = rms error in angle

   λ = wavelength

   $\mathscr{R}_o$ = energy ratio with uniform illumination

   $\mathscr{L}_o$ = rms aperture area width

   K = difference slope

   $K_o = \mathscr{L}_o/\lambda$ = maximum possible value of K

   η_a = aperture efficiency

   $\mathscr{R}_m$ = energy ratio for on-axis target

   θ_3 = half-power beamwidth

   k_m = normalized monopulse slope

   $\mathscr{L}_s$ = rms aperture illumination power width

   $\mathscr{R}$ = received energy ratio

   L_p = beamshape loss in scanning radar⎱ ⎰one-way and two-way values denoted by
   k_p = slope factor for scanning radar⎰ ⎱subscripts

   $(S/N)_m$ = signal-to-noise power ratio for on-axis target

   L_k = crossover loss

   k_s = conical-scan slope

Table 2.9

NORMALIZED TRACKING ERROR FOR NONOPTIMUM PROCESSING

Monopulse

$$\frac{\sigma_\theta}{\theta_3} = \frac{\sqrt{L_x L_{mv} L_c}}{k_m \sqrt{\mathcal{R}_m}} = \frac{\sqrt{(S/N)+1}}{k_m(S/N)} \times \frac{\sqrt{L_{mv} L_c}}{\sqrt{2n L_m}}$$

Monopulse, wideband IF $(L_m = B_n\tau > 2)$

$$\frac{\sigma_\theta}{\theta_3} = \frac{\sqrt{(S/N)+1}}{k_m(S/N)} \times \frac{\sqrt{L_{mv} L_c}}{\sqrt{2n B_n\tau}}$$

Monopulse, narrowband IF $(L_m = L_{mv},\ B_n\tau < 1,\ B_n\tau < 2B_v\tau)$

$$\frac{\sigma_\theta}{\theta_3} = \frac{\sqrt{(S/N)+1}}{k_m(S/N)} \times \frac{\sqrt{L_c}}{\sqrt{2n}}$$

Conical Scan (in terms of beam axis $\mathcal{R}_m$)

$$\frac{\sigma_\theta}{\theta_3} = \frac{\sqrt{2 L_k L_x L_{mv} L_c}}{k_s \sqrt{\mathcal{R}_m}}$$

Conical Scan [in terms of beam axis $(S/N)_m$]

$$\frac{\sigma_\theta}{\theta_3} = \frac{\sqrt{2(S/N)_m + L_k}}{k_s(S/N)_m} \times \frac{\sqrt{L_k L_{mv} L_c}}{\sqrt{2n L_m}}$$

Conical Scan (in terms of tracking axis S/N)

$$\frac{\sigma_\theta}{\theta_3} = \frac{\sqrt{2(S/N)+1}}{k_s(S/N)} \times \frac{\sqrt{L_{mv} L_c}}{\sqrt{2n L_m}}$$

Search Radar (normally $L_m \cong L_{mv}$, since optimum gates cannot be used)

$$\frac{\sigma_\theta}{\theta_3} = \frac{\sqrt{(S/N)_m + L_p}}{k_p(S/N)_m} \times \frac{\sqrt{L_p L_{mv} L_c}}{\sqrt{2n L_m}} \simeq \frac{0.5\sqrt{(S/N)_m + L_p}}{(S/N)_m} \times \sqrt{\frac{L_c}{n}}$$

Further Definitions

> L_x = detector loss
>
> L_{mv} = video filter loss
>
> L_c = collapsing loss
>
> (S/N) = IF signal-to-noise ratio
>
> L_m = IF filter loss
>
> B_n = IF filter noise bandwidth
>
> τ = pulsewidth
>
> n = number of pulses integrated
>
> B_v = video filter noise bandwidth

(f) Estimate processing losses L_x, L_{mv}, and L_c from Sec. 2.6, and apply according to Table 2.9. If off-axis measurement is involved, apply Fig. 2.12 and Table 2.10.

(g) If the final error σ_θ exceeds $\theta_3/6$, the target will get beyond the linear slope region, and tracking will not be possible.

Table 2.10

ERROR FOR OFF-AXIS TARGETS

Monopulse, target off beam axis ($L_\theta = G_m/G(\theta)$ *for one-way,* $[G_m/G(\theta)]^2$ *for two-way*)

$$\frac{\sigma_\theta}{\theta_3} = \frac{\sqrt{L_\theta[1 + (k_m\theta/\theta_3)^2]}}{k_m\sqrt{\mathscr{R}_m}} = \frac{\sqrt{1 + (k_m\theta/\theta_3)^2}}{k_m\sqrt{\mathscr{R}}} \tag{2.60}$$

Monopulse, beam at angle θ from antenna axis

$$\sigma_{(\sin\theta)} = \frac{1}{K\sqrt{\mathscr{R}_o}}$$

$$\sigma_\theta = \frac{1}{K\cos\theta\sqrt{\mathscr{R}_o}}$$

$$\frac{\sigma_\theta}{\theta_3} = \frac{1}{k_m\cos\theta\sqrt{\mathscr{R}_m}}\qquad$$ where θ_3 is evaluated for a beam on the antenna axis and $\mathscr{R}_m$ is the energy ratio from a target at θ

$$\frac{\sigma_\theta}{\theta'_3} = \frac{1}{k_m\sqrt{\mathscr{R}_m}}\qquad$$ where θ'_3 and $\mathscr{R}_m$ are both evaluated for the actual beam at angle θ from the antenna axis

---

Summary of Procedure: Search Radar

(a) If radar parameters are known, find the on-axis ratio $(S/N)_m$ and the number of hits per scan from Eqs. (1.34) and (1.26). If the starting point is a known probability of detection (blip/scan ratio), estimate the corresponding ratios $n(S/N)_m$ and $\mathscr{R}$ from detection curves (e.g., Barton, 1964, Figs. 1.9 and 1.12).

(b) Use the approximation, Eq. (2.45), to calculate error, unless detailed knowledge of illumination function justifies more accurate evaluation of $\mathscr{L}_s$ or slope k_p (Sec. 2.5).

(c) Estimate processing losses L_x, L_{mv}, and L_c from Sec. 2.6 and apply according to Table 2.9.

(d) In the absence of other data, assume $\sigma_\theta = \theta_3/10$, from Eq. (2.45) with $n(S/N)_m = 25$.

Summary of Procedure: Conical-Scan Tracker

(a) Find $\mathscr{R}_m$ from Eq. (1.27), using on-axis antenna gains, or from given $(S/N)_m$ using Eqs. (1.30)–(1.33).

(b) From known offset angle θ_k or crossover loss L_k, find the effective slope $k_s/\sqrt{L_k}$ from Fig. 2.9, for one-way or two-way operation (as applicable). Calculate error from Eq. (2.46).

(c) Estimate processing losses L_x, L_{mv}, and L_c from Sec. 2.6, and apply according to Table 2.9.

(d) If the final error σ_θ exceeds $\theta_3/6$, the target will get beyond the linear slope region, and tracking will not be possible.

In discussing measurement of range, or time delay, we shall make use of the basic concepts presented in Chap. 1, and develop from these a series of equations which parallel those given in Chap. 2 for angular measurement. For simplicity, we will write all equations in terms of time delay and time functions, rather than range. Conversion from time to range is made, for the round-trip case of radar echoes or transponder beacons, by the simple equation

$$R = \frac{t_d c}{2}. \tag{3.1}$$

The same constant $c/2$ relates range error σ_r to time-delay error σ_t.

The velocity of light in a vacuum, c, has been measured to be

$$c = 2.997925 \times 10^8 \text{ m/sec}$$
$$= 9.835692 \times 10^8 \text{ ft/sec}$$
$$= 1.618750 \times 10^5 \text{ naut mi/sec.}$$

Thus, the conversion can also be written as

$$\frac{t_d}{R} = \frac{\sigma_t}{\sigma_r} = \frac{2}{c} = 6.671281 \times 10^{-9} \text{ sec/m}$$
$$= 2.033410 \times 10^{-9} \text{ sec/ft}$$
$$= 12.35521 \times 10^{-6} \text{ sec/naut mi.}$$

Corrections to be applied for paths through the atmosphere are discussed in App. D.

The factors which determine the ability of a radar system to measure time delay between transmission and reception of a particular signal, in a background of thermal noise, are:

1. The bandwidth of the signal, and the spectral density over that band;

2. The relative response of the radar receiver to the various spectral components of the signal;

3. The ratio of the total signal energy, received during the period of measurement, to the receiver noise spectral density; and

4. The method used to process the signal and to extract the "center of gravity," leading edge, or other recognizable feature of the received signal.

In this chapter, we summarize the relationships between signal and receiver bandwidths, spectral distributions, frequency response, signal duration, signal-to-noise ratio, and time-delay error, for commonly-used signal waveforms and measurement systems. The discussion will cover initially the case of a single, band-limited pulse whose energy is large relative to the noise spectral density. Linear processing on the signal envelope will be assumed to recover all the available range information. Later sections will extend the theory to trains of pulses, considering nonlinear effects and practical losses. It will be shown that the theory applies with equal force to continuous differentiating filters ("slope-reversal" method) and to time discriminators operating around a predetermined point in the range sweep ("split-gate" method).

3.1 Radar Signal and Filter Parameters

The theory of time-delay estimation for the case where matched filters are used has been discussed in many of the basic radar references (see, for instance, Woodward, 1953; Skolnik, 1962). In extending this theory to the general case of practical (mismatched) filters, we must introduce separate notations for the input signal, the filter response functions, and the output signal. Although we will develop the procedure for error calculation along the same lines as used in Chap. 2 for angle measurement, the expressions will be complicated by the presence of the separate terms describing the input signal. Table 3.1 summarizes the analogies and differences between angle and time-delay measurements, and between the general case and the matched-filter case for time-delay.

A direct analogy between angle and delay measurements can be drawn in the special case in which a uniform signal spectrum extends over the receiver bandpass. Here, the receiver weighting functions describe the output: $H(f) = A_x(f)/A_m$ becomes analogous to $g(x)$; $a_x(t)$ is analogous to $F(\theta)$; and $\beta_x = \beta_{h1}$ is analogous to $\mathscr{L}_\theta/\lambda$. A different analogy connects angular measurement in the case of a uniformly scanning beam pattern $F(\theta)$ to delay measurement on the corresponding waveform $a(t)$. We will develop below the expressions for the general case, and will apply these to several specific examples.

Figure 3.1 shows a simplified model of the system considered. In an actual system, the received signal would be amplified at IF and a coherent phase detector or an envelope detector would be used to recover the modulation envelope. The system can be divided into two main portions: the elements which generate a marker at some recognizable point in the received pulse, and those which measure the delay of this marker relative to the corresponding point in the transmitted pulse. In this chapter, we are concerned with the first portion, in which the accuracy limit is set by input noise. Since we are not concerned with absolute voltage levels,

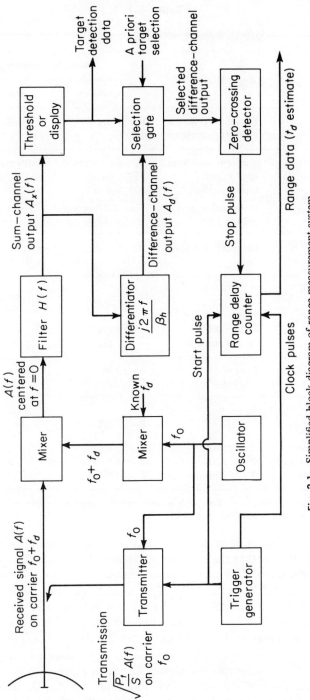

Fig. 3.1   Simplified block diagram of range measurement system.

Table 3.1

COMPARISON OF MEASUREMENT FUNCTIONS

	Angle Measurement	*Time-Delay Measurement*	
		General Case	*Matched Filter*
Coordinates	Distance x across aperture; angle θ	Frequency f; time t	Frequency f; time t
Input signal functions	Uniform wavefront over x; point source in angle	Voltage spectrum $A(f)$; waveform $a(t)$	Voltage spectrum $A(f)$; waveform $a(t)$
Weighting by receiver	Illumination $g(x)$; pattern $F(\theta)$	Transfer function $H(f)$; weighting function $h(t)$	$H(f) = A^*(f)/A_m$ $h(-t) = a^*(t)/A_m$
Output signal functions	Identical to $g(x)$, $F(\theta)$	$A_x(f) = H(f)A(f)$ $a_x(t) = a(t)\otimes h(t)$	$A_x(f) = \lvert A(f)\rvert^2/A_m$ $a_x(t) = a(t)\otimes a^*(-t)/A_m$
Resolution cell (3-db) widths:			
Signal	Source width $\simeq 0$	τ_{3a}	τ_{3a}
Receiver	Beamwidth θ_3	τ_{3h}	$\tau_{3h} = \tau_{3a}$
Output	Beamwidth θ_3	τ_{3x}	$\tau_{3x} \equiv \tau_o$
Derivative weighting	$g_d(x) =$ $(j2\pi x/\mathscr{L}_s)g(x)$	$H_d(f) =$ $(j2\pi f/\beta_h)H(f)$	$H_d(f) =$ $(j2\pi f/\beta_a)A^*(f)/A_m$
Rms widths:*			
Signal power	∞	$\beta_a \leq 1.63/\tau_{3a}$	$\beta_a \simeq 1.63/\tau_o$
Signal voltage	∞	$\beta_{a1} \simeq 1.63/\tau_{3a}$	–
Weighting power	$\mathscr{L}_s \leq 1.63\lambda/\theta_3$	$\beta_h \leq 1.63/\tau_{3h}$	$\beta_h = \beta_a$
Weighting voltage	$\mathscr{L}_\theta \simeq 1.63\lambda/\theta_3$	$\beta_{h1} \simeq 1.63/\tau_{3h}$	–
Output voltage	$\mathscr{L}_\theta \simeq 1.63\lambda/\theta_3$	$\beta_x \simeq 1.63/\tau_{3x}$	$\beta_x = \beta_a$

*Relationships between rms bandwidths and signal durations apply to monochromatic signals (no phase modulation). See Sec. 3.3.

we may consider the receiver to have unity gain at the center of its passband, and to be adjusted so that the sum and difference channels have equal noise outputs to their detectors. A properly phased, coherent detector will then produce a signal output which follows the peaks of the modulated input carrier, and the output signal-to-noise power ratio from the detector will be twice that at the input. In an envelope detector, the same effect will occur at high S/N ratio, since the quadrature component of noise will not affect the detected output. In the equations for noise error, we will use the ratio $2S/N$, or the doubled energy ratio $\mathscr{R} = 2E/N_o$, which accounts for this rejection of quadrature noise.

The general form of the signal and filter functions can be seen in Fig. 3.2. The signal and sum-channel functions have even symmetry about $f = 0$ and $t = 0$.

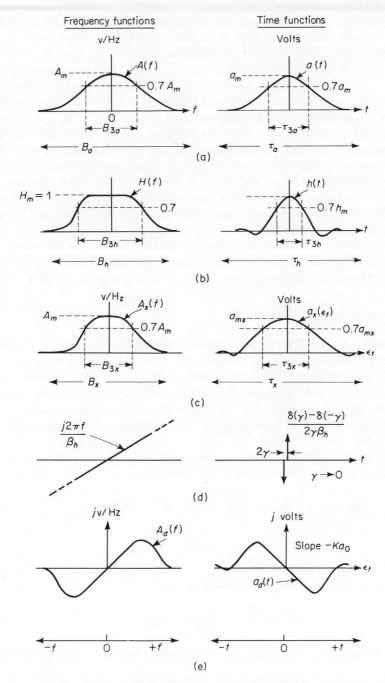

Fig. 3.2 Signal and filter functions: (a) input signal envelope; (b) filter function; (c) sum-channel output signal; (d) ideal differentiator; (e) difference-channel output signal.

The differentiated output has odd symmetry. We introduce a time-error variable ϵ_t to denote time relative to the peak of the filter output, which is delayed somewhat beyond the range delay t_d. Thus, the expressions given in Chap. 1 for response functions in terms of t_d will be written here in terms of ϵ_t. In the later discussion of the split-gate circuit, ϵ_t will denote the displacement of the gate relative to the signal.

In describing the total bandwidths and durations, B_a and τ_a, B_h and τ_h, or B_x and τ_x, we must note that a band-limited signal, in principle, extends over infinite time. Similarly, a time-limited signal extends over infinite bandwidth. For practical measurement purposes, however, we can choose to ignore that portion of the spectrum or waveform which falls far below maximum amplitude (e.g., 0.01, or perhaps 0.001 in extreme cases). Because the total widths of the functions cannot be described exactly, we will make use of half-power (3-db) widths, denoted by the subscript 3, and equivalent (noise) widths, denoted by subscript n. The equivalent widths are defined below.

Initially, the functions will be written as complex modulation functions, with $f = 0$ representing the carrier. When we discuss applications, $a(t)$ will usually be a real waveform, an exception being made for uniform-spectrum chirp signals.

Description of Input Signal

The received signal from a point target in free space is a replica of the transmission, greatly attenuated, delayed in time, and shifted in frequency by the Doppler effect. We assume here that the Doppler shift can be removed, along with the carrier frequency, in the input mixer, to leave the low-frequency modulation function which describes the signal envelope and phase. Initially, we will consider a single, nonrepetitive waveform of finite energy $E_1 \gg N_o$, where N_o is the spectral density of input noise.

The functions which describe the signal were given in Table 3.1. The noise bandwidth and equivalent signal duration are defined by

$$B_{na} \equiv \frac{1}{A_m^2} \int_{B_a} |A(f)|^2 \, df \quad \text{(Hz)}, \tag{3.2}$$

$$\tau_{na} \equiv \frac{1}{a_m^2} \int_{\tau_a} |a(t)|^2 \, dt \quad \text{(sec)}. \tag{3.3}$$

The signal energy, for a carrier modulated by $a(t)$, is

$$E_1 = \frac{A_m^2 B_{na}}{2} = \frac{a_m^2 \tau_{na}}{2} \quad \text{(W-sec)}. \tag{3.4}$$

The "peak" signal power (mean-square voltage at the maximum of the signal envelope) is

$$S_a = \frac{a_m^2}{2} = \frac{E_1}{\tau_{na}} \quad \text{(W)}. \tag{3.5}$$

Detectability and measurement capability for the signal are dependent upon the energy ratio $\mathscr{R}_1 = 2E_1/N_o$.

Description of Receiving Filter

The response of the receiving filter in the sum channel of the ranging system is described by a transfer function in the frequency domain and an impulse response or weighting function in the time domain (see Table 3.1). We will refer our filter functions to maxima at $f = 0$ and $t = 0$, although these reference points may be shifted to other values by appropriate circuit elements. The noise bandwidth and the equivalent duration of the weighting function are defined by

$$B_{nh} \equiv \int_{B_h} |H(f)|^2 \, df \quad (\text{Hz}), \tag{3.6}$$

$$\tau_{nh} \equiv \frac{1}{h_m^2} \int_{\tau_h} |h(t)|^2 \, dt \quad (\text{sec}). \tag{3.7}$$

Unity gain of the filter is provided at low frequencies:

$$H_m = H(0) = 1,$$

$$\int_{\tau_h} h(t) \, dt = H_m = 1,$$

$$h_m^2 = \frac{B_{nh}}{\tau_{nh}} H_m^2 = \frac{B_{nh}}{\tau_{nh}}. \tag{3.8}$$

This last relationship follows from Eqs. (3.6) and (3.7), using Parseval's theorem.

Description of Filter Output

The signal output of the receiving filter is described by the spectrum A_x and the waveform a_x, from Table 3.1. The waveform is the *response function* $\psi(\epsilon_t, 0)$ of the system for zero frequency offset (see Chap. 1), the Doppler shift having been removed:

$$a_x(\epsilon_t) = \psi(\epsilon_t, 0) = \int_{-\infty}^{\infty} H(f)A(f) \exp(j2\pi f \epsilon_t) \, df$$

$$= \int_{-\infty}^{\infty} h(\epsilon_t - t)a(t) \, dt. \tag{3.9}$$

The maximum output voltage, at the point which defines $\epsilon_t = 0$, is given by

$$a_{mx} = a_x(0) = \int_{-\infty}^{\infty} H(f)A(f) \, df = \int_{-\infty}^{\infty} h(-t)a(t) \, dt. \tag{3.10}$$

The maximum signal power at the output is

$$S_x = \frac{a_{mx}^2}{2}, \tag{3.11}$$

and the average output noise power is

$$N = N_o B_{nh}. \tag{3.12}$$

The calculation of IF signal-to-noise ratio from the radar equation gives S_x/N directly when a filter matching loss is included as in Eqs. (1.34) and (1.35). For simplicity, we will drop the subscript and use S/N to refer to this output ratio, unless another point of reference is specified.

Filter Efficiency

The parameters for the special case of the matched filter are shown in Table 3.2. The transfer function of this filter is the complex conjugate of the signal spectrum, and leads to the largest possible value of S/N at the output. By using this ratio as a reference, the efficiency of any other type of filter may be defined, for the single-pulse case, as

$$\eta_f \equiv \frac{2(S_x/N)}{\mathscr{R}_1} = \frac{\psi^2(0,0)/N}{\psi_o^2(0,0)/N_o B_{na}} = \frac{a_{mx}^2/N_o B_{nh}}{a_o^2/N_o B_{na}} = \frac{a_{mx}^2}{2E_1 B_{nh}}$$

$$= \frac{\left| \int_{B_x} A(f)H(f)\,df \right|^2}{\int_{B_a} |A(f)|^2\,df \int_{B_h} |H(f)|^2\,df}. \tag{3.13}$$

Table 3.2

SPECIAL CASE: THE MATCHED FILTER

$$H(f) = \frac{A^*(f)}{A_m} \qquad A_x(f) = \frac{|A(f)|^2}{A_m}$$

$$B_a = B_h = B_x$$

$$\tau_x = 2\tau_a = 2\tau_h$$

$$\tau_{3x} \equiv \tau_o$$

$$\tau_{na} = \tau_{nh}$$

$$a_{mx} \equiv a_o = A_m B_{na} = \frac{2E_1}{A_m}$$

$$S_x = \frac{a_o^2}{2} = E_1 B_{na}$$

$$N = N_o B_{na}$$

$$\frac{2S_x}{N} = \frac{2E_1}{N_o} = \mathscr{R}_1$$

$$\eta_f = 1$$

The reciprocal of η_f is the *filter matching loss*, L_m, used in Chap. 1. The filter efficiency, in range measurement, is analogous to aperture efficiency η_a in Eq. (2.3), for cases in which a uniform signal spectrum exists over $B_a = B_h$. This corresponds to the appearance of a uniform wavefront across the aperture of an antenna.

Error Sensitivity

Time-delay error sensitivity of the system is defined in terms of derivatives of the difference-channel output, as in Table 3.3. In our model, we have combined all the filtering into a single, sum-channel filter $H(f)$, and have obtained the difference signal by ideal differentiation of the sum-channel output, $a_x(t)$. The normalization factor $1/\beta_h$ has been included to make the rms noise voltage out of the difference channel equal to that leaving the sum channel, and to express it in the dimension of volts. The slope factors have been defined in a way analogous to the angular measurement factors, using the same symbols.

Table 3.3

DIFFERENCE-CHANNEL DESCRIPTION

$$A_d(f) = \frac{j2\pi f}{\beta_h} A_x(f) = \begin{cases} \text{amplitude spectrum of} \\ \quad \text{difference signal} \end{cases} \quad \text{(V/Hz)}$$

$$a_d(\epsilon_t) = \frac{1}{\beta_h} \frac{da_x}{d\epsilon_t} = \begin{cases} \text{voltage waveform of} \\ \quad \text{difference signal} \end{cases} \quad \text{(V)}$$

$$N_d = N = N_o \, B_{nh} = \text{Mean square output noise} \quad \text{(W)}$$

$$\sqrt{\frac{B_{nh}}{B_{na}}} = \begin{cases} \text{Ratio of actual output noise voltage to} \\ \quad \text{matched-filter value} \end{cases}$$

$$K \equiv \frac{-\sqrt{B_{na}/B_{nh}}}{a_o} \frac{da_d}{d\epsilon_t}\Big|_{\epsilon_t=0} = \text{Relative difference slope} \ (s^{-1})$$

$$K_o \equiv \text{Maximum possible value of } K \text{ for given signal}$$

$$K_r \equiv \frac{K}{K_o} = \text{Difference slope ratio}$$

$$= \frac{(\sigma_t)_{min}}{\sigma_{t1}} \text{ for given signal of energy } E_1$$

Three "rms bandwidth" quantities may be defined to describe the signals and the filter. These, in turn, are related to the second derivatives of three waveforms which can be generated by elements of the system, and can be used to calculate error sensitivity.

$$\beta_a \equiv \left[\frac{\displaystyle\int_{B_a} (2\pi f)^2 \, |A(f)|^2 \, df}{\displaystyle\int_{B_a} |A(f)|^2 \, df} \right]^{1/2} = \begin{cases} \text{rms width of signal} \\ \text{energy spectrum, as} \\ \text{in Woodward (1953)} \end{cases} \quad (3.14)$$

$$\beta_a^2 = -\frac{1}{a_o}\frac{d^2 a_x}{d\epsilon_t^2}\Big|_{\epsilon_t=0} = \begin{cases}\text{normalized second derivative of} \\ \text{the actual signal passed through} \\ \text{a matched filter}\end{cases} \quad (3.15)$$

$$\beta_h \equiv \left[\frac{\int_{B_h}(2\pi f)^2\,|H(f)|^2\,df}{\int_{B_h}|H(f)|^2\,df}\right]^{1/2} = \begin{cases}\text{rms width of squared} \\ \text{filter transfer function} \\ \text{(analogous to }\mathscr{L}_s\text{)}\end{cases} \quad (3.16)$$

$$\beta_h^2 = \frac{-1}{a_{mx}}\frac{d^2 a_x}{d\epsilon_t^2}\Big|_{\epsilon_t=0} = \begin{cases}\text{normalized second derivative of} \\ \text{output when the input is a signal} \\ \text{matched to the filter}\end{cases} \quad (3.17)$$

$$\beta_x \equiv \left[\frac{\int_{B_x}(2\pi f)^2\,|A(f)H(f)|\,df}{\int_{B_x}|A(f)H(f)|\,df}\right]^{1/2} = \begin{cases}\text{rms width of actual} \\ \text{output voltage spectrum}\end{cases} \quad (3.18)$$

$$\beta_x^2 = \frac{-1}{a_{mx}}\frac{d^2 a_x}{d\epsilon_t^2}\Big|_{\epsilon_t=0} = \begin{cases}\text{normalized second derivative} \\ \text{of the actual signal output} \\ \text{of the filter}\end{cases} \quad (3.19)$$

The term β_h is also the normalization factor for the difference channel, giving the ratio of rms differentiator output noise to input noise.

The relative difference slope K, which is a measure of error sensitivity, was defined in Table 3.3. It represents the ratio of difference-channel voltage slope to rms noise, normalized to the voltage signal-to-noise ratio of a matched filter (see Table 3.2). This is equivalent to expressing the voltage slope as a fraction of a_o per unit of angular error, and the noise voltage as a fraction of $\sqrt{N_o B_{na}}$. The value of K can be expressed directly in terms of rms bandwidths and filter efficiency.

$$\boxed{\begin{array}{c}\textit{Relative Difference Slope} \\[4pt] K = \dfrac{-\sqrt{B_{na}/B_{nh}}}{a_o\beta_h}\dfrac{d^2 a_x}{d\epsilon_t^2}\Big|_{\epsilon_t=0} = \dfrac{a_{mx}}{a_o}\sqrt{\dfrac{B_{na}}{B_{nh}}}\dfrac{\beta_x^2}{\beta_h} = \dfrac{\beta_x^2}{\beta_h}\sqrt{\eta_f}\end{array}} \quad (3.20)$$

The difference slope ratio can also be expressed in terms of rms bandwidths and efficiency,

$$K_r \equiv \frac{K}{K_o} = \frac{\beta_x^2}{\beta_a\beta_h}\sqrt{\eta_f}, \quad (3.21)$$

where the ideal (matched-filter) difference slope is simply $K_o = \beta_a$. These expressions are useful in cases where the signal and filter functions are known in the frequency domain. Values of β_a and β_h may be obtained directly from tables given below and in the appendices. Both β_x and η_f must be calculated from each combination of signal spectrum and filter transfer function, or from the known output waveform using Eq. (3.19).

3.2 Expressions for Noise Error

Ideal Case (Matched Filter)

The expression derived by Woodward and given in Chap. 1 as Eq. (1.19) defines the minimum error in time-delay measurement for a given signal spectrum. It applies when delay is measured to the peak output from a matched filter.

$$\boxed{\begin{array}{c} \textit{Ideal Minimum Error} \\ (\textit{Matched Filter}) \\[6pt] (\sigma_t)_{min} = \dfrac{1}{\beta_a \sqrt{\mathscr{R}}} = \dfrac{1}{K_o \sqrt{\mathscr{R}}} \end{array}} \tag{3.22}$$

Here, $\mathscr{R}$ is the ratio of twice the total signal energy entering the receiver to noise power per unit bandwidth, and β_a is defined as in Eq. (3.14). Since the matched filter integrates all the received energy, over as many pulses as are received in the measurement period, this expression is applicable to either single-pulse or multiple-pulse measurements provided all pulses have the same spectrum. Values

Table 3.4

SIGNAL AND FILTER PARAMETERS

Spectrum $A(f)$, or Transfer Function $H(f)$	$B_a\tau_{3a}$, $B_h\tau_{3h}$	B_{3a}/B_a, B_{3h}/B_h	B_{na}/B_a, B_{nh}/B_h	β_a/B_a, β_h/B_h	β_{a1}/B_a, β_{h1}/B_h	$\beta_a\tau_{3a}$, $\beta_h\tau_{3h}$	$\beta_{a1}\tau_{3a}$, $\beta_{h1}\tau_{3h}$
Rectangular, $A(f) = 1$	0.886	1.000	1.000	1.81	1.81	1.607	1.607
Triangular, $A(f) = 1 - |2f/B|$	1.276	0.293	0.333	0.994	1.28	1.268	1.636
Parabolic, $A(f) = 1 - 2f^2/B^2$	0.972	0.765	0.718	1.53	1.66	1.481	1.613
Parabolic, $A(f) = 1 - 4f^2/B^2$	1.155	0.541	0.533	1.188	1.407	1.372	1.624
Cosine, $A(f) = \cos(\pi f/B)$	1.189	0.500	0.500	1.136	1.370	1.350	1.629
Cosine^2, $A(f) = \cos^2(\pi f/B)$	1.441	0.367	0.375	0.89	1.134	1.283	1.636
Cosine^4, $A(f) = \cos^4(\pi f/B)$	1.853	0.263	0.313	0.669	0.886	1.240	1.645
Gaussian, $A(f) = \exp(-f^2/2\sigma_a^2)$ (assume $B = 6\sigma_a$)	* 1.59	* 0.278	* 0.295	* 0.74	* 1.048	1.177 1.177	1.665 1.665

*For Gaussian spectrum, $\tau_{3}\sigma_a = \dfrac{\sqrt{\ln 2}}{\pi}$, $B_3/\sigma_a = 2\sqrt{\ln 2}$, $B_n/\sigma_a = \sqrt{\pi}$, $\beta/\sigma_a = \sqrt{2}\pi$, $\beta_1/\sigma_a = 2\pi$.

of β_a for different signal spectra are given in Table 3.4, and a discussion of the special case of the rectangular pulse will be found in Sec. 3.4 below.

Mismatched Filter

When the receiving filter is not matched to the signal spectrum, the signal-to-noise ratio observed at the filter output, prior to the differentiator, will be decreased and the noise error will be increased. For a single pulse with energy ratio $\mathcal{R}_1$, the error is

$$\boxed{\begin{array}{c} \textit{Time-Delay Error for Mismatched Filter} \\[2mm] \sigma_{t1} = \dfrac{1}{K\sqrt{\mathcal{R}_1}} = \dfrac{\sqrt{\eta_f}}{K\sqrt{2S/N}} = \dfrac{1}{K\sqrt{2L_m S/N}} \end{array}} \tag{3.23}$$

If we express the error slope in terms of the spectral response parameters of the sum-channel filter preceding the differentiator, using Eq. (3.20), we have

$$\sigma_{t1} = \frac{\beta_h}{\beta_x^2 \sqrt{\eta_f \mathcal{R}_1}} = \frac{\beta_h}{\beta_x^2 \sqrt{2S/N}}. \tag{3.24}$$

The efficiency η_f and the rms bandwidth β_x must be found for each signal-filter combination, according to Eqs. (3.13) and (3.19), while β_h is a characteristic of the filter alone, given by Eq. (3.16). For most signal-filter combinations, an approximation may be used to find β_x from the separate signal and filter parameters β_{a1} and β_{h1} given in Table 3.4.

$$\frac{1}{\beta_x^2} \cong \frac{1}{\beta_{a1}^2} + \frac{1}{\beta_{h1}^2} \tag{3.25}$$

$$\beta_{a1} \equiv \left[\frac{\int_{B_a} (2\pi f)^2 |A(f)| \, df}{\int_{B_a} |A(f)| \, df} \right]^{1/2} = \begin{cases} \text{rms width of signal} \\ \text{amplitude spectrum} \end{cases} \tag{3.26}$$

$$\beta_{a1}^2 = -\frac{1}{a_m} \frac{d^2 a}{dt^2} \Big|_{t=0} = \begin{cases} \text{normalized second derivative of} \\ \text{actual signal passed through phase-} \\ \text{compensated, broadband filter} \end{cases} \tag{3.27}$$

$$\beta_{h1} \equiv \left[\frac{\int_{B_h} (2\pi f)^2 |H(f)| \, df}{\int_{B_h} |H(f)| \, df} \right]^{1/2} = \begin{cases} \text{rms width of filter} \\ \text{transfer function} \end{cases} \tag{3.28}$$

$$\beta_{h1}^2 = -\frac{1}{h_m}\frac{d^2h}{dt^2}\bigg|_{t=0} = \begin{cases} \text{normalized second derivative of} \\ \text{impulse response of actual filter} \end{cases} \tag{3.29}$$

3.3 Error Normalized to Pulse Width

Normalization of error and difference slope to the half-power width of the output pulse leads to simplified expressions for "monochromatic" signals (those which contain no intrapulse frequency modulation or coding). This is because the half-power width of such waveforms is closely related to the second derivative at the center of the pulse:

$$\beta_{a1}\tau_{3a} \cong \beta_{h1}\tau_{3h} \cong \beta_x\tau_{3x} \cong 1.63. \tag{3.30}$$

This relationship also applies, within limits, to the case of the rectangular pulse, described in Sec. 3.4. Thus, for the matched-filter case, we have $\beta_a = \beta_x$, $\tau_o = \tau_{3x}$, and

$$\boxed{\begin{array}{c} \textit{Ideal Error Normalized} \\ \textit{to Output Pulse Width} \\[2mm] \dfrac{(\sigma_t)_{min}}{\tau_o} = \dfrac{1}{\beta_a\tau_o\sqrt{\mathscr{R}}} \cong \dfrac{1}{1.63\sqrt{\mathscr{R}}} \end{array}} \tag{3.31}$$

For the general case of the mismatched filter, we have

$$\frac{\sigma_{t1}}{\tau_{3x}} = \frac{1}{K\tau_{3x}\sqrt{\mathscr{R}_1}} \cong \frac{(\beta_h/\beta_x)}{1.63\sqrt{\eta_f\mathscr{R}_1}}. \tag{3.32}$$

The error may also be expressed as a fraction of the input pulse width τ_{3a} by introducing a "pulse broadening factor" $K_h = \tau_{3x}/\tau_{3a}$. This leads to

$$\boxed{\begin{array}{c} \textit{Error Normalized to Input} \\ \textit{Pulse Width} \\[2mm] \dfrac{\sigma_{t1}}{\tau_{3a}} = \dfrac{1}{K\tau_{3a}\sqrt{\mathscr{R}_1}} \cong \dfrac{K_h\beta_h/\beta_x}{1.63\sqrt{\eta_f\mathscr{R}_1}} \end{array}} \tag{3.33}$$

The normalized difference slope $K\tau_{3a}$ is a convenient measure of system performance, which will be computed and plotted below for cases of special interest. Use of this parameter is to be recommended both because it is dimensionless and because it usually has a value only slightly greater than unity for near-optimum systems. Thus, if we have no precise information about signal and filter shape below the 3-db points, a good estimate of error can be made by substituting $K\tau_{3a} = 1$ in any of the error equations.

In the general case, with known signal and filter functions, we may use Eqs. (3.25) and (3.30) to express the normalized slope as a function of filter efficiency and the ratios of parameters listed in Table 3.4:

$$K\tau_{3a} \cong 1.63 \frac{r}{1 + r^2} \frac{\beta_{h1}}{\beta_h} \sqrt{\eta_f}. \tag{3.34}$$

Here, the ratio $r \equiv \beta_{h1}/\beta_{a1}$ and the factor η_f are measures of filter-to-signal matching, while β_{h1}/β_h is a function of filter shape alone (note that $1 \leq \beta_{h1}/\beta_h \leq \sqrt{2}$). The pulse broadening factor is also a function of the ratio r:

$$K_h \cong \frac{\beta_{a1}}{\beta_x} = \sqrt{\frac{1 + r^2}{r^2}}. \tag{3.35}$$

Special Case : Gaussian Spectrum and Filter

The parameters of a monochromatic Gaussian pulse passed through a Gaussian filter are given in Table 3.5. The bandwidth ratio $r = \sigma_h/\sigma_a$ determines the

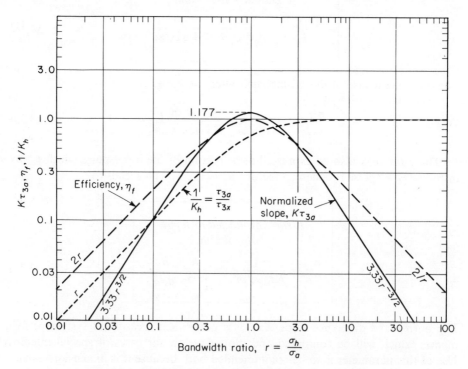

Fig. 3.3   Slope and efficiency for Gaussian pulse and filter.

efficiency, the pulse broadening factor, and the normalized error slope. These factors are plotted in Fig. 3.3. We see that the normalized slope is reduced, for large mismatch, by the $\frac{3}{2}$ power of r or $1/r$, whichever is greater than unity. We can attribute a factor $\sqrt{r}$ or $\sqrt{1/r}$ to the mismatch of the sum-channel filter, resulting in reduced S/N ratio, and the remaining r or $1/r$ to the mismatch of the first derivative response.

Table 3.5

PARAMETERS FOR GAUSSIAN PULSE SPECTRUM AND
GAUSSIAN FILTER

$$\left. \begin{aligned} A(f) &= A_m \exp\left(-\frac{f^2}{2\sigma_a^2}\right) = A_m \exp(-2\pi^2\sigma_t^2 f^2) \\ a(t) &= a_m \exp\left(-\frac{t^2}{2\sigma_t^2}\right) = a_m \exp(-2\pi^2\sigma_a^2 t^2) \end{aligned} \right\} \quad \text{where } \sigma_t = \frac{1}{2\pi\sigma_a}$$

$$H(f) = \exp\left(-\frac{f^2}{2\sigma_h^2}\right) = \exp(-2\pi^2\sigma_c^2 f^2) \quad \text{where } \sigma_c = \frac{1}{2\pi\sigma_h}$$

$$A_x(f) = A_m \exp\left(-\frac{f^2}{2\sigma_x^2}\right) = A_m \exp\left(-\frac{f^2(\sigma_a^2 + \sigma_h^2)}{2\sigma_a^2\sigma_h^2}\right)$$

$$a_x(t) = a_{mx} \exp\left(-\frac{t^2}{2\sigma_o^2}\right) = a_{mx} \exp(-2\pi^2\sigma_x^2 t^2)$$

$$\sigma_a\sigma_t = \sigma_x\sigma_o = \frac{1}{2\pi}$$

$$r \equiv \frac{\beta_{h1}}{\beta_{a1}} = \frac{\sigma_h}{\sigma_a}$$

$$\frac{\sigma_x}{\sigma_a} = \frac{B_{3x}}{B_{3a}} = \frac{\tau_{3a}}{\tau_{3x}} = \frac{1}{K_h} = \sqrt{\frac{r^2}{1+r^2}}$$

$$B_{3h}\tau_{3a} = \frac{2r\ln 2}{\pi} = 0.44\,r$$

$$\eta_f = \frac{2r}{1+r^2}$$

$$K = 4\pi\sigma_a\left(\frac{r}{1+r^2}\right)^{3/2} = \frac{4\sqrt{\ln 2}}{\tau_{3a}}\left(\frac{r}{1+r^2}\right)^{3/2} = \frac{3.33}{\tau_{3a}}\left(\frac{r}{1+r^2}\right)^{3/2}$$

$$K_o = \beta_a = \sqrt{2}\,\pi\sigma_a = \frac{\sqrt{2\ln 2}}{\tau_{3a}} = \frac{1.177}{\tau_{3a}}$$

$$K_r \equiv \frac{K}{K_o} = \left(\frac{2r}{1+r^2}\right)^{3/2}$$

$$\sigma_{t1} = \frac{1}{K\sqrt{\mathscr{R}_1}} = \frac{\tau_{3a}}{3.33}\sqrt{\frac{(1+r^2)^3}{r^3\mathscr{R}_1}}$$

$$(\sigma_t)_{min} = \frac{\tau_{3a}}{1.177\sqrt{\mathscr{R}}} \quad \text{for } r = 1$$

Special Case : Gaussian Spectrum, Rectangular Filter

To illustrate the accuracy of the approximations in Eqs. (3.34) and (3.35), we will compute exact and approximate performance of the rectangular filter on a Gaussian pulse. The efficiency and slope for this case can be expressed in terms of the error function, erf x, which is defined by

$$\text{erf } x = \frac{2}{\sqrt{\pi}} \int_0^x \exp(-t^2)\, dt = \frac{1}{\sqrt{\pi}} \int_{-x}^x \exp(-t^2)\, dt.$$

We let $x = B_h/\sqrt{8}\,\sigma_a$, so that $\exp(-x^2)$ represents the relative spectral density $A(f)/A_m$ at each edge of the filter passband ($f = \pm B_h/2$). Then, from Eq. (3.13),

$$\eta_f = \frac{\left[\dfrac{1}{A_m} \displaystyle\int_{-B_h/2}^{B_h/2} A(f)\, df\right]^2}{\dfrac{B_h}{A_m^2} \displaystyle\int_{-\infty}^{\infty} |A(f)|^2\, df} = \sqrt{\frac{\pi}{2}}\,\frac{(\text{erf } x)^2}{x}.$$

Also, $\beta_{h1} = \beta_h = \pi B_h/\sqrt{3}$, $r = x\sqrt{2/3}$, and

$$K_h = \sqrt{\frac{1.5 + x^2}{x^2}},$$

$$K\tau_{3a} = \sqrt{6\sqrt{\pi/2}\ln 2}\; x^{-3/2}\left[\text{erf } x - \frac{2}{\sqrt{\pi}}\,x\,e^{-x^2}\right].$$

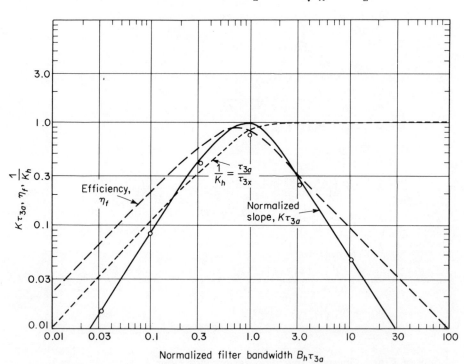

Fig. 3.4 Slope and efficiency for Gaussian pulse, rectangular filter. Circles indicate values of $K\tau_{3a}$ calculated from Eq. (3.34).

These factors are plotted in Fig. 3.4, along with the approximate values, as functions of the normalized bandwidth $B_h \tau_{3a} = 0.75x$. Note the similarity in shape of the curves, when compared to Fig. 3.3. Optimum $K\tau_{3a}$ is at a bandwidth slightly higher than that which gives highest efficiency. The approximate values are in excellent agreement with exact values at all points except near the peak of the curve, where they underestimate the actual slope. When we consider that the rectangular filter is as far removed from the Gaussian as we can get in shape, it is apparent that the approximations based on Eq. (3.25) are good enough for engineering purposes for most combinations of filter shape and pulse waveform.

Special Case : Uniform Spectrum, Pulse Compression

The case of uniform spectral density, $A(f) = A_m$ over B_a, is of interest because it represents the idealized "chirp" and many other pulse-compression signals, and because it is analogous to the antenna case. Table 3.6 gives the special relationships for this case, in which we assume that the filter response does not extend beyond the signal spectrum, and that the input waveform is essentially uniform over τ_a.

Table 3.6

SPECIAL RELATIONSHIPS FOR UNIFORM SPECTRUM,
RECTANGULAR PULSE, $B_a \tau_a \gg 1$

$A(f) = A_m$ over frequency interval $B_a = B_{3a} = B_{na} \geq B_h$

$|a(t)| = a_m$ over time interval $\tau_a = \tau_{3a} = \tau_{na} \leq \tau_h$

$A_x(f) = A_m H(f)$, where the receiving filter is matched to signal phase and has tapered amplitude weighting

$E_1 = \dfrac{A_m^2 B_a}{2} = \dfrac{a_m^2 \tau_a}{2}$

$\beta_a = \beta_{a1} = \dfrac{\pi B_a}{\sqrt{3}} = K_o = \dfrac{1.607}{\tau_o}$

$\beta_x = \beta_{h1}$

$\eta_f = \dfrac{1}{B_a B_{nh}} \left[\int_{B_h} |H(f)| \, df \right]^2$

$(\sigma_t)_{min} = \dfrac{\sqrt{3}}{\pi B_a \sqrt{\mathscr{R}}} = \dfrac{\tau_o}{1.607 \sqrt{\mathscr{R}}}$ (matched filter, $B_a \tau_o = 0.886$)

$\sigma_{t1} = \dfrac{\beta_h}{\beta_{h1}^2 \sqrt{\eta_f \mathscr{R}_1}} = \dfrac{\tau_a(\beta_h/\beta_{h1})}{1.607 K_c \sqrt{\eta_f \mathscr{R}_1}} = \dfrac{\tau_a(\beta_h/\beta_{h1}) K_h}{1.81 D \sqrt{\eta_f \mathscr{R}_1}}$

$K = \dfrac{\left| \int_{B_h} (2\pi f)^2 H(f) \, df \right|}{\left[B_a \int_{B_h} (2\pi f)^2 |H(f)|^2 \, df \right]^{1/2}}$ [see Eq. (2.20)]

$= \dfrac{\beta_{h1}^2 \sqrt{\eta_f}}{\beta_h}$

The only dissimilarity between this case and the antenna lies in the definition of η_f and τ_o, where we have considered signal energy extending over B_a, which may exceed the filter bandwidth B_h. In the antenna case, we considered only that portion of the incident wave which reached the aperture.

For a pulse-compression signal, we define the *compression ratio* K_c as the ratio of input pulse duration to output duration at the -3-db point:

$$K_c \equiv \frac{\tau_a}{\tau_{3x}}.$$

For a matched filter, $\tau_{3x} = \tau_o = 0.886/B_a$, and $(K_c)_{max} = 1.13 B_a \tau_a$. Another term used to describe such signals is the *dispersion factor* $D = B_a \tau_a$ (also known as the *time-bandwidth product*). This is approximately equal to the compression ratio which is attainable with the matched filter, and serves as a more consistent description of the signal, being independent of the receiver processing.

When filter shaping is used to reduce sidelobes, the output pulse is broadened. For the pulse compression case, we may define the pulse broadening factor as was done by Cook (1967):

$$K_h \equiv \frac{\tau_{3x}}{\tau_o} = 1.13 \, B_a \tau_{3x} \geq 1. \tag{3.36}$$

As before, the broadening factor is equal to the ratio of rms bandwidth of the signal input spectrum to that at the filter output:

$$K_h \simeq \frac{\beta_{a1}}{\beta_x} = \frac{\pi B_a}{\sqrt{3} \, \beta_{h1}}. \tag{3.37}$$

This factor may be found from the data in Table 3.4 and in the appendices. Figures 2.2–2.4 may be used to find the various slope and efficiency factors as functions of the time-sidelobe level, by replacing $\mathscr{L}_\theta/\lambda$ with β_{h1} or β_x, $\mathscr{L}_s/\lambda$ with β_h, η_a with $\eta_f B_a/B_h$, and w with B_h. The compression ratio when shaping is used is

$$K_c = \frac{\tau_a}{\tau_{3x}} = 1.13 \frac{B_a \tau_a}{K_h} = 1.13 \frac{D}{K_h}. \tag{3.38}$$

These results are based on the assumption of known signal frequency, or Doppler shift. A brief discussion of combined range-Doppler errors for linear-FM signals is given in Sec. 4.7.

3.4 The Rectangular Pulse

Special treatment of the rectangular pulse is required because its rms bandwidth β_a is infinite. We will describe the limit to range accuracy with optimum processing, as derived by Manasse (1955), and relate this to the accuracy for the band-limited rectangular pulse, as described by Skolnik (1960). A further limit, imposed by noise ambiguity at low energy ratios, has been discussed by Woodward (1953), and will also be described here.

Manasse's Expression

Through the method of inverse probability, Manasse derived the following limit for time-delay error on a rectangular pulse of width τ.

$$\boxed{\begin{array}{c} \textit{Ideal Error Limit on Rectangular Pulse} \\[1em] (\sigma_t)_{min} = \dfrac{\tau\sqrt{2}}{\mathcal{R}} \qquad (B_a\tau \to \infty,\ \mathcal{R} \gg 1) \end{array}} \qquad (3.39)$$

Although the presence of $\mathcal{R}$ to the first power in the denominator appears unusual, in the light of the usual square-root relationship, it can be explained qualitatively as follows. If we use a matched filter and differentiator, the noise bandwidth of the combination will be infinite. We must include a band-limiting filter of width B_h to avoid infinite noise output, and must adjust B_h to avoid ambiguous zero-crossings in the region near the maximum of the matched-filter response. When $\mathcal{R}$ is increased, we obtain benefit in two ways, because the error with a given B_h is decreased, and we are also permitted to use a larger B_h without introducing noise ambiguities. The resulting block diagram of an optimum estimator for the rectangular pulse is shown in Fig. 3.5. We use an "$\mathcal{R}$ estimator" in the sum channel

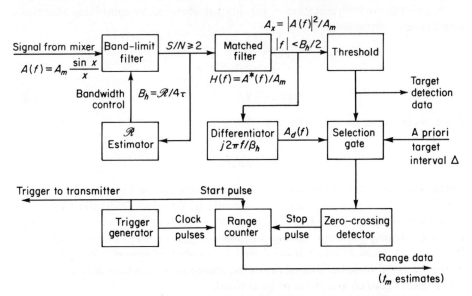

Fig. 3.5 Simplified block diagram of adaptive estimator for rectangular pulse.

to control the band-limiting filter in accordance with observed energy ratio. The zero-crossing selection gate is only permitted to pass a signal when the sum-channel voltage exceeds a given threshold during the interval Δ in which the signal is known, *a priori*, to lie.

Skolnik's Expression

The accuracy attainable on a band-limited rectangular pulse was calculated by Skolnik (1960), who showed that the rms bandwidth β_a could be approximated closely by

$$\beta_a \cong \sqrt{\frac{2B_a}{\tau}} \qquad (B_a\tau \gg 1). \tag{3.40}$$

Using this approximation, the error limit is

> *Ideal Error on Band-Limited Rectangular Pulse*
>
> $$\sigma_t \cong \sqrt{\frac{\tau}{2B_a\mathscr{R}}} \qquad (\text{Matched filter, } B_h = B_a) \tag{3.41}$$

The normalized error σ_t/τ is plotted in Fig. 3.6 as a function of $\mathscr{R}$ for different values of $B_a\tau$ and $\beta_a\tau$. Manasse's limit is also shown as a shaded diagonal line, indicating that B_h must be restricted to values outside the shaded region.

A quantitative interpretation of this limit is obtained by combining Manasse's and Skolnik's equations:

$$\frac{\tau\sqrt{2}}{\mathscr{R}} \leq \sqrt{\frac{\tau}{2B_h\mathscr{R}}},$$

$$\mathscr{R} \geq 4B_h\tau. \tag{3.42}$$

For a single pulse, these lead to

$$\frac{S}{N} = \frac{\mathscr{R}_1}{2B_h\tau} \geq 2,$$

where S/N refers to the signal-to-noise power ratio at the input to the matched filter in Fig. 3.5, and to the ratio of differentiated signal to differentiated noise at the zero-crossing detector. The criterion $S/N > 2$ is necessary to ensure that only one zero crossing will occur during the interval $\pm 3\sigma_t$ centered on the signal peak. When energy from n pulses is combined in a coherent system, $\mathscr{R}/2B_h\tau = nS/N$ is still the signal-to-noise ratio at the differentiator output, which must exceed two if ambiguous zero crossings are to be avoided.

Figure 3.6 also indicates the signal bandwidth which is needed to take advantage

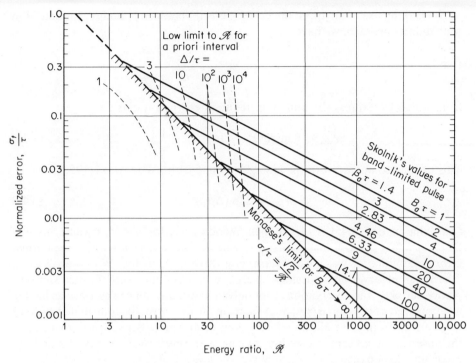

Fig. 3.6   Measurement accuracy vs energy ratio (rectangular pulse).

of Manasse's $1/\mathscr{R}$ relationship for error. If any element in the transmitter, propagation path, or receiver places a limit on $B_a\tau$, the error will follow Skolnik's $1/\sqrt{\mathscr{R}}$ curves for values of $\mathscr{R}$ beyond $4B_a\tau$.

Noise Ambiguity

Also shown on the plot are dashed curves representing the minimum $\mathscr{R}$ needed to avoid noise ambiguity caused by false alarms in the *a priori* interval Δ. When our target information is indefinite at the beginning of the measurement period, we must be prepared to find zero crossings over a relatively large interval Δ, extending for many pulse widths each side of the true position. This makes necessary a larger energy ratio if we are to avoid false alarms far outside the normal distribution of error. The average contribution to the total squared error caused by a false alarm is $\Delta^2/6$, when both targets and false alarms are uniformly distributed over the interval Δ. The component of mean-squared error for a probability P_n of a false alarm within Δ is

$$\sigma^2 = P_n\frac{\Delta^2}{6} = p_n\frac{\Delta}{\tau}\frac{\Delta^2}{6}. \tag{3.43}$$

Here, p_n is the false-alarm probability per resolvable range element τ.

By making use of an approximate relationship between p_n and $\mathscr{R}$, for detection probabilities near 0.5, we have

$$p_n \cong \exp\left(-\mathscr{R}/2\right) = \frac{6\tau\sigma^2}{\Delta^3}, \qquad (3.44)$$

from which we express the minimum required energy ratio as

$$\mathscr{R}_{min} = 6\ln(\Delta/\tau) - 4\ln(\sigma/\tau) - 3.6. \qquad (3.45)$$

This may be compared with the ambiguity limit set by Woodward (1953):

$$\mathscr{R}_{min} = 2\ln\left(\Delta\beta_a\mathscr{R}_{min}\right). \qquad (3.46)$$

Our value in Eq. (3.45) is higher than Woodward's, for $\sigma < \tau$, because his was based on a fifty percent uncertainty as to whether the reading lay within the normal error distribution or was a false alarm elsewhere in Δ. Below Woodward's $\mathscr{R}_{min}$, "the radar observation is useless—except in combination with further signals." Above our $\mathscr{R}_{min}$, the observation is not only of value, but can be described as having a normalized rms error $\sigma/\tau < 1$. A portion of this error is contributed by very large errors from false alarms of very low probability. Between the two limits, the observation has some value, but may have an rms error in excess of τ and approaching $\Delta/\sqrt{6}$.

Rectangular Pulse with Rectangular Filter

For moderate bandwidths of the filter ($B_h\tau < 2$), the performance of a rectangular filter is comparable to that of the band-limited matched filter. In this region, the output pulse is described with reasonable accuracy by the quadratic term in the Taylor expansion about the pulse center, and the relationships of Sec. 3.3 can be applied. The difference slope K, representing the second derivative at the pulse center, provides a measure of system accuracy, and the approximation $\beta_x\tau_{3x} \cong 1.63$ [see Eq. (3.30)] is valid. Figure 3.7 shows the degree to which these expressions and parameters can be used for the rectangular pulse with the rectangular filter and ideal differentiator. As $B_h\tau$ is increased beyond 2, the central peak of the pulse splits to create two peaks with a central dip, and two displaced zero crossings are observed (Fig. 3.8). Beyond this point, the simple differentiator and zero-crossing detector are no longer adequate for range estimation, except in an approximate sense.

When using Fig. 3.7 for $B_h\tau < 2$, we must be careful to distinguish between the total energy of the rectangular pulse and that which lies within the filter passband. The fraction of energy within the filter is indicated by $\mathscr{R}'/\mathscr{R}$ in the figure. It is the smaller energy $\mathscr{R}'$ which is used in Skolnik's expression for accuracy on the band-limited pulse, while the full energy $\mathscr{R}$ is used with the normalized slope $K\tau$. When

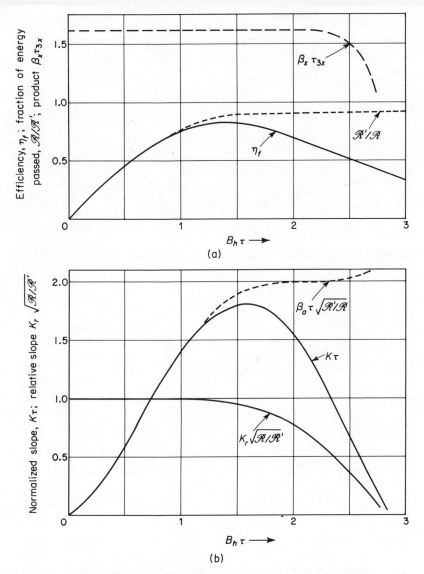

Fig. 3.7 Performance of rectangular filter on rectangular pulse: (a) efficiency and product $B_x\tau_{3x}$ vs $B_h\tau$; (b) slope parameters vs $B_h\tau$.

the rms bandwidth β_a is scaled down in accordance with this ratio, we find that the slope $K\tau$ for the rectangular filter lies within five percent of the matched-filter value, when $B_h\tau \leq 1.5$, and departs only by twenty-five percent for $B_h\tau = 2$ [Fig. 3.7(b)]. As a result, unless we wish to take advantage of the range information contained in the spectral sidelobes of the pulse, the use of a matched filter will not prove especially beneficial when compared to the simple bandpass filter.

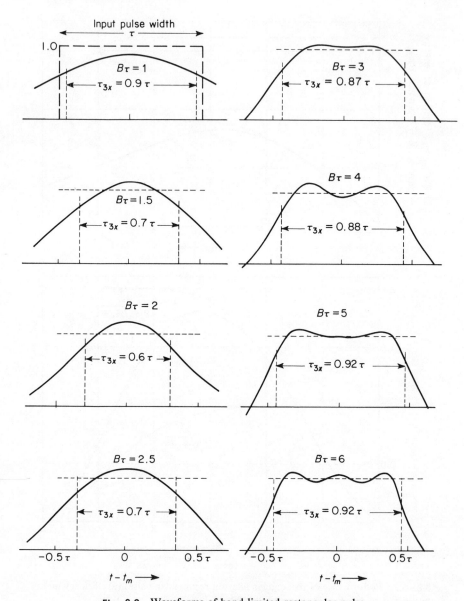

Fig. 3.8 Waveforms of band-limited rectangular pulse.

3.5 The Split-Gate Time Discriminator

Precision tracking radars normally use some form of split-gate circuit as a range or time discriminator. A simplified block diagram is shown in Fig. 3.9, and the waveforms are shown in Fig. 3.10. We see that the curve for discriminator voltage

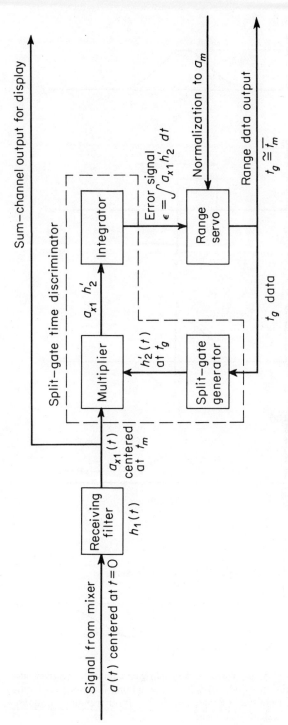

Fig. 3.9 Simplified block diagram of split-gate range tracker.

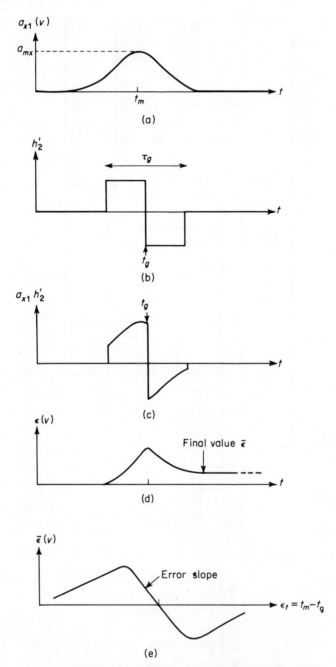

Fig. 3.10 Waveforms in split-gate circuit: (a) receiver output envelope; (b) split-gate weighting function; (c) multiplier output; (d) integrator output; (e) discriminator response to gate displacement.

vs displacement (or time-delay error) has the same form as the output waveform $a_d(\epsilon_t)$ from a differentiating filter, described in the preceding sections. Therefore, if we may continue to neglect the effects of nonlinearities prior to the gate, we may replace the split-gate discriminator with an equivalent linear filter, as in Fig. 3.11. Here, we have divided the discriminator transfer function into an ideal differentiator $j2\pi f/\beta_h$, preceded by a filter function $H_2(f)$ in cascade with the actual sum-channel filter $H_1(f)$. These functions are shown in Fig. 3.12, and described in Table 3.7. It is apparent that the rectangular split-gate function is matched to a triangular input pulse, $a(t)$, whose spectrum has the form $(\sin^2 x)/x^2$, but not to rectangular or other pulse shapes. When the gate width τ_g is equal to or greater than the total pulse width, the system is already subject to an excessive amount of filtering, and the sum-channel filter should be kept wide enough to avoid further degradation (see Sec. 3.7 for a discussion of nonlinear effects on this allocation of filtering).

Obviously, the split-gate input to the multiplier need not be of the rectangular form shown in Fig. 3.10. By choosing a weighting function $h_2(t)$ which better matches the input waveform, and using its first derivative as the multiplier input $h'_2(t)$, we can approach the optimum range estimator. The theory developed in Secs. 3.1–3.4 for linear filters can be applied here to evaluate gating waveforms and to describe the single-pulse errors in range estimation for the split-gate discriminator. The error in the smoothed estimate $t_g \cong \bar{t}_o$ will depend also on the period over which the servo averages the data. This effect will be discussed in Sec. 3.7 as being equivalent to linear filtering over n independent pulse measurements.

Because of the importance of the rectangular-gated time discriminator in practical radar systems, its output characteristics will be explored further here, for pulses of rectangular and Gaussian shape passed through various receiving filters.

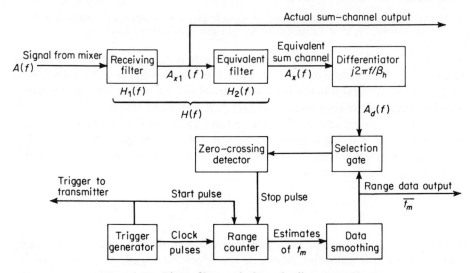

Fig. 3.11   Linear-filter equivalent of split-gate tracker.

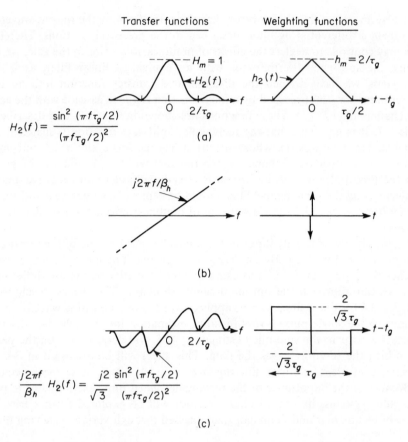

Fig. 3.12 Filter functions equivalent to split-gate discriminator: (a) equivalent sum-channel; (b) ideal differentiator; (c) split-gate discriminator function.

We may dispose first of the cases in which the gate is narrow, relative to the reciprocal of filter bandwidth ($B_{h1}\tau_g \ll 1$). For all practical purposes, the narrow split gate serves as an ideal differentiator for outputs of such a filter, and performance may be calculated without regard to exact gate width and shape.

A more interesting case is when the receiver bandwidth is relatively wide, such that most of the filtering is provided by the width of the gate ($B_{h1}\tau_g \gg 1$). The system performance is then found directly from the time derivative used in Table 3.3 to define the difference slope:

$$K\tau_{3a} = \frac{-\tau_{3a}}{a_o}\sqrt{\frac{B_{na}}{B_{nh}}}\frac{da_d}{d\epsilon_t}\bigg|_{\epsilon_t=0}$$

Since a_d is the convolution of the gate function with the filter output (essentially with the input, for wideband filters), its time derivative is the convolution of the first derivative of the gate function with the signal (Fig. 3.10). This is simply the

Table 3.7

RECTANGULAR SPLIT-GATE FUNCTIONS

$$H_2(f) = \frac{\sin^2(\pi f \tau_g/2)}{(\pi f \tau_g/2)^2} = \text{Equivalent sum-channel transfer function}$$

$$h_2(t) = \frac{2}{\tau_g}\left(1 - \left|\frac{2t}{\tau_g}\right|\right) = \text{Equivalent sum-channel filter weighting function}$$

$$H'(f) = \frac{j2\pi f}{\beta_h}H_2(f) = \frac{j2}{\sqrt{3}}\frac{\sin^2(\pi f \tau_g/2)}{\pi f \tau_g/2} = \begin{array}{l}\text{Transfer function of split-gate}\\ \text{circuit}\end{array}$$

$$\left.\begin{array}{ll} h'(t) = \dfrac{2}{\sqrt{3}\,\tau_g}; & -\dfrac{\tau_g}{2} < (t - t_g) < 0 \\[2ex] \quad\ = \dfrac{-2}{\sqrt{3}\,\tau_g}; & 0 < (t - t_g) < \dfrac{\tau_g}{2} \end{array}\right\} \begin{array}{l}\text{Weighting function of split-gate}\\ \text{circuit}\end{array}$$

$$B_{nh} = \frac{4}{3\tau_g} = \text{Noise bandwidth of split-gate circuit and of equivalent filter } H_2$$

$$\beta_h = \frac{\sqrt{12}}{\tau_g} = \text{Rms bandwidth of } |H_2(f)|^2$$

$$\tau_{nh} = \tau_g/3 = \text{Equivalent duration of } h_2(t)$$

$$\tau_{nd} = \tau_g = \text{Equivalent duration of split-gate weighting function}$$

---

difference between the signal amplitude at the center of the gate and the average amplitude at the two ends of the gate, multiplied by a constant factor:

$$\frac{da_d}{d\epsilon_t} = \frac{-4a_{mx}}{\sqrt{3}\,\tau_g}\left[1 - \frac{1}{2a_{mx}}a_{x1}(\tau_g/2) - \frac{1}{2a_{mx}}a_{x1}(-\tau_g/2)\right]. \tag{3.47}$$

For symmetrical waveforms, $a_x(\tau_g/2) = a_x(-\tau_g/2)$, and we can write

$$K\tau_{3a} = 2\tau_{3a}\frac{a_{mx}}{a_o}\sqrt{\frac{B_{na}}{\tau_g}}\left[1 - \frac{a_{x1}(\tau_g/2)}{a_{mx}}\right]$$

$$\cong \frac{2\tau_{3a}}{\sqrt{\tau_g\tau_{na}}}\left[1 - \frac{a(\tau_g/2)}{a_m}\right] \qquad (B_{nh} \gg B_{na}). \tag{3.48}$$

When the gate width exceeds the total signal width, the term in brackets is unity. Then, for the rectangular pulse, we have $\tau_{na} = \tau_{3a} = \tau$, and

$$K\tau = 2\sqrt{\tau/\tau_g} \qquad \text{(rectangular pulse)}. \tag{3.49}$$

Under the same conditions, the Gaussian pulse has $\tau_{na} = 1.06\tau_{3a}$, and

$$K\tau_{3a} = 1.94\sqrt{\tau_{3a}/\tau_g} \qquad \text{(Gaussian pulse)}. \tag{3.50}$$

Similarly, the $(\sin x)/x$ pulse has $\tau_{na} = 1.15\tau_{3a}$, and

$$K\tau_{3a} = 1.86\sqrt{\tau_{3a}/\tau_g} \qquad \left(\frac{\sin x}{x}\text{ pulse}\right). \tag{3.51}$$

The performance of other pulse shapes with a broadband filter will lie between these extreme cases.

When the gate width is less than the total signal width, we must reduce the slope by the factor $[1 - a(\tau_g/2)/a_m]$ from Eq. (3.48). This leads to zero slope for the rectangular pulse when $\tau_g > \tau$. The gate will be free to wander over the flat top of the pulse, with a restoring signal being generated only when the edge of the gate overlaps the edge of the pulse. Results for the Gaussian pulse are less extreme, but there is still a degradation of performance when the gate width drops below the optimum value ($\tau_g/\tau_{3a} \cong 2$; see Fig. 3.13). Also shown on the figure are curves for rectangular and Gaussian pulses with matched receiving filters followed by split gates. These will be found in agreement with results published previously (Barton, 1964, p. 363; M_r in the earlier book is identical to our $K\tau_{3a}$).

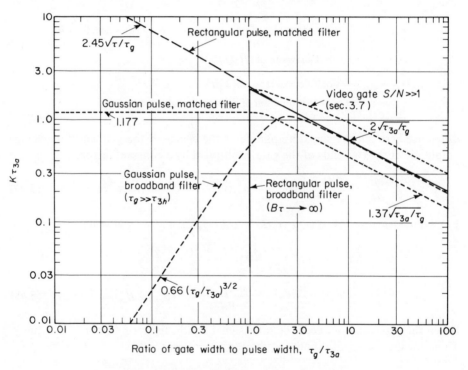

Fig. 3.13 Normalized slope for split-gate discriminator.

3.6 The Leading-Edge Tracker

The leading-edge tracker is a device which measures range to the leading edge of a near-rectangular pulse, ignoring the flat top and the trailing edge. It is used primarily to select the shortest-range target from a group or a mass of distributed echoes, as in a formation of aircraft or an aircraft dropping confusion reflectors. Analysis of this tracker is simple when the signal bandwidth exceeds the bandwidth of the receiving filter. The simplified block diagram is shown in Fig. 3.14. The pulse is passed through the filter $H_1(f)$, which introduces a sloped leading edge of rise time $t_d' \cong 1/B_h$. This filtered signal $a_1(t)$ is differentiated once to produce a short pulse waveform $a_x(t)$, of width approximately t_d'. Then a second differentiator is used to locate the center of the short pulse. (A split-gate discriminator may replace the second differentiator if $\tau_g \cong t_d'$). To establish the performance of the system, we adjust the gain of each differentiator to make the output noise equal that at the filter output, and then apply the defining equation for difference slope K, from Table 3.3:

$$B_{nh} = \int_{B_h} |H_1(f)|^2 \, df = \frac{1}{\beta_4^4} \int_{B_h} (2\pi f)^4 |H_1(f)|^2 \, df,$$

$$\beta_4 \equiv \left[\frac{1}{B_{nh}} \int_{B_h} (2\pi f)^4 |H_1(f)|^2 \, df \right]^{1/4}, \tag{3.52}$$

$$a_x(t) = \frac{1}{\beta_4} \frac{da_1(t)}{dt} = \frac{a_m}{\beta_4} h_1(t), \tag{3.53}$$

$$K = \frac{-1}{\beta_4 a_o} \sqrt{\frac{B_{na}}{B_{nh}}} \frac{d^2 a_x}{d\epsilon_t^2}\bigg|_{\epsilon_t=0} = \frac{-a_m}{\beta_4^2 a_o} \sqrt{\frac{B_{na}}{B_{nh}}} \frac{d^2 h_1}{d\epsilon_t^2}\bigg|_{\epsilon_t=0}$$

$$= \frac{a_m h_m}{a_o} \sqrt{\frac{B_{na}}{B_{nh}}} \frac{\beta_{h1}^2}{\beta_4^2}. \tag{3.54}$$

From Sec. 3.1, $h_m^2 = B_{nh}/\tau_{nh}$. Also, for the rectangular pulse, $B_{na} = 1/\tau$, and for $B_h \gg 1/\tau$ we have $a_m = a_o$. Thus,

$$K = \frac{1}{\sqrt{\tau \tau_{nh}}} \frac{\beta_{h1}^2}{\beta_4^2},$$

$$K\tau = \sqrt{\frac{\tau}{\tau_{nh}}} \frac{\beta_{h1}^2}{\beta_4^2}. \tag{3.55}$$

If the receiving filter is rectangular and of width B_h, the equivalent duration $\tau_{nh} = 1/B_h$ and the rms durations are $\beta_{h1} = \pi B_h/\sqrt{3}$ and $\beta_4 = \pi B_h/\sqrt{5}$, to give

$$K\tau = \sqrt{\frac{5 B_h \tau}{9}}. \tag{3.56}$$

This result is poorer than the ideal performance on the band-limited rectangular pulse [Eq. (3.41)] by a factor of about two, reflecting the nonoptimum weighting and the loss in information from the trailing edge of the pulse.

For a Gaussian receiving filter of noise bandwidth B_{nh}, we have $\tau_{nh} = 0.5/B_{nh}$, $\beta_{h1}^2 = 4\pi B_{nh}^2$, and $\beta_4^2 = \sqrt{12}\pi B_{nh}^2$, giving

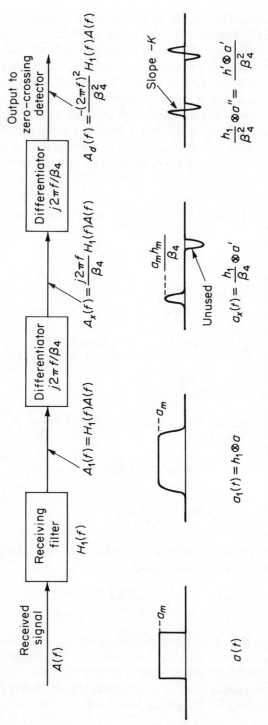

Fig. 3.14 Block diagram and waveforms of leading-edge tracker.

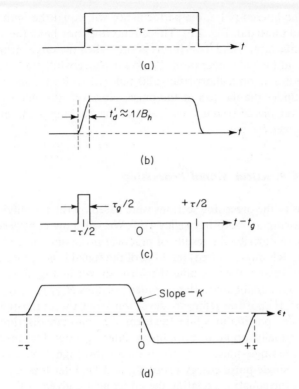

Fig. 3.15 Separated-gate waveforms and response: (a) input impulse $a(t)$; (b) filtered pulse $a_1(t)$; (c) separated-gate weighting function; (d) discriminator response.

$$K\tau = \sqrt{\frac{8B_{nh}\tau}{3}} = 1.63\sqrt{B_{nh}\tau}. \qquad (3.57)$$

Since the input signal must actually be of restricted bandwidth B_{na}, the above expressions are only true when B_{na} exceeds B_{nh} by a significant amount (or when $B_{na} \geq B_{nh}$, for rectangular spectrum and filter). When a $(\sin x)/x$ signal spectrum is restricted by a broadband Gaussian passband in the transmitter, the peak amplitude of the first derivative $a_x(t)$ will be reduced by $\sqrt{2}$ for $B_{nh} = B_{na}$, and $K\tau$ will reach a maximum value $1.15\sqrt{B_{na}\tau}$. As with other trackers, the filtering action of a split gate may be included in $H_1(f)$, generally with some decrease in performance as compared to the ideal differentiator.

Separated-Gate Discriminator

A variation of the leading-edge tracker takes the form of a discriminator with a pair of narrow gates, separated to cover the leading and trailing edges of the band-limited rectangular pulse (Fig. 3.15). To the extent that these gates match the first derivative of the filtered pulse, the performance will equal the optimum predicted by Skolnik (and plotted in Fig. 3.6). This circuit does not have the ability to select the shortest-range target, and it performs poorly when the target echo is broadened to extend beyond the gate separation. Its performance while tracking is equivalent to that of a split gate on a short triangular pulse of width τ_g, and no use is made of the energy under the flat top of the pulse. However, this energy does serve to provide broad regions of response each side of the tracking point, making it easier to acquire a target.

3.7 Effects of Practical Signal Processing

The derivations of the preceding sections were based on the simplifying assumption of linear processing at "zero-frequency IF." We will now consider the modifications necessary to describe the results of practical processing procedures employed when optimum (RF phase-locked) tracking of the signal is not practical. The actual receiver and processor often assume the form shown in Fig. 3.16. The receiving filter $H(f)$ has been divided into IF and video portions $H_1(f)$ and $H_2(f)$, where frequency f in the IF portion refers to deviation from the center-frequency of the filter, f_o. The video portion may take the form of a split-gate discriminator, but the description is equally applicable to linear filtering after the envelope detector in cases where the single-pulse S/N ratio is reasonably high. Considerations of integration time t_o, single-pulse energy ratio $\mathcal{R}_1$, and IF filter loss L_m were discussed in Sec. 1.4, and the equations relating these factors are given as Eqs. (1.23)–(1.33). These are equally applicable to range-measurement problems. Curves of L_m vs. $B_{nh}\tau_{3a}$ are given in Fig. 3.17 for several cases of interest.

Detector Loss, L_x

The loss in the envelope detector will depend upon the signal-to-noise ratio S/N at the output of the IF filter. The expression for this loss is identical to that for the monopulse angle detector:

$$L_x = \frac{(S/N) + 1}{(S/N)} = \frac{S + N}{S} \qquad \text{[See Eq. (2.53)].}$$

When this loss is applied to reduce the effective value of energy ratio $\mathcal{R}$ in the n-pulse measurement case, we can obtain from Eq. (3.23) the following:

$$\boxed{\begin{array}{c} \textit{Normalized Error for n-Pulse} \\ \textit{Measurement} \\[2mm] \dfrac{\sigma_t}{\tau_{3a}} = \dfrac{\sqrt{L_x}}{K\tau_{3a}\sqrt{\mathcal{R}}} = \dfrac{\sqrt{(S/N) + 1}}{K\tau_{3a}(S/N)\sqrt{2nL_m}} \end{array}} \qquad (3.58)$$

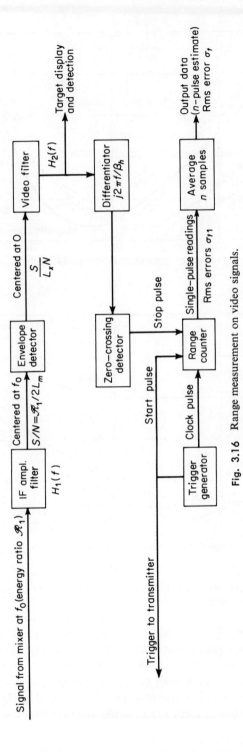

Fig. 3.16   Range measurement on video signals.

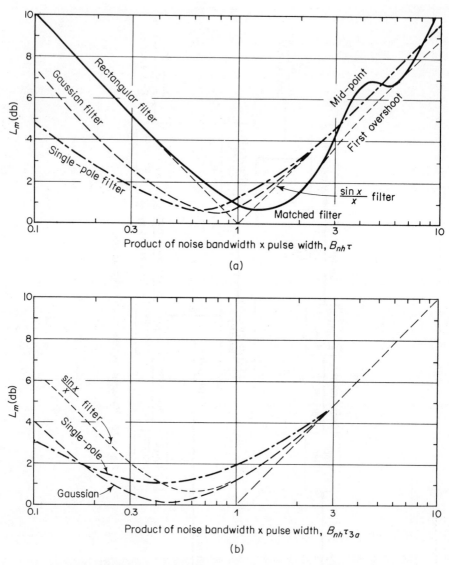

Fig. 3.17 IF filter loss: (a) rectangular pulse with different filters; (b) Gaussian pulse with different filters.

The effect of L_m is merely to decrease (S/N) at the envelope detector, and hence to increase L_x when the signal is not sufficiently strong. Any further degradation in efficiency of measurement, owing to filter mismatch, is taken into account when $K\tau_{3a}$ is calculated from the overall filter response $H(f) = H_1(f)H_2(f)$. For this reason, calculation of a video matching loss L_{mv}, needed in the analysis of angle error, is not required here, except as a possible step in evaluation of $K\tau_{3a}$.

Zero-Crossing Ambiguity

In the earlier analysis, it was assumed that the S/N ratio was sufficient to ensure only one zero crossing of the differentiated signal within the *a priori* target interval. This condition is met for targets which are detectable when a matched filter precedes the differentiator, subject only to the band-limit restriction on rectangular pulses (Sec. 3.4). However, when detection depends on noncoherent integration of many pulses, and when output data is obtained by averaging the many zero-crossing times, false zero-crossings caused by noise can occur at times widely removed from the true signal peak. The additional component of rms error is given by Eq. (3.43), and this component increases very rapidly as the single-pulse S/N ratio approaches unity. This is the primary reason for use of the split-gate type of tracker on noncoherent radars which must make range measurements near the maximum range of detection.

Beamshape Loss, L_p

In search radar, when the integration or measurement time t_o is computed between one-way half-power points of the beam pattern, the actual energy available for measurement is reduced by the beamshape loss factor L_p. Combining this with the detector loss, which is also increased as a result of L_p, we obtain the following:

$$\boxed{\begin{aligned} &\textit{Normalized Range Error for Search Radar}\\[6pt] &\frac{\sigma_t}{\tau_{3a}} = \frac{\sqrt{L_x L_p}}{K\tau_{3a}\sqrt{\mathscr{R}}} = \frac{\sqrt{(S/N)_m + L_p}}{K\tau_{3a}(S/N)_m}\sqrt{\frac{L_p}{2nL_m}}\end{aligned}} \qquad (3.59)$$

The term $(S/N)_m$ refers to the beam-center value of the single-pulse S/N ratio.

Collapsing Loss for Wide Split Gate

The losses in angular measurement, caused by excessive gate width, were listed in Table 2.7 under the heading of "collapsing loss." For the coherent system, the loss was

$$L_c = \tau_g/\tau,$$

which gives the same increase in error as we have computed for the wide gate in Eq. (3.49). Thus, we may regard the effect of noise in the extended gate either as a reduction in normalized slope $K\tau_{3a}$, or as a reduction in effective energy ratio by the loss L_c. When a linear envelope detector is used, Table 2.7 gives, for $S/N \gg 1$,

$$L_c = 1 + \frac{\tau_g - \tau}{\tau}\left(2 - \frac{\pi}{2}\right).$$

This factor should replace τ_g/τ in Eq. (3.49) when this type of detector is used. The error increases less rapidly for increasing gate width, as shown by the upper line in Fig. 3.13. However, when the S/N ratio approaches unity, the loss caused by the wide gate reverts to its normal value τ_g/τ. In a tracking radar which integrates over many pulses, the tracking threshold is set by the inability of the range gate to maintain lock on signals below noise level, where the full value of collapsing loss combines with the detector loss to produce rapid degradation in performance.

IF Split Gate

To realize the performance of an idealized split gate (Fig. 3.12) without excessive detector loss, the gate function $h_2'(t)$ may be introduced in a broadband IF stage as a phase reversal, and followed by narrow-band filtering and phase-sensitive detection. The reference signal for detection is derived from a range-sum channel, gated by a single (even) waveform of width $\tau_g = \tau$ and filtered in a circuit similar to the narrow-band portion of the difference channel. The two channels are equivalent to the sum and difference channels of the monopulse angle tracker. When proper time alignment has been established with the signal, the sum channel represents a matched filter for a single received pulse ($L_m = 1$), and L_x is as small as possible for noncoherent reception of the signal. When a number of coherent pulses are received, the narrow-band filters which follow range gating may be used for coherent integration and Doppler resolution, providing the most efficient possible use of received energy.

The IF gate circuit also lends itself to separated-gate operation (Fig. 3.15), providing the full performance predicted by Skolnik for the band-limited rectangular pulse, and not requiring the construction of $(\sin x)/x$ filters in the frequency domain. In either the split-gate or separated-gate configuration, the IF circuit has an advantage over the video discriminator when near-rectangular pulses are received at low single-pulse energy ratios. The advantage arises from the ability of the broadband IF gate circuits to use effectively the high-frequency content of the signal spectrum, while the post-gate, narrow-band filter minimizes detector loss. The same advantages may be obtained in a linear processor which has a matched filter prior to the detector and differentiator.

3.8 Summary of Range Noise Error

The basic equations for thermal noise in range measurement are repeated in Table 3.8. These equations omit the effects of nonlinearities and detector loss, which are included in the expressions of Table 3.9. Procedures for obtaining estimates of single-pulse time-delay error are summarized in Table 3.10. The energy ratio $\mathcal{R}_1$, calculated from a radar equation such as Eq. (1.29), is used in each case. If the matching of the IF filter to the pulse spectrum is good enough to ensure $S/N > 4$,

Table 3.8

BASIC ERROR EQUATIONS

Ideal

$$(\sigma_t)_{min} = \frac{1}{\beta_a \sqrt{\mathscr{R}}} = \frac{1}{K_o \sqrt{\mathscr{R}}} \qquad (3.22)$$

Mismatched Filter

$$\sigma_{t1} = \frac{1}{K\sqrt{\mathscr{R}_1}} = \frac{1}{K\sqrt{2L_m S/N}} \qquad (3.23)$$

Ideal (normalized to 3-db pulse width out of matched filter)

$$\frac{(\sigma_t)_{min}}{\tau_o} = \frac{1}{\beta_a \tau_o \sqrt{\mathscr{R}}} \cong \frac{1}{1.63 \sqrt{\mathscr{R}}} \qquad (3.31)$$

Mismatched Filter (normalized to 3-db input pulse width)

$$\frac{\sigma_{t1}}{\tau_{a3}} = \frac{1}{K\tau_{3a}\sqrt{\mathscr{R}_1}} \cong \frac{K_h \beta_h/\beta_x}{1.63 \sqrt{\eta_f \mathscr{R}_1}} \qquad (3.33)$$

(for values of normalized slope $K\tau_{3a}$, see Figs. 3.3, 3.4, 3.7, and 3.13)

Ideal (special case of rectangular pulse)

$$\frac{(\sigma_t)_{min}}{\tau} = \frac{\sqrt{2}}{\mathscr{R}} \qquad (K\tau = \sqrt{\mathscr{R}/2}) \qquad (3.39)$$

Ideal (band-limited rectangular pulse)

$$\frac{(\sigma_t)_{min}}{\tau} \cong \frac{1}{\sqrt{2B_h \tau \mathscr{R}}} \qquad (K\tau \cong \sqrt{2B_h \tau}) \qquad (3.41)$$

Brief Definitions of Symbols

B_h = total filter bandwidth
K = relative difference slope (Table 3.3)
K_o = value of K for matched filter
K_h = pulse broadening factor, τ_{3x}/τ_{3a}
$\mathscr{R}$ = $2E/N_o$ = total signal energy ratio
$\mathscr{R}_1$ = single-pulse energy ratio
β_a = rms signal bandwidth, Eq. (3.14)
β_h = rms filter bandwidth, Eq. (3.16)
β_x = rms output signal bandwidth, Eq. (3.18)
η_f = filter efficiency, Eq. (3.13)
σ_{t1} = single-pulse rms error
$(\sigma_t)_{min}$ = minimum error (matched filter)
σ_t = n-pulse rms error
τ = width of rectangular pulse
τ_{3a} = 3-db width of signal pulse
τ_o = 3-db width of output pulse from matched filter
τ_{3x} = 3-db width of output pulse from filter

Table 3.9

NONOPTIMUM PROCESSING OF n PULSES

General Expression with Processing Loss L

$$\frac{\sigma_t}{\tau_{3a}} = \frac{\sqrt{L}}{K\tau_{3a}\sqrt{\mathscr{R}}}$$

Noncoherent n-Pulse Measurement (detector loss L_x)

$$\frac{\sigma_t}{\tau_{3a}} = \frac{\sqrt{L_x}}{K\tau_{3a}\sqrt{\mathscr{R}}} = \frac{\sqrt{(S/N)+1}}{K\tau_{3a}(S/N)\sqrt{2nL_m}} \qquad (3.58)$$

Search-Radar Measurement (beamshape loss L_p)

$$\frac{\sigma_t}{\tau_{3a}} = \frac{\sqrt{L_xL_p}}{K\tau_{3a}\sqrt{\mathscr{R}}} = \frac{\sqrt{(S/N)_m + L_p}}{K\tau_{3a}(S/N)_m}\sqrt{\frac{L_p}{2nL_m}} \qquad (3.59)$$

Relationship of Energy Ratio to S/N (Power) Ratio

$$\mathscr{R} = n\mathscr{R}_1 = 2nL_mS/N \qquad (1.30)$$

$$n = f_r t_o = \frac{f_r}{2\beta_n} \qquad \text{(tracker with servo bandwidth } \beta_n\text{)}$$

$$= \frac{f_r\theta_3}{\omega} \qquad \text{(search radar scanning at } \omega \text{ rad/sec)}$$

---

the errors can be found to within about ten percent accuracy. Extension of these procedures to other cases will involve the following additional steps:

(a) Multiple-pulse systems, $S/N > 4$, including coherent systems with $(S/N)_f > 4$.

Use procedures in Table 3.10, and then calculate output error $\sigma_t = \sigma_{t1}/\sqrt{n}$, where n is the number of pulses integrated. This is equivalent to use of $\mathscr{R} = n\mathscr{R}_1$ in each expression.

(b) Noncoherent systems, $S/N < 4$.

Calculate as in (a), but reduce $\mathscr{R}$ by the loss of the envelope detector:

$$L_x = \frac{(S/N) + 1}{S/N}.$$

If a differentiator is used to provide data from an *a priori* target interval $\Delta \gg \sigma_t$, the error component contributed by noise ambiguities must be calculated from Eq. (3.43) and added in an rms fashion to σ_t. The false-alarm probability can be found from standard curves of the Rice distribution (e.g., Barton, 1964, Fig. 1.8).

(c) Coherent systems, $(S/N)_f < 4$.

Calculate as in (a) but reduce $\mathscr{R}$ by the detector loss

$$L_x = \frac{(S/N)_f + 1}{(S/N)_f}.$$

Table 3.10

PROCEDURE FOR SINGLE-PULSE TIME-DELAY ERROR ($\mathcal{R} = \mathcal{R}_1 \gg 1$)

Type of Circuit	Smooth Waveform, Monochromatic Signal		Rectangular Pulse, Band-Limited to B_a, B_h		Pulse Compression, Uniform Spectrum over B_a	
	Find:	*Using:*	*Find:*	*Using:*	*Find:*	*Using:*
Matched Filter, Differentiator	$K_o = \beta_a$	Eq. (3.14), Table 3.4, App. A, B	$(\sigma_t)_{min}$	Eq. (3.39)	β_a	$\beta_a = \pi B_a/\sqrt{3}$
			σ_{t1}	Eq. (3.41)	$(\sigma_t)_{min}$	Eq. (3.22)
			Select larger	Fig. 3.6	τ_o	$\tau_o = 0.886/B_a$
	$(\sigma_t)_{min}$	Eq. (3.22)			$(\sigma_t)_{min}$	Eq. (3.31)
Mismatched Filter, Differentiator	$\beta_h, \beta_{h1}, \beta_{a1}$	Eq. (3.16), (3.26)–(3.29) App. A, B and Eq. (3.30)	$K\tau$	Fig. 3.7	$\beta_h, \beta_{h1} = \beta_x$	Eqs. (3.16), (3.28) App. A, B, and Eq. (3.30)
	$K\tau_{3a}$	Eq. (3.34) or Figs. 3.3–3.4	σ_{t1}/τ	Eq. (3.33)	K	Eq. (3.20)
					σ_{t1}	Eq. (3.23)
	σ_{t1}/τ_{3a}	Eq. (3.33)				
Filter and Split Gate	$a_x(t)$	Eq. (3.9)	$K\tau$	Eqs. (3.46)–(3.49) or Fig. 3.13	$a_x(t)$	Eq. (3.9) or App. A, B
	$K\tau_{3a}$	Eqs. (3.46)–(3.51) or Fig. 3.13			$K\tau_{3x}$	Eqs. (3.46)–(3.51)*
	σ_{t1}/τ_{3a}	Eq. (3.33)	σ_{t1}/τ	Eq. (3.33)	σ_{t1}/τ_{3x}	Eq. (3.33)*

* τ_{3x} replaces τ_{3a}.

(d) Search Radar.

Follow steps (a), (b), or (c), but further reduce $\mathcal{R}$ and S/N by the beam-shape loss L_p, as given in Table 2.4.

In the absence of specific information about the pulse shape, spectrum, or filter characteristics, a rough estimate of the error may be made by assuming $K\tau_{3a} \cong 1$ in Eq. (3.58):

$$\frac{\sigma_t}{\tau_{3a}} \cong \sqrt{\frac{L_x}{\mathcal{R}}} = \frac{\sqrt{(S/N) + 1}}{(S/N)\sqrt{2nL_m}}.$$

The measurement of signal frequency, or Doppler shift, has assumed great importance in many modern radar systems. We will describe the frequency measurement capabilities of the radar in terms of the signal and filter functions defined in the previous chapter (see Table 3.1), adding those new parameters needed to calculate frequency error.

When a target is illuminated by a wave transmitted at frequency f_o, the received echo will be shifted by an amount f_d, according to the Doppler equation.*

$$f_d = f_o \left[\frac{c - v_r}{c + v_r} - 1 \right]$$
$$= -\frac{2 f_o v_r}{c} \left[1 - \frac{v_r}{c} + \frac{v_r^2}{c^2} - \cdots \right], \qquad (4.1)$$

$$f_d \cong -\frac{2 f_o v_r}{c} = -\frac{2 v_r}{\lambda}. \qquad (4.2)$$

Here, v_r is the time derivative of range (positive for outbound targets), c is the velocity of light, and $\lambda = c/f_o$ is the wavelength of the transmission. Solving for v_r, we have

$$v_r = -\frac{f_d c}{2 f_o} \left[1 - \frac{f_d}{2 f_o} + \frac{f_d^2}{4 f_o^2} - \cdots \right], \qquad (4.3)$$

$$v_r \cong -\frac{f_d c}{2 f_o} = -\frac{f_d \lambda}{2}. \qquad (4.4)$$

The relative contribution of the second-order term is less than thirty parts per million in f_d or v_r, for objects controlled by the earth's gravitational field ($v_r < 10^4$ m/sec). The approximations of Eqs. (4.2) and (4.4) are normally adequate for objects within the earth's atmosphere.

In this chapter, we summarize the relationships among the signal and receiver bandwidths,

* For signals with large bandwidth-time products, second-order effects may have to be considered.

4

Doppler
Measurement
in
Noise

spectral distributions, frequency response, signal duration, signal-to-noise ratio, and Doppler error, for commonly used signal waveforms and measurement systems. Measurements on the spectral envelope of the signal will be described first, and the theory will then be extended to measurements based on the fine structure of a repetitive, coherent signal.

4.1 Spectral Envelope and Fine Structure

The signal and receiver descriptions developed in Chap. 3 for range measurement will provide the basis for analysis of Doppler measurement as well. However, we must first clarify the time and frequency scales within which the measurement is conducted. If a single pulse is transmitted, it will be characterized by the time and frequency functions listed in Table 3.1, and illustrated in Fig. 3.2(a–c). Measurement of frequency can be performed on such a pulse, and results from a train of similar pulses can be combined by averaging without regard to the relative phase of successive pulses. We refer to such a process as measurement on the spectral envelope, or *noncoherent* measurement.

In other cases, it is possible to maintain a phase reference over a train of pulses, either by deriving all pulses from a stable, continuous oscillator or by heterodyning

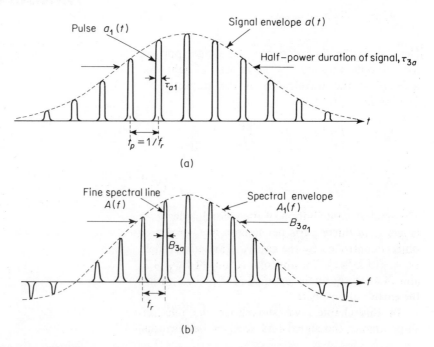

(a)

(b)

Fig. 4.1   Waveform and spectrum of coherent pulse train: (a) waveform of pulse train; (b) spectrum of pulse train.

the received signal with a *coherent oscillator* which is locked to the phase of each transmitted pulse. The resulting signal waveform and heterodyned spectrum will appear as in Fig. 4.1, with a fine line structure $A(f)$ beneath the spectral envelope $A_1(f)$. Measurements can now be made within the width of the fine lines, provided that the ambiguities can be resolved to choose the proper line in the received spectrum. The time function $a(t)$ may be the two-way pattern of a scanning antenna (in search), or a time function describing the stability of the transmitter-target-receiver system (in tracking).

In analysis of accuracy, the only difference between the noncoherent and coherent cases will be the scale and shape of the time and frequency functions. The noncoherent case will be described in terms of the width and shape of an individual pulse, typically on a scale of microseconds (with a spectral envelope width in MHz), while the coherent time function may extend over a period in the order of a second (fine line width in Hz). We will discuss the noncoherent case first, before proceeding to the coherent pulse train. The additional subscripts used in Fig. 4.1 to denote the single-pulse functions $a_1(t)$ and $A_1(f)$ will be used in the later discussion to avoid confusion with the fine-line functions, but are omitted in the general analysis which follows.

4.2 Frequency Estimator Model

Response Function

In Chap. 1, we wrote an expression for the response function ψ of a receiving system with a given signal in terms of time and frequency errors relative to the point to which the system was tuned:

$$\psi(\epsilon_t, \epsilon_f) = \int_{-\infty}^{\infty} H(f)A(f - \epsilon_f) \exp(j2\pi f \epsilon_t)\, df$$

$$= \int_{-\infty}^{\infty} h(\epsilon_t - t)a(t) \exp(j2\pi\epsilon_f t)\, dt. \tag{4.5}$$

The signal and filter functions are those defined in Table 3.1. Let us assume here that the time delay of the signal is known, and that we wish to measure the Doppler shift f_d which has been added to the transmission frequency f_o by target motion. We tune our receiving system to a point in time $t_m = t_d$ and in frequency to $f_o + f_m \cong f_o + f_d$, and then vary f_m to find the peak of the response function at $f_d - f_m = \epsilon_f \cong 0$. The precision with which this peak can be located will depend on the shape of the response function along the frequency axis ($\epsilon_t = 0$):

$$\psi(0, \epsilon_f) = \int_{-\infty}^{\infty} H(f)A(f - \epsilon_f)\, df$$

$$= \int_{-\infty}^{\infty} h(-t)a(t) \exp(j2\pi\epsilon_f t)\, dt \qquad \text{(volts)}. \tag{4.6}$$

Noise will displace the apparent peak of this function by some amount, and this noise error will be reduced if the response function has a sharp peak at $\epsilon_f = 0$.

It has been established (see Chap. 1) that the ideal response function ψ_o is generated when the receiving system is matched to the signal. As a standard for comparison, then, we will use the shape of ψ_o along the frequency axis:

$$\psi_o(0, \epsilon_f) = \frac{1}{C} \int_{-\infty}^{\infty} A^*(f) A(f - \epsilon_f)\, df$$

$$= \frac{1}{C} \int_{-\infty}^{\infty} |a(t)|^2 \exp(j2\pi\epsilon_f t)\, dt. \tag{4.7}$$

Here, C is a constant with dimensions of volt-seconds, which sets the gain of the receiving system to both signals and noise. The peak of the ideal response function has an amplitude

$$\psi_o(0, 0) = \frac{1}{C} \int_{-\infty}^{\infty} |A(f)|^2\, df = \frac{1}{C} \int_{-\infty}^{\infty} |a(t)|^2\, dt = \frac{2E_1}{C} \quad \text{(volts)}. \tag{4.8}$$

Correlator Configuration

The system shown in Fig. 4.2 has been chosen to illustrate the analogies between frequency and time-delay estimation, rather than as an example of a practical frequency estimator. An approximate match to a signal of known delay is obtained by multiplying the signal by a function similar to the signal envelope, and integrating the product in a narrowband filter centered at $f_o + f_m$. If the multiplier output is applied to a bank of closely spaced filters, the response function can be inspected over a range of frequencies, and the filter giving the maximum output can be selected to indicate f_d. To aid in identifying this filter, a difference-channel signal is generated by a second multiplier, which applies a linear-odd time function. The filter pair having the smallest ratio of Δ/Σ, when the Σ voltage exceeds a given threshold, is the best estimate of f_d. Granularity of the estimate can be reduced to any value by decreasing the width v and the spacing of the filters and increasing their number, until an accuracy limit is reached which depends upon input noise. In a practical correlator, the input signal may be heterodyned from $f_o + f_d$ to an intermediate frequency $f_c + f_d$, and the filter banks operated at $f_c \pm (f_d)_{max}$.

There is a close analogy between this Doppler estimator and the stacked-beam angle estimator which uses sums and differences of adjacent beams, taken from a beam-forming matrix. In the angle case, the weighting functions in time are replaced by spatial weighting functions across the aperture, and the corresponding far-field patterns are analogous to the filter transfer functions here.

Sum-Channel Description

Figure 4.3 shows the time functions for the correlator of Fig. 4.2, and the corresponding frequency transforms. Although this set of functions is similar in many

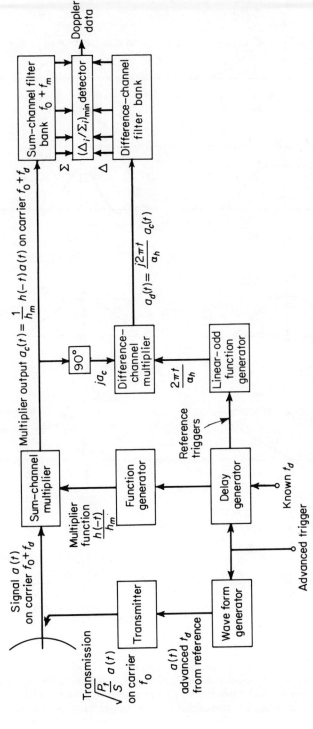

Fig. 4.2 Idealized Doppler estimator (correlator configuration). Both filter banks consist of narrowband filters of width ν covering all frequencies from $f_o + (f_d)_{min}$ to $f_o + (f_d)_{max}$.

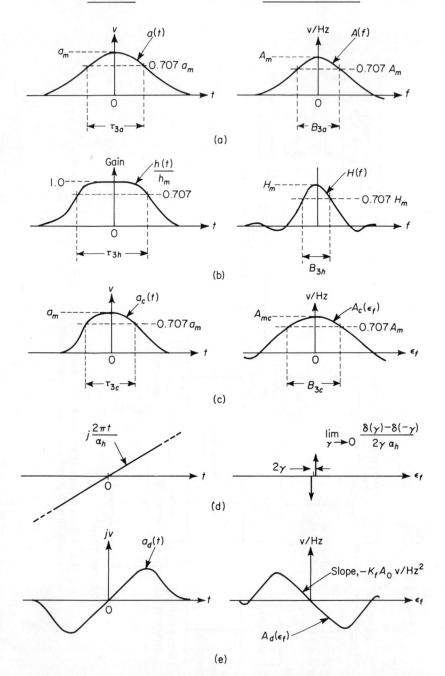

Fig. 4.3 Signal and correlator functions: (a) input signal envelope; (b) multiplier weighting function; (c) sum-channel multiplier output [the curve for $A_c(\epsilon_f)$ also represents $\sum (\epsilon_f)$ in volts.]; (d) difference-channel multiplier function; (e) difference-channel multiplier output [the curve for $A_d(\epsilon_f)$ also represents $\Delta(\epsilon_f)$ in volts].

ways to that illustrated in Fig. 3.2 for time-delay estimation, we note that in the correlator it is the sum-channel *waveform $a_c(t)$* which can be expressed as a simple product of signal and weighting functions. The corresponding spectrum $A_c(\epsilon_f)$ is the convolution of signal and filter frequency functions, and hence is broader than either of these. The relationships listed in Table 4.1 show the following properties of the correlator sum channel:

1. The multiplier function is a time-reversed, dimensionless image of the weighting function $h(t)$, arbitrarily scaled here to unit maximum amplitude.

2. Except for the scaling factor $1/h_m$, the spectrum $A_c(\epsilon_f)$ at the multiplier output represents the desired response function, viewed along the frequency axis.

3. Output envelope voltages taken from the narrowband filters represent samples of the voltage spectral density $A_c(\epsilon_f)$ at discrete points, and are analogous to discrete samples of the waveform $a_x(t)$ in the time-delay estimator.

4. The over-all (dimensionless) voltage gain of the system is v/h_m relative to the response function ψ, but this does not affect the signal-to-noise ratio or other performance factors.

Table 4.1

SUM-CHANNEL DESCRIPTION

Coordinates and Constants

t = time, measured relative to the center of signal and weighting functions, in seconds

f = frequency, measured relative to the center of the signal spectrum and filter transfer function, in Hz

$\epsilon_f = f_d - f_m$ = Doppler error, in Hz

v = width of each narrowband filter, in Hz

Functional Relationships

$a_c(t) = (1/h_m)\, h(-t)a(t)$ = multiplier output envelope, in volts

$A_c(\epsilon_f) = \dfrac{1}{h_m} \displaystyle\int_{-\infty}^{\infty} H(f)A(f - \epsilon_f)\, df = \dfrac{1}{h_m}\, \psi(0, \epsilon_f)$

$\qquad$ = multiplier output spectrum, in volt-seconds

$\Sigma(\epsilon_f) = v\, A_c(\epsilon_f) = \dfrac{v}{h_m}\, \psi(0, \epsilon_f)$ = sum-channel output envelope, in volts

$S_x = \frac{1}{2}[\Sigma(0)]^2$ = average output power at $\epsilon_f = 0$, in watts

$N_o\tau_{nh}$ = noise energy density out of each filter, in watt-seconds^2

$vN_o\tau_{nh}$ = noise energy out of each filter, in watt-seconds

$N = v^2 N_o\tau_{nh}$ = average noise power out of each filter, in watts

$\dfrac{S_x}{N} = \dfrac{1}{2N}[\Sigma(0)]^2 = \dfrac{A_{mc}^2}{2N_o\tau_{nh}}$ = signal-to-noise ratio at output, for $\epsilon_f = 0$

Correlator Efficiency

The correlator system constitutes a matched filter when the multiplying function is

$$\frac{h(-t)}{h_m} = \frac{a^*(t)}{a_m}.$$

The system parameters for this case are shown in Table 4.2, and are analogous to those given in Table 3.2 for the frequency-domain matched filter. When other multiplying functions are used, the efficiency is given as follows:

$$\eta_f \equiv \frac{2(S_x/N)}{\mathscr{R}_1} = \frac{A_{mc}^2/N_o\tau_{nh}}{A_o^2/N_o\tau_{na}} = \frac{A_{mc}^2}{2E_1\tau_{nh}}$$

$$= \frac{\left|\int_{-\infty}^{\infty} h(-t)a(t)\,dt\right|^2}{\int_{\tau_h} |h(t)|^2\,dt \int_{\tau_a} |a(t)|^2\,dt}$$

$$= \frac{\left|\int_{-\infty}^{\infty} H(f)A(f)\,df\right|^2}{\int_{B_h} |H(f)|^2\,df \int_{B_a} |A(f)|^2\,df}. \tag{4.9}$$

Table 4.2

MATCHED-FILTER CORRELATOR

(The constant in Eq. (4.8) is set to $C = a_m/h_m$ V-sec.)

$$h(-t) = \frac{h_m}{a_m} a^*(t) \qquad \tau_{nh} = \tau_{na}$$

$$H(f) = \frac{h_m}{a_m} A^*(f) \qquad B_{nh} = B_{na}$$

$$\tau_a = \tau_h = \tau_c$$

$$B_a = B_h$$

$$B_{3c} \equiv B_o$$

$$a_c(t) = \frac{|a(t)|^2}{a_m}$$

$$A_c(\epsilon_f) = \frac{1}{a_m} \int_{-\infty}^{\infty} A^*(f)A(f - \epsilon_f)\,df = \frac{1}{h_m} \psi_o(0, \epsilon_f)$$

$$A_{mc} = A_o = \frac{1}{a_m} \int_{-\infty}^{\infty} |A(f)|^2\,df = \frac{1}{a_m} \int_{-\infty}^{\infty} |a(t)|^2\,dt$$

$$= \frac{1}{h_m} \psi_o(0, 0) = \frac{2E_1}{a_m} = a_m\tau_{na}$$

$$\Sigma_o = vA_o$$

$$S_x = \frac{v^2 A_o^2}{2} = v^2 E_1\tau_{na}$$

$$N = v^2 N_o\tau_{na}$$

$$\frac{2S_x}{N} = \frac{2E_1}{N_o} = \mathscr{R}_1$$

$$\eta_f = 1$$

Error Sensitivity

Frequency error sensitivity of the correlator is given by derivatives of the difference-channel output (Table 4.3). Following the practice used in Chap. 3, we have combined into the sum-channel weighting function $h(t)$ all the filter action

Table 4.3

DIFFERENCE-CHANNEL DESCRIPTION

$a_d(t) = \dfrac{j2\pi t}{\alpha_h} a_c(t)$ = difference-channel multiplier output

$A_d(\epsilon_f) = \dfrac{1}{\alpha_h} \dfrac{\partial A_c(\epsilon_f)}{\partial \epsilon_f}$ = difference-channel multiplier output voltage spectrum, in volt-seconds

$\Delta(\epsilon_f) = vA_d(\epsilon_f)$ = difference-channel filter output in volts

$N_d = N = v^2 N_o \tau_{nh}$ = average output noise in watts

$\sqrt{\tau_{nh}/\tau_{na}}$ = ratio of output noise to matched-filter value

$K_f \equiv -\dfrac{\sqrt{\tau_{na}/\tau_{nh}}}{\Sigma_o} \dfrac{\partial \Delta(\epsilon_f)}{\partial \epsilon_f}\bigg|_{\epsilon_f=0} = -\dfrac{\sqrt{\tau_{na}/\tau_{nh}}}{A_o} \dfrac{\partial A_d(\epsilon_f)}{\partial \epsilon_f}\bigg|_{\epsilon_f=0}$

    = relative difference slope in relative amplitude per hertz

K_{of} = maximum possible difference slope for given signal

$K_{rf} \equiv K_f/K_{of} = \dfrac{(\sigma_f)_{min}}{\sigma_f}$ = difference slope ratio

$h_d(t) = \dfrac{j2\pi t}{\alpha_h} h(t)$ = weighting function from input through difference channel multiplier

$H_d(f) = \dfrac{1}{\alpha_h} \dfrac{dH(f)}{df}$ = transfer function from input through difference channel multiplier

and have obtained the difference signal by ideal differentiation of the multiplier output spectrum $A_c(f)$. The normalizing factor α_h, defined in Table 4.4, is included to make the noise outputs of the sum and difference channels equal and to maintain the dimension of volts in both channels. The relative difference slope is again defined as the ratio of difference-channel voltage slope to rms noise, divided by the ideal sum-channel signal-to-noise voltage ratio.

In frequency estimation, the signals and weighting function may be described by "rms time duration" parameters, and these are in turn related to the second derivatives (with respect to frequency) of the corresponding voltage spectra or transfer functions (Table 4.4). Using these definitions, we can express the difference slope K_f as a function of rms time durations and filter efficiency factor.

$$\boxed{\begin{array}{c} \textit{Relative Difference Slope} \\[6pt] K_f = -\dfrac{\sqrt{\tau_{na}/\tau_{nh}}}{A_o \alpha_h} \dfrac{\partial^2 A_c}{\partial \epsilon_f^2}\bigg|_{\epsilon_f=0} = \dfrac{A_{mc}}{A_o} \sqrt{\dfrac{\tau_{na}}{\tau_{nh}}} \dfrac{\alpha_c^2}{\alpha_h} = \dfrac{\alpha_c^2}{\alpha_h} \sqrt{\eta_f} \end{array}} \qquad (4.10)$$

Table 4.4

RMS TIME DURATION PARAMETERS

$$\alpha_a \equiv \left[\frac{\displaystyle\int_{\tau_a} (2\pi t)^2 \, |a(t)|^2 \, dt}{\displaystyle\int_{\tau_a} |a(t)|^2 \, dt} \right]^{1/2} = \begin{cases} \text{rms duration of signal power, as used by Woodward} \\ (1953) \end{cases}$$

$$\alpha_h \equiv \left[\frac{\displaystyle\int_{\tau_h} (2\pi t)^2 \, |h(t)|^2 \, dt}{\displaystyle\int_{\tau_h} |h(t)|^2 \, dt} \right]^{1/2} = \begin{cases} \text{rms duration of squared filter weighting function, also} \\ \text{equal to differentiator gain to filter noise output} \end{cases}$$

$$\alpha_c \equiv \left[\frac{\displaystyle\int_{\tau_c} (2\pi t)^2 \, |a(t)h(t)| \, dt}{\displaystyle\int_{\tau_c} |a(t)h(t)| \, dt} \right]^{1/2} = \text{rms duration of actual output signal voltage}$$

Note: A physical interpretation of rms duration for each function can be found as the curvature of the corresponding voltage spectral density.

$$\alpha_a^2 = -\frac{1}{A_{mc}} \frac{\partial^2 A_c}{\partial \epsilon_f^2} \bigg|_{\substack{\epsilon_f = 0 \\ h(t) = h_m a(-t)/a_m}} = \begin{cases} \text{normalized second derivative of spectrum of } |a(t)|^2, \text{ or} \\ \text{of correlator output when } h(t) \text{ is matched to } a(t) \end{cases}$$

$$\alpha_h^2 = -\frac{1}{A_{mc}} \frac{\partial^2 A_c}{\partial \epsilon_f^2} \bigg|_{\substack{\epsilon_f = 0 \\ a(t) = a_m h(-t)/h_m}} = \begin{cases} \text{normalized second derivative of correlator output when} \\ a(t) \text{ is matched to } h(t) \end{cases}$$

$$\alpha_c^2 = -\frac{1}{A_{mc}} \frac{\partial^2 A_c}{\partial \epsilon_f^2} \bigg|_{\epsilon_f = 0} = \text{normalized second derivative of actual output voltage spectrum}$$

The difference slope ratio is

$$K_{rf} \equiv \frac{K_f}{K_{of}} = \frac{\alpha_c^2}{\alpha_a \alpha_h} \sqrt{\eta_f}, \tag{4.11}$$

where the ideal (matched-filter) difference slope is $\alpha_a = K_{of}$. While α_a and α_h may be found separately for the signal and the filter (using tables given below and in the appendices), α_c and η_f must be calculated for each combination of signal waveform and filter weighting function.

Depending on the form of these functions, it may prove convenient to find the difference slope K_f from either rms time durations or derivatives of the frequency functions. For example, if the signal spectrum or filter transfer function can be approximated by a function whose first or second derivatives are constant over each segment of the frequency band, then the integrals in the frequency domain are more easily evaluated. A number of equivalent expressions for K_f are given in Table 4.5, and several of these will be used in the later analysis.

A special property of the rms duration of a time function is its relationship to the central curvature of the corresponding voltage spectrum (see note in Table 4.4). As will be seen below, this property sets an upper limit on the product $K_f B_{3a}$ for practical waveforms and filters, and permits us to express in convenient form the error as a fraction of signal bandwidth.

Table 4.5

ALTERNATE EXPRESSIONS FOR K_f

Basic definition: $K_f \equiv -\dfrac{\sqrt{\tau_{na}/\tau_{nh}}}{\Sigma_o}\dfrac{\partial\Delta(\epsilon_f)}{\partial\epsilon_f}\Bigg|_{\epsilon_f=0}$ (A)

$K_f = -\dfrac{\sqrt{\tau_{na}/\tau_{nh}}}{A_o}\dfrac{\partial A_d(\epsilon_f)}{\partial\epsilon_f}\Bigg|_{\epsilon_f=0} = \dfrac{A_{mc}}{A_o}\sqrt{\dfrac{\tau_{na}}{\tau_{nh}}}\dfrac{\alpha_c^2}{\alpha_h} = \dfrac{\alpha_c^2}{\alpha_h}\sqrt{\eta_f}$ (B)

If all filtering is lumped into the sum-channel function,

$A_d(\epsilon_f) = \dfrac{1}{\alpha_n}\dfrac{\partial A_c}{\partial\epsilon_f} = \dfrac{1}{\alpha_h h_m}\displaystyle\int_{-\infty}^{\infty} H(f)\dfrac{\partial A(f-\epsilon_f)}{\partial\epsilon_f}\,df$ (C)

$= \dfrac{1}{h_m}\displaystyle\int_{-\infty}^{\infty} H_d(f+\epsilon_f)A(f)df,$ (D)

$\dfrac{\partial A_d}{\partial\epsilon_f}\Bigg|_{\epsilon_f=0} = \dfrac{1}{h_m}\displaystyle\int_{-\infty}^{\infty}\dfrac{\partial H_d(f+\epsilon_f)}{\partial\epsilon_f}A(f)df,$ (E)

$K_f = -\dfrac{\sqrt{\tau_{na}/\tau_{nh}}}{A_o h_m}\displaystyle\int_{-\infty}^{\infty}\dfrac{\partial H_d(f+\epsilon_f)}{\partial\epsilon_f}A(f)\,df.$ (F)

Substituting $A_0 = A_m\sqrt{B_{na}\tau_{na}}$, $h_m = H_m\sqrt{B_{nh}/\tau_{nh}}$,

$K_f = -\dfrac{1}{\sqrt{B_{na}B_{nh}}}\displaystyle\int_{-\infty}^{\infty}\dfrac{1}{H_m}\dfrac{\partial H_d(f+\epsilon_f)}{\partial\epsilon_f}\dfrac{1}{A_m}A(f)\,df$ (G)

$= -\dfrac{\displaystyle\int_{-\infty}^{\infty}\dfrac{\partial H_d(f+\epsilon_f)}{\partial\epsilon_f}A(f)df}{\left[\displaystyle\int_{B_a}|A(f)|^2\,df\displaystyle\int_{B_h}|H(f)|^2\,df\right]^{1/2}},$ (H)

$K_f = \dfrac{\dfrac{1}{\alpha_h}\displaystyle\int_{-\infty}^{\infty}\dfrac{\partial^2 H(f+\epsilon_f)}{\partial\epsilon_f^2}A(f)\,df}{\left[\displaystyle\int_{B_a}|A(f)|^2\,df\displaystyle\int_{B_h}|H(f)|^2\,df\right]^{1/2}}$ (I)

$= \dfrac{\displaystyle\int_{\tau_c}(2\pi t)^2\,|h(t)a(t)|\,dt}{\left[\displaystyle\int_{\tau_h}(2\pi t)^2\,|h(t)|^2\,dt\displaystyle\int_{\tau_a}|a(t)|^2\,dt\right]^{1/2}}.$ (J)

4.3 Expressions for Noise Error (Single Pulse)

Ideal Case (Matched Filter)

The expression derived by Woodward, and given in Chap. 1 as Eq. (1.20), defines the minimum error in frequency measurement for a given signal waveform. It applies when frequency is measured by determining the maximum output of a matched filter as it is tuned over the signal spectrum. In the correlator of Fig. 4.2, it describes the extent to which noise will shift the apparent null in the difference-channel narrowband filter bank, when the weighting function is a time-reversed image of the signal waveform.

$$\boxed{\begin{array}{c} \textit{Ideal Minimum Error} \\ (\textit{Matched Filter}) \\[2mm] (\sigma_f)_{min} = \dfrac{1}{\alpha_a\sqrt{\mathscr{R}_1}} = \dfrac{1}{K_{of}\sqrt{\mathscr{R}_1}} \end{array}} \qquad (4.12)$$

Here, $\mathscr{R}_1$ is the ratio of total energy in the signal pulse to noise power per hertz, and α_a is as defined in Table 4.4 for the pulse envelope waveform $a(t)$. Values of α_a for several waveforms are given in Table 4.6.

Mismatched Filter

When the receiver weighting function $h(t)$ is not matched to the signal waveform, there will be a decrease in signal-to-noise ratio in the sum channel, and a decrease in the difference slope as well. The error for a single pulse will be:

$$\boxed{\begin{array}{c} \textit{Frequency Error for Mismatched Filter} \\[2mm] \sigma_{f1} = \dfrac{1}{K_f\sqrt{\mathscr{R}_1}} = \dfrac{\sqrt{\eta_f}}{K_f\sqrt{2S/N}} = \dfrac{1}{K_f\sqrt{2L_m S/N}} \end{array}} \qquad (4.13)$$

This can be expressed in terms of the rms durations of signal and weighting functions:

$$\sigma_{f1} = \frac{\alpha_h}{\alpha_c^2\sqrt{\eta_f\,\mathscr{R}_1}} = \frac{\alpha_h}{\alpha_c^2\sqrt{2S/N}}. \qquad (4.14)$$

For most signals and weighting functions, the output signal duration to the filter banks may be found approximately from

$$\frac{1}{\alpha_c^2} \simeq \frac{1}{\alpha_{a1}^2} + \frac{1}{\alpha_{h1}^2}, \qquad (4.15)$$

where

$$\alpha_{a1} \equiv \left[\frac{\displaystyle\int_{\tau_a} (2\pi t)^2\,|a(t)|\,dt}{\displaystyle\int_{\tau_a} |a(t)|\,dt}\right]^{1/2} = \left\{\begin{array}{l} \text{rms duration of signal voltage} \\ \text{waveform} \end{array}\right. \qquad (4.16)$$

$$\alpha_{a1}^2 = -\frac{1}{A_m}\frac{\partial^2 A}{\partial f^2}\bigg|_{f=0} = \left\{\begin{array}{l} \text{normalized second derivative of the} \\ \text{signal voltage spectrum} \end{array}\right. \qquad (4.17)$$

$$\alpha_{h1} \equiv \left[\frac{\int_{\tau_h} (2\pi t)^2 \, |h(t)| \, dt}{\int_{\tau_h} |h(t)| \, dt}\right]^{1/2} = \text{rms duration of weighting function} \qquad (4.18)$$

$$\alpha_{h1}^2 = -\frac{1}{H_m} \frac{\partial^2 H}{\partial f^2}\bigg|_{f=0} = \text{second derivative of filter transfer function} \qquad (4.19)$$

These functions are also tabulated in Table 4.6 for the waveforms listed there.

Table 4.6

SIGNAL AND FILTER PARAMETERS

Waveform or Weighting Function	τB_{3a}, τB_{3h}	τ_{3a}/τ, τ_{3h}/τ	τ_{na}/τ, τ_{nh}/τ	α_a/τ_a, α_h/τ_h	α_{a1}/τ_a, α_{h1}/τ_h	$\alpha_a B_{3a}$, $\alpha_h B_{3h}$	$\alpha_{a1}B_{3a}$, $\alpha_{h1}B_{3h}$
Rectangular, $a(t) = 1$	0.886	1.000	1.000	1.81	1.81	1.607	1.607
Triangular, $a(t) = 1 - |2t/\tau|$	1.276	0.293	0.333	0.994	1.28	1.268	1.636
Parabolic, $a(t) = 1 - 2t^2/\tau^2$	0.972	0.765	0.718	1.53	1.66	1.481	1.613
Parabolic, $a(t) = 1 - 4t^2/\tau^2$	1.155	0.541	0.533	1.188	1.407	1.372	1.624
Cosine, $a(t) = \cos(\pi t/\tau)$	1.189	0.500	0.500	1.136	1.370	1.350	1.629
Cosine², $a(t) = \cos^2(\pi t/\tau)$	1.441	0.367	0.375	0.89	1.134	1.283	1.636
Cosine⁴, $a(t) = \cos^4(\pi t/\tau)$	1.853	0.263	0.313	0.669	0.886	1.240	1.645
Gaussian,							
$a(t) = \exp(-t^2/2\sigma_t^2)$	*	*	*	*	*	1.177	1.665
(let $\tau = 6\sigma_t$)	1.59	0.278	0.295	0.74	1.048	1.177	1.665

*For Gaussian pulse, $B_3\sigma_t = \sqrt{\ln 2}/\pi$, $\tau_3/\sigma_t = 2\sqrt{\ln 2}$, $\tau_n/\sigma_t = \sqrt{\pi}$, $\alpha/\sigma_t = \sqrt{2\pi}$, $\alpha_1/\sigma_t = 2\pi$.

4.4 Error Normalized to Signal Bandwidth

For monochromatic signals whose spectral width is set by the waveform rather than by phase modulation, the following relationships connect the rms durations to the half-power spectral widths:

$$\alpha_{a1} B_{3a} \cong \alpha_{h1} B_{3h} \cong \alpha_c B_{3c} \cong 1.63. \qquad (4.20)$$

Using these, we can write simplified expressions for the frequency error in terms of spectral width at the sum-channel multiplier output. For the matched filter,

$$\boxed{\begin{array}{c} \textit{Ideal Error Normalized} \\ \textit{to Output Spectrum Width} \\ \dfrac{(\sigma_f)_{min}}{B_o} = \dfrac{1}{\alpha_a B_o \sqrt{\mathscr{R}_1}} \cong \dfrac{1}{1.63\sqrt{\mathscr{R}_1}} \end{array}} \qquad (4.21)$$

For the mismatched filter, we have

$$\frac{\sigma_{f1}}{B_{3c}} = \frac{1}{K_f B_{3c} \sqrt{\mathcal{R}_1}} \cong \frac{\alpha_h/\alpha_c}{1.63\sqrt{\eta_f \mathcal{R}_1}}. \tag{4.22}$$

Because the output spectral width will depend both on the signal spectrum and the filter transfer function, we will describe the broadening of the output spectrum relative to the signal by a factor $K_s \equiv B_{3c}/B_{3a}$, and calculate this factor for each combination of signal and filter. The error may then be expressed in terms of the input spectrum width:

$$\boxed{\begin{array}{c} \textit{Error Normalized} \\ \textit{to Input Spectrum Width} \\[2mm] \dfrac{\sigma_{f1}}{B_{3a}} = \dfrac{1}{K_f B_{3a} \sqrt{\mathcal{R}_1}} \cong \dfrac{K_s \alpha_h/\alpha_c}{1.63\sqrt{\eta_f \mathcal{R}_1}} \end{array}} \tag{4.23}$$

The normalized slope $K_f B_{3a}$ is a convenient measure of performance for a frequency estimator, and we will compute and plot this below for several special cases of interest.

In general, Eqs. (4.10) and (4.20) may be combined to express the normalized slope as a function of correlator efficiency and the ratios of parameters given in Table 4.6:

$$K_f B_{3a} \cong 1.63 \frac{p}{1+p^2} \frac{\alpha_{h1}}{\alpha_h} \sqrt{\eta_f}, \tag{4.24}$$

where $p \equiv \alpha_{h1}/\alpha_{a1}$ is a measure of correlator-to-signal matching (as is also η_f), and α_{h1}/α_h is a function of the multiplier function alone. The spectrum broadening factor is also a function of the duration ratio p, as follows:

$$K_s \cong \frac{\alpha_{a1}}{\alpha_c} = \sqrt{\frac{1+p^2}{p^2}}. \tag{4.25}$$

Special Case : Gaussian Signal and Multiplier Functions

The parameters of a monochromatic Gaussian pulse, multiplied by a Gaussian function, are given in Table 4.7. The duration ratio $p = \sigma_c/\sigma_t$ determines efficiency, spectrum broadening factor, and normalized difference slope, as shown in Fig. 4.4. The situation is analogous to that of delay measurement on a Gaussian pulse with a Gaussian filter (Table 3.5 and Fig. 3.3).

Table 4.7

PARAMETERS FOR GAUSSIAN PULSE AND GAUSSIAN FILTER

$$a(t) = a_m \exp\left(-\frac{t^2}{2\sigma_t^2}\right) = a_m \exp\left(-2\pi^2\sigma_a^2 t^2\right)$$

$$A(f) = A_m \exp\left(-\frac{f^2}{2\sigma_a^2}\right) = A_m \exp\left(-2\pi^2\sigma_t^2 f^2\right)$$

$$\frac{h(t)}{h_m} = \exp\left(-\frac{t^2}{2\sigma_c^2}\right) = \exp\left(-2\pi^2\sigma_h^2 t^2\right)$$

$$a_c(t) = a_m \exp\left(-\frac{t^2}{2\sigma_0^2}\right) = a_m \exp\left(-t^2\frac{\sigma_t^2 + \sigma_c^2}{2\sigma_t^2\sigma_c^2}\right)$$

$$A_c(f) = A_{mc} \exp\left(-2\pi^2 f^2 \frac{\sigma_t^2\sigma_c^2}{\sigma_t^2 + \sigma_c^2}\right)$$

$$p \equiv \frac{\alpha_{h1}}{\alpha_{a1}} = \frac{\sigma_c}{\sigma_t}$$

$$\frac{\sigma_0}{\sigma_t} = \frac{\tau_{3c}}{\tau_{3a}} = \frac{B_{3a}}{B_{3c}} = \frac{1}{K_s} = \sqrt{\frac{p^2}{1 + p^2}}$$

$$\tau_{3h}B_{3a} = \frac{2p\ln 2}{\pi} = 0.44p$$

$$\eta_f = \frac{2p}{1 + p^2}$$

$$K_f = 4\pi\sigma_t\left(\frac{p}{1 + p^2}\right)^{3/2} = \frac{4\sqrt{\ln 2}}{B_{3a}}\left(\frac{p}{1 + p^2}\right)^{3/2} = \frac{3.33}{B_{3a}}\left(\frac{p}{1 + p^2}\right)^{3/2}$$

$$K_{of} = \alpha_a = \sqrt{2}\,\pi\sigma_t = \frac{\sqrt{2\ln 2}}{B_{3a}} = \frac{1.177}{B_{3a}}$$

$$K_{rf} = \frac{K_f}{K_{of}} = \left(\frac{2p}{1 + p^2}\right)^{3/2}$$

$$\sigma_{f1} = \frac{1}{K_f\sqrt{\mathcal{R}_1}} = \frac{B_{3a}}{3.33}\sqrt{\frac{(1 + p^2)^3}{p^3\mathcal{R}_1}}$$

$$(\sigma_f)_{min} = \frac{B_{3a}}{1.177\sqrt{\mathcal{R}}} \qquad \text{for } p = 1$$

Special Case: Gaussian Signal, Rectangular Gate

A practical approximation of the correlator system takes the form of a rectangular range gate in place of the linear multiplier in the RF or IF stage of the receiver. The performance curves are of the same form as for time-delay measurement on a Gaussian signal passed through a rectangular filter (Fig. 3.4). If we let τ_g denote the gate width, and $y = \tau_g/(\sqrt{8}\,\sigma_t)$, then $\exp(-y^2)$ represents the relative signal amplitude $a(t)/a_m$ at each end of a centered gate. From Eq. (4.9), we have

$$\eta_f = \frac{\left|\dfrac{1}{a_m}\displaystyle\int_{-\tau_g/2}^{\tau_g/2} a(t)\,dt\right|^2}{\dfrac{\tau_g}{a_m^2}\displaystyle\int_{\tau_a} |a(t)|^2\,dt} = \sqrt{\frac{\pi}{2}}\frac{(\operatorname{erf} y)^2}{y}. \qquad (4.26)$$

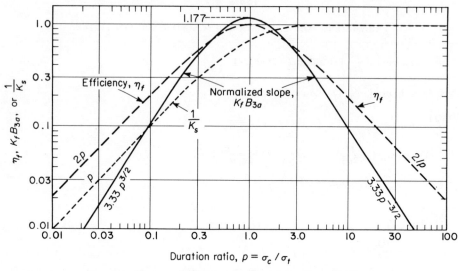

Fig. 4.4 Slope and efficiency for Gaussian pulse and filter.

Also, $\alpha_{h1} = \alpha_h = \pi\tau_g/\sqrt{3}$, $p = y\sqrt{2/3}$, and $K_s = \sqrt{(1.5 + y^2)}/y$.

$$K_f B_{3a} = \sqrt{6\sqrt{\pi/2}\ln 2}\; y^{-3/2}\left[\operatorname{erf} y - \frac{2}{\sqrt{\pi}}ye^{-y^2}\right]. \tag{4.27}$$

Figure 4.5 shows these factors, and also shows how well the points calculated from the approximation of Eq. (4.20) agree with the exact curves.

Special Case : Rectangular Pulse, Rectangular Gate

Rectangular pulses and gates are often approximated in practical radar, and the gate may be wider or narrower than the received pulse. Table 4.8 lists the performance parameters for both situations. The normalized difference slope $K_f B_{3a}$ is not plotted, since it can be represented on a log-log scale by two straight-line segments of slopes $\frac{3}{2}$ and $-\frac{3}{2}$ relative to the ratio τ_g/τ_a, intersecting at the matched-filter value $K_f B_{3a} = 1.607$.

Special Case : Pulse Compression Waveform

A long rectangular pulse is often coded with internal phase modulation to provide better range resolution. If the receiving (compression) filter is matched to the signal pulse over its duration τ_a, the frequency-measurement properties of the system are identical to those of the uncoded (monochromatic) pulse of the same duration. This ideal performance is given by Eq. (4.12), with $\alpha_a = \pi\tau_a/\sqrt{3}$. Reduction of time sidelobes requires that amplitude weighting be applied to the receiving

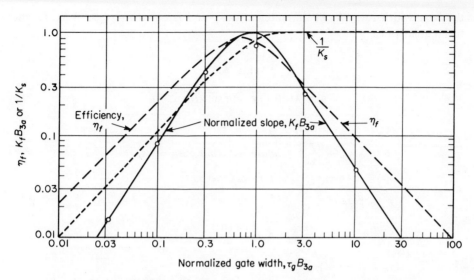

Fig. 4.5 Slope and efficiency for Gaussian pulse, rectangular gate. The solid curve represents the exact value of $K_f B_{3a}$. Circles along this curve show values of $K_f B_{3a}$ calculated from Eq. (4.20).

filter. This weighting reduces the error sensitivity of the system in accordance with Eqs. (4.14) and (4.15), by reducing the rms duration of the weighting function and the multiplier output. Note the output waveform of the equivalent correlator, $a_c(t)$, is not compressed in time, but its spectrum is compressed in frequency as a result of removal of phase modulation in the multiplier.

By substituting in Eq. (4.10) the appropriate expressions from Table 4.4, with $|a(t)| = 1$ over the duration τ_a, we obtain:

$$K_f = \frac{\int_{\tau_h} (2\pi t)^2 \, |h(t)| \, dt}{\left[\tau_a \int_{\tau_h} (2\pi t)^2 \, |h(t)|^2 \, dt \right]^{1/2}} \qquad (\tau_h \leq \tau_a). \qquad (4.28)$$

This will be recognized as the same form used previously to calculate angle difference slope in Eq. (2.14) and range difference slope for the uniform-spectrum case (Table 3.6). Furthermore, for the linear FM (chirp) signal, the weighting applied to $h(t)$ is identical to that on $H(f)$, so that the slope is related to time-side-lobe level in accordance with Figs. 2.2–2.4. When the error is normalized to the total signal spectrum width B_a, we have

$$\frac{\sigma_{f1}}{B_a} = \frac{K_s(\alpha_h/\alpha_{h1})}{1.81 D \sqrt{\eta_f \mathcal{R}_1}}, \qquad (4.29)$$

where the dispersion factor $D \equiv B_a \tau_a$, and the spectral broadening factor for pulse compression is defined as $K_s \equiv B_{3c}/B_o$. This broadening factor is analogous to the beamwidth ratio K_θ in the antenna case, and is equal to $\pi\tau_a/\sqrt{3}\,\alpha_{h1}$ for the rectangular pulse.

Table 4.8

RECTANGULAR PULSE AND GATE

$$a(t) = a_m \quad \text{over } \tau_a \leq \tau_g$$

$$A(f) = A_m \frac{\sin \pi f \tau_a}{\pi f \tau_a} \quad \text{where } A_m = a_m \tau_a,\ B_{na} = 1/\tau_a$$

$$h(t) = 1 \quad \text{over } \tau_g, \qquad a_c(t) = a_m \quad \text{over } \tau_a$$

$$E_1 = \frac{a_m^2 \tau_a}{2}$$

$$\alpha_a = \alpha_{a1} = \alpha_c = \frac{\pi\tau_a}{\sqrt{3}} = K_{of} = \frac{1.607}{B_o}$$

$$\alpha_h = \alpha_{h1} = \frac{\pi\tau_g}{\sqrt{3}}$$

$$\eta_f = \frac{a_m^2 \tau_a^2}{\tau_g a_m^2 \tau_a} = \tau_a/\tau_g$$

$$(\sigma_f)_{min} = \frac{\sqrt{3}}{\pi\tau_a\sqrt{\mathscr{R}}} = \frac{B_o}{1.607\sqrt{\mathscr{R}}} \qquad \text{(matched filter, } \tau_a B_o = 0.886)$$

$$\sigma_{f1} = \frac{\alpha_h}{\alpha_a^2\sqrt{\eta_f \mathscr{R}_1}} = \frac{\sqrt{3\tau_g^3}}{\pi\sqrt{\tau_a^5 \mathscr{R}_1}}$$

$$K_f B_{3a} = 1.607(\tau_a/\tau_g)^{3/2}$$

Similarly, when $\tau_a \geq \tau_g$, $a_c(t) = a_m$ over τ_g

$$\alpha_c = \alpha_h = \alpha_{h1} = \frac{\pi\tau_g}{\sqrt{3}}$$

$$\eta_f = \tau_g/\tau_a$$

$$K_f B_{3a} = 1.607(\tau_g/\tau_a)^{3/2}$$

$$\sigma_{f1} = \frac{\alpha_h}{\alpha_h^2\sqrt{\eta_f \mathscr{R}_1}} = \frac{\sqrt{3\tau_a}}{\pi\sqrt{\tau_g^3 \mathscr{R}_1}}$$

4.5 The Frequency Discriminator

Simplified Model

The preceding discussion of correlator characteristics lays the groundwork for consideration of the more often used frequency discriminator. A simplified block diagram of such a system is shown in Fig. 4.6, and the frequency functions are sketched in Fig. 4.7. Basically, two receiving filters are used, with a small displacement between their center frequencies. When their outputs are subtracted, a difference-channel output is generated whose voltage vs frequency-error (Δ vs ϵ_f)

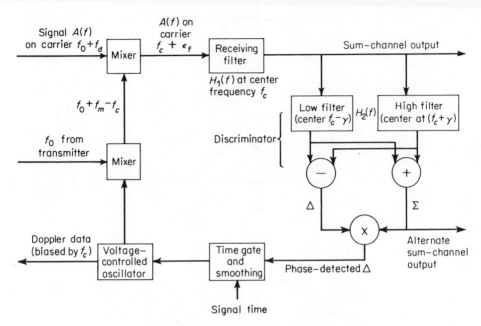

Fig. 4.6 Simplified block diagram of Doppler tracker.

characteristic resembles that of the correlator (Fig. 4.3). Thus the Doppler shift of signals whose spectra are centered between the two filter frequencies may be measured with an accuracy determined by the difference slope K_f of the equivalent correlator. If each filter is matched to the signal, and the displacement is small relative to the width of the signal spectrum, ideal performance will be obtained for signals near the midpoint. Either tracking of the signal or *a priori* estimation of the signal frequency will be required to permit the system to be tuned to the approximate signal frequency.

Equivalent Correlator

For any receiver-discriminator characteristic curve (Δ vs ϵ_f), we may substitute the equivalent correlator system of Fig. 4.8, using an ideal differentiator and a combined weighting function $h(t)$ representing the convolution of h_1 with h_2. The function $h_1(t)$ is intended to describe the actual sum-channel weighting, but the difference-channel output is affected only by the convolution of h_1 with h_2, the latter representing the weighting function of the discriminator filter pair. The functions shown in Fig. 4.9 represent the weighting and filter response for an idealized triangular discriminator characteristic. If we assume that the receiving filter H_1 is broadband, relative to H_2 and to the signal, then the system will have the characteristics given in Table 4.9. In this case, the discontinuities in the transfer function lead to weighting functions of infinite duration. The difference slope

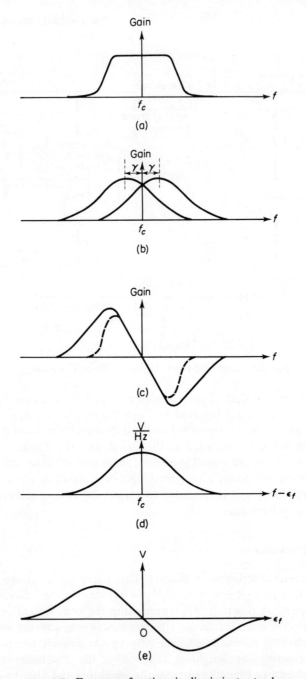

Fig. 4.7 Frequency functions in discriminator-tracker: (a) receiving filter $H_1(f)$; (b) discriminator filters $H_2(f)$; (c) discriminator frequency response $H_d(f)$ [solid curve, without $H_1(f)$; broken curve, with $H_1(f)$]; (d) signal input $A(f)$; (e) difference-channel output vs error, Δ vs ϵ_f.

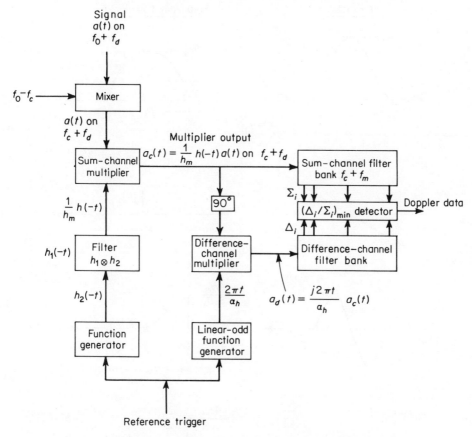

Fig. 4.8   Correlator equivalent of frequency discriminator Doppler tracker.

K_f is found most conveniently from Eq. (G) of Table 4.5, since the derivative of H_d is very easy to convolve with the signal spectrum. The results are plotted in Fig. 4.10 for a rectangular pulsed signal. Also plotted are similar curves for an idealized rectangular discriminator, consisting of two adjacent rectangular filters whose outputs are subtracted. The curves for the gated systems will be discussed later.

Effect of Excessive Discriminator Bandwidth

The fall-off of normalized slopes for $B_h \tau \gg 1$ in Fig. 4.10 reflects the failure to use information received over the entire pulse duration. The equivalent correlator has a weighting function which samples the pulse near its center, rejecting most of the available energy. In practice, the discriminator bandwidth may have to be quite wide to accommodate unknown frequency shifts, but its output can be

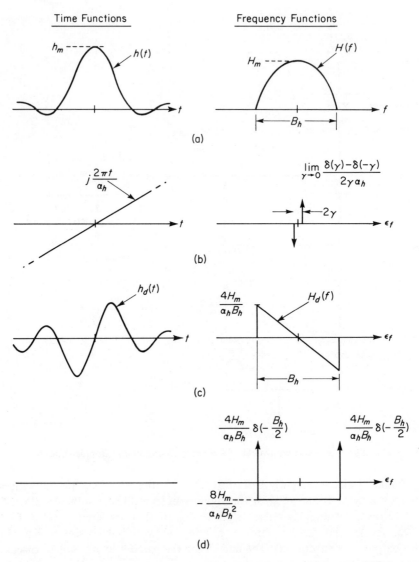

Fig. 4.9   Triangular discriminator functions: (a) sum-channel filter; (b) difference-channel multiplier; (c) equivalent difference-channel filter; (d) derivative of $H_d(f)$, used in Eq. (G) of Table 4.5.

averaged or filtered over the entire pulse width to recover some of the lost information (see Sec. 4.7 below).

Gated Receiver Preceding Discriminator

The effect of gating in a wideband stage prior to the receiving filter is to eliminate noise which originates before and after the gate in time. When the gate width

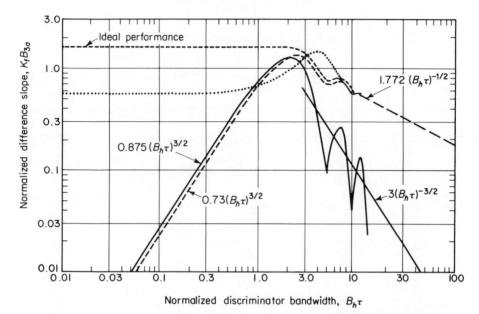

Fig. 4.10 Sensitivity of idealized discriminators with rectangular pulsed signal. Solid curve represents the triangular discriminator; long dashes, rectangular discriminator; short dashes, gated rectangular discriminator ($\tau_g = \tau$); dotted line, gated rectangular discriminator ($\tau_g = 2\tau$).

equals the pulse width, the bandwidth of succeeding stages can be reduced to achieve an ideal, matched system ($K_f B_{3a} = 1.607$). As the gate is widened, the normalized slope is reduced:

$$K_f B_{3a} \cong 1.607 \, (\tau/\tau_g)^{3/2} \qquad \text{for } \tau_g > \tau, \, B_h\tau < 1. \tag{4.30}$$

Thus, when the gate is twice the width of the pulse, the slope for low bandwidths is reduced to about 0.57, as shown by the dotted curve in Fig. 4.10. For the idealized rectangular discriminator, the slope increases with B_h, reaching a peak value almost as high as for the matched-filter case when $B_h\tau \cong 4$.

Curves for the gated discriminator can be calculated most easily by combining frequency- and time-domain integrals in the expression for K_f, using the numerator from Eq. (I) of Table 4.5 and the denominator from Eq. (J):

$$K_f = \frac{-\int_{-\infty}^{\infty} \dfrac{\partial^2 H(f + \epsilon_f)}{\partial \epsilon_f^2} A(f) \, df}{\left[\int_{-\infty}^{\infty} |a(t)|^2 \, dt \int_{-\tau_g/2}^{\tau_g/2} (2\pi t)^2 \, |h(t)|^2 \, dt \right]^{1/2}}. \tag{4.31}$$

Reference to Table 4.5 will show that the dimensions and constants in this form are consistent with the definition of K_f. The filter functions $H(f)$ and $h(t)$ apply to

Table 4.9

TRIANGULAR DISCRIMINATOR PARAMETERS

$h_2(t) = h_m \dfrac{3}{x^2} \left[\dfrac{\sin x}{x} - \cos x \right]$ = equivalent sum-channel weighting function, where $x = \pi B_h t$

$H_2(f) = H_m \left[1 - \left(\dfrac{2f}{B_h} \right)^2 \right]$ = equivalent sum-channel transfer function, for region $|f| \leqslant B_h/2$

$h_d(t) = \dfrac{j2\pi t}{\alpha_h} h_2(t)$ = weighting function of discriminator

$H_d(f) = \dfrac{1}{\alpha_h} \dfrac{\partial H_2(f)}{\partial f} = \dfrac{H_m}{\alpha_h} \dfrac{8f}{B_h^2}$ = transfer function of discriminator, for $|f| \leqslant B_h/2$

$\tau_{nh} = \dfrac{6}{5 B_h}$ = equivalent duration of sum-channel function $h_2(t)$

$\alpha_h = \dfrac{\sqrt{10}}{B_h}$ = rms duration of sum-channel function $|h_2(t)|^2$

$B_{nh} = \dfrac{8 B_h}{15}$ = noise bandwidth of sum-channel function $H_2(f)$

$K_f = \dfrac{1}{A_m} \sqrt{\dfrac{3}{B_{na} B_h}} \left[\dfrac{2}{B_h} \displaystyle\int_{B_h} A(f)\, df - A(B_h/2) - A(-B_h/2) \right]$

For a pulsed signal, $A(f) = A_m (\sin \pi f \tau)/\pi f \tau$, with $B_{na} = 1/\tau$,

$K_f = \sqrt{\dfrac{12\tau}{B_h}} \left[\dfrac{\text{Si}(z)}{z} - \dfrac{\sin z}{z} \right]$ \quad where $z = \pi B_h \tau/2$

$K_f B_{3a} = 1.96 (B_h \tau)^{-3/2} [\text{Si}(z) - \sin z]$

$\cong 3.05 (B_h \tau)^{-3/2}$ \quad for $B_h \tau \gg 1$

$\cong 0.875 (B_h \tau)^{3/2}$ \quad for $B_h \tau \ll 1$

the combination of receiving filter and discriminator which follows the gate, and the effect of the gate on output noise is given by the placing of finite limits on the second integral of the denominator. As long as the signal lies entirely inside the gate, its output is not affected by the gate, and integrals over infinite limits may be applied. For the rectangular pulse and gate, the result can be expressed as a normalized difference slope given by

$$K_f B_{3a} = \dfrac{1.97[1 - (\sin z)/z]}{\sqrt{z[\text{Si}(2y) - 0.5\,\text{Si}(4y) - (\sin y)^4/y]}}, \qquad (4.32)$$

where $z = \pi B_h \tau/2$ and $y = \pi B_h \tau_g/4$.

In all cases plotted in Fig. 4.10, the oscillation of the functions above $B_h \tau = 1$ can be attributed to the sharp discontinuities in the idealized discriminator response. Practical circuits do not exhibit these discontinuities, and their response functions will be smoother than those plotted, although similar in general form.

4.6 Application to Coherent Pulse Trains

Long-Duration Signals

In the preceding analysis, frequency measurements on a single-pulse signal have been described, and the error has been related to the duration of that pulse and the width of its spectrum. Nothing in the analysis has placed a limit on the duration of signals considered, and we may apply all the results to a "pulse" formed by keying a CW transmitter for any period, or by scanning the antenna of a CW system past a stable target. As long as the match between bandwidths of the signal and the receiver-discriminator (or between the durations of the signal and the weighting function of the correlator) is considered, the expressions for difference slope and noise are directly applicable to signals of any duration. The signal frequency must remain constant or follow a known function over this duration, of course, to permit the receiving system to form the appropriate sharply-peaked response function in the frequency domain.

Pulse Trains

In cases where a phase reference can be maintained over a train of pulses, the duration of the entire train can be taken as the effective signal duration for purposes of frequency measurement. The receiving system must be matched to the fine-line structure within the spectral envelope, shown in Fig. 4.1. One practical method for realizing such a filter is to apply a range gate prior to fine-line filtering in the receiver (Fig. 4.11). This gate, repeating in time with each pulse of the received train, may be considered as part of a "spectral envelope" filter which optimizes the receiver for reception of each individual pulse. Imperfection in this match will degrade the signal-to-noise ratio at the input to the fine-line filter, but will not otherwise affect the frequency error of the estimator.

The equivalent correlator configuration of Fig. 4.12 shows how the two filter functions can be separated for simplicity in analysis. The mixer, shaped gate, and ideal narrowband filter of width f_r will convert the received pulse train to a low-frequency modulation envelope on the carrier at $f_c + f_d - if_r$. At this point, the ratio of signal power to the noise power within f_r is equal to the S/N ratio at the peak of each filtered pulse in Fig. 4.11, modified by integration over the gate τ_g. As the amplitude of the pulse train varies (with antenna scan angle or target fluctuation), the modulation envelope reproduces the changes, and the fine-line correlator sees this envelope as its input waveform $a(t)$. Frequency measurements are made within the interval $f_c \pm f_r/2$, and the central line of a signal whose Doppler shift exceeds $f_r/2$ is brought into this interval by proper selection of the frequency offset if_r applied in the first mixer. An error in selection of the integer i causes an ambiguity in Doppler measurement by some multiple of f_r, and if the error is large it may also reduce the ratio $(S/N)_1$ which is available to the estimator.

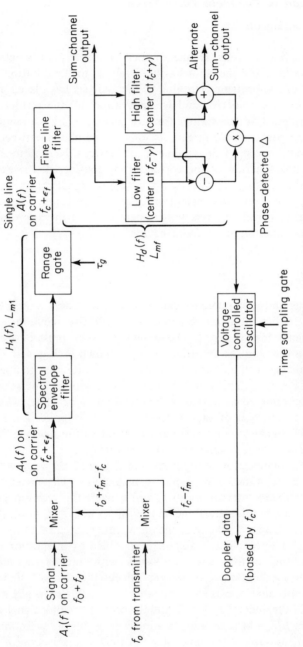

Fig. 4.11 Simplified block diagram of fine-line tracker.

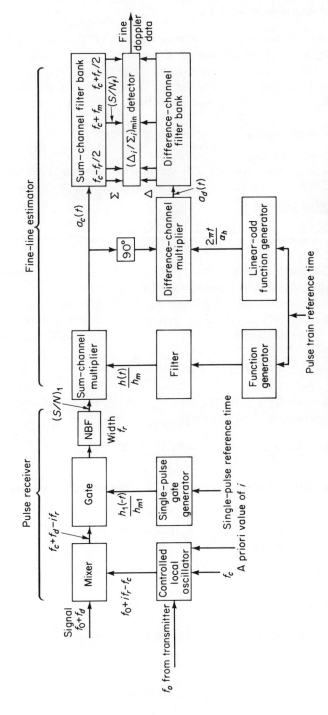

Fig. 4.12 Correlator equivalent to fine-line frequency tracker.

Evaluation of Signal-to-Noise Ratio

Since we will apply an equation similar to Eq. (4.13) to find the frequency error caused by noise, we must use a value of $\mathscr{R}$ appropriate to the entire pulse train, and must include loss factors to express the effects of the two filter processes—one applied to each individual pulse and one to the envelope of the train. For an observation which includes the entire pulse train, the energy ratio is given by

$$\mathscr{R} = \frac{1}{N_o} \int_{-\infty}^{\infty} \left[a_1(t) \frac{a(t)}{a_m} \right]^2 dt, \tag{4.33}$$

where $a_1(t)$, the single-pulse waveform, is repeated at time intervals t_p and modulated by the normalized function $a(t)/a_m$. The single-pulse energy ratio reaches a peak value, when $a(t) = a_m$, of

$$\mathscr{R}_m = \frac{1}{N_o} \int_{\tau_{a1}} |a_1(t)|^2 \, dt = \frac{2S\tau}{N_o} = 2L_{m1}(S/N)_m, \tag{4.34}$$

where L_{m1} is the matching loss of the pulse receiver, $(S/N)_m$ is the maximum of $(S/N)_1$ during the train, and $S\tau$ is the energy of a single pulse.

In search radar, we consider the "time-on-target" t_o to be the period during which the target lies between the 3-db points of the one-way antenna pattern. The number of pulses exchanged per scan is $n = f_r t_o$, and the total energy is

$$\mathscr{R} = \frac{n\mathscr{R}_m}{L_p} = \frac{2nL_{m1}(S/N)_m}{L_p}, \tag{4.35}$$

where L_p is the beamshape loss. The loss L_{m1} reduces the effective signal energy ratio available to the estimator, which is

$$\mathscr{R}_f = \frac{\mathscr{R}}{L_{m1}} = \frac{n\mathscr{R}_m}{L_{m1}L_p} = \frac{2n(S/N)_m}{L_p}. \tag{4.36}$$

Finally, we may define a signal-to-noise ratio at the sum-channel output of the estimator.

$$\left(\frac{S}{N} \right)_f \equiv \frac{[\Sigma(0)]^2}{2N_f} = \frac{\eta_f \mathscr{R}_f}{2} = \frac{n(S/N)_m}{L_p L_{mf}}, \tag{4.37}$$

where $L_{mf} = 1/\eta_f$ is the loss in matching the fine-line filter to the individual spectral line, found from Eq. (4.9).

In tracking radar, t_o is the period over which a measurement is conducted, and all pulses in the train are received with equal gain ($L_p = 1.0$). If a tracking servo or data output filter of noise bandwidth β_n (one-sided) is used, $t_o = 1/(2\beta_n)$, $n = f_r/(2\beta_n)$, and the estimator has

$$\mathscr{R}_f = \frac{n\mathscr{R}_1}{L_{m1}} = \left(\frac{S}{N} \right)_1 \left(\frac{f_r}{\beta_n} \right), \tag{4.38}$$

$$\left(\frac{S}{N}\right)_f = \frac{n\mathscr{R}_1}{2L_{m1}L_{mf}} = \frac{(S/N)_1(f_r/\beta_n)}{2L_{mf}}.$$ (4.39)

The measurement or observation period t_o in a tracking system often exceeds the time over which signal coherence can be maintained (i.e., radar or target instabilities may spread the fine-line width B_{na} beyond $2\beta_n = 1/t_o$). The total energy is still given by Eq. (4.33), but η_f will be reduced by the appearance of amplitude and phase instability terms in the signal functions $a(t)$ and $A(f)$, and by the inability of the fine-line filter to match to these unknown functions. In practice, then, the fine-line filter bandwidth must be broadened to pass the spectrum of the unstable signal, and the extended observation time provides only the advantage of simple averaging of errors, reducing the output error in proportion to the square root of the observation time or energy. When the fine-line filter is matched to the signal line, Eq. (4.39) becomes

$$\left(\frac{S}{N}\right)_f = \left(\frac{S}{N}\right)_1\left(\frac{f_r}{B_{na}}\right).$$

Error Expressions for Coherent Pulse Train

Where the pulse train maintains its coherence over the observation, the frequency error is found from the equations derived in Secs. 4.2–4.5 above, by inserting the fine-line signal and filter functions along with the effective energy ratio $\mathscr{R}_f$. The ideal error expression of Eq. (4.12) is changed only by substitution of total energy $\mathscr{R}$ for the single-pulse energy $\mathscr{R}_1$, provided that the pulse-receiving filter is matched to the individual pulse, and the estimating filter to the pulse-train envelope.

$$\boxed{\begin{array}{c} \textit{Ideal Minimum Error} \\ \textit{(Matched Filters)} \\[6pt] (\sigma_f)_{min} = \dfrac{1}{\alpha_a\sqrt{\mathscr{R}}} = \dfrac{1}{K_{of}\sqrt{\mathscr{R}}} \end{array}}$$ (4.40)

For mismatched filters, we modify the earlier expressions slightly to identify the separate losses introduced by the pulse-receiving filter (L_{m1}) and by the fine-line estimating filter $(L_{mf} = 1/\eta_f)$.

$$\boxed{\begin{array}{c} \textit{Frequency Error for Mismatched Filters} \\[6pt] \sigma_f = \dfrac{1}{K_f\sqrt{\mathscr{R}_f}} = \dfrac{\sqrt{L_{m1}}}{K_f\sqrt{\mathscr{R}}} = \dfrac{\sqrt{\eta_f}}{K_f\sqrt{2(S/N)_f}} \end{array}}$$ (4.41)

This error may also be expressed in terms of rms durations of the pulse train and the fine-line filter functions, as was done in Eq. (4.14) for the single pulse:

$$\sigma_f = \frac{\alpha_h}{\alpha_c^2\sqrt{\eta_f\mathscr{R}}} = \frac{\alpha_h}{\alpha_c^2\sqrt{2(S/N)_f}}. \qquad (4.42)$$

When normalized to the 3-db width of the individual spectral lines, we have

> *Error Normalized to Width*
> *of Signal Lines*
>
> $$\frac{\sigma_f}{B_{3a}} = \frac{1}{K_f B_{3a}\sqrt{\mathscr{R}_f}} \simeq \frac{K_s\alpha_h/\alpha_c}{1.63\sqrt{2(S/N)_f}} \qquad (4.43)$$

The curves of Figs. 4.4 and 4.10 apply to frequency discriminators used in fine-line measurement as well as for single pulses. For most cases, however, the Gaussian functions will better describe the pulse-train envelope in scanning, the broadening of spectral lines caused by system instability, and the filter functions found in practical discriminators.

Scanning Radar

The parameters listed in Table 4.7 for the Gaussian signal can be applied directly to coherent measurement in a stable system which scans past the target. The pulse-train amplitude for the two-way (echo) case is

$$a(t) = a_m \exp\left[\frac{-t^2}{2\sigma_t^2}\right] = a_m \exp\left[\frac{-2.77t^2}{t_o^2}\right] \quad \text{(V)}, \qquad (4.44)$$

where the time-on-target $t_o = \sqrt{2}\,\tau_{3a} = 2.35\sigma_t$ is measured between 3-db points on the one-way pattern. The filter-to-signal duration ratio p is related to t_o by

$$p = \sigma_c/\sigma_t = \tau_{3h}/\tau_{3a} = \sqrt{2}\,\tau_{3h}/t_o. \qquad (4.45)$$

Using Eq. (4.36) for energy ratio available to the estimator, we can write the following for frequency error:

$$\sigma_f = \frac{\sqrt{L_p}}{K_f\sqrt{2n(S/N)_m}} = \frac{\sqrt{L_p}}{5.32t_o\sqrt{2n(S/N)_m}}\left[\frac{1+p^2}{p}\right]^{3/2} \qquad (4.46)$$

For the optimum fine-line filter, $\tau_{3h} = t_o/\sqrt{2}$, $p = 1$, and

$$(\sigma_f)_{opt} = \frac{1}{K_{of}\sqrt{\mathscr{R}_f}} = \frac{\sqrt{L_p}}{1.88t_o\sqrt{2n(S/N)_m}}. \qquad (4.47)$$

The curves of Fig. 4.5 can be applied to the coherent case where a pulse receiver is gated on during the period τ_g which surrounds the passage of the beam over the target. In terms of t_o, we have $y = 0.832\tau_g/t_o$, and $\tau_g B_{3a} = 0.625\tau_g/t_o$. The optimum performance for frequency measurement requires that the gate width be about 1.5 times the time-on-target. This is longer than the optimum integration time for detection, because the measurement of frequency places a premium on long-duration observations.

Resolution of Doppler Ambiguity

Precise measurements on a received spectral line are of little value unless the line can be identified with the corresponding transmitted line (i.e., unless the integer i can be chosen correctly). There are three methods for making the choice of i:

1. The signal frequency may be known *a priori*, on the basis of external measurements or physical constraints, to lie within a particular ambiguity interval f_r in frequency;

2. The frequency may be measured noncoherently over the pulse train, in addition to the coherent measurement, with sufficient accuracy to identify the proper interval; or

3. The frequency may be computed from range rate, as measured over the period of the pulse train.

In use of the second option, we find in Sec. 4.7 that the optimum accuracy of noncoherent measurements, averaged over a uniform train of n pulses of width τ, is

$$\sigma_{fn} = \frac{\sqrt{3}}{\pi\tau\sqrt{\mathscr{R}}} = \frac{\sqrt{3}}{\pi\tau\sqrt{n\mathscr{R}_1}}.$$

To resolve the ambiguity with high probability of success (e.g., 99.7 percent), we must have $\sigma_{fn} \ll f_r$. This requires that

$$\frac{\pi\tau f_r\sqrt{\mathscr{R}}}{\sqrt{3}} \geq 6 \qquad \text{or} \qquad \tau f_r\sqrt{\mathscr{R}} \geq 3.3.$$

This is the same constraint derived by Manasse (1955, p. 19) in his discussion of optimum frequency measurements on pulse trains. Manasse also shows that failure to meet this constraint imposes a limitation on frequency accuracy which corresponds to the noncoherent value σ_{fn}. For pulse trains with small duty factor $D_u = \tau f_r$, this is a difficult requirement to meet. The energy ratio $\mathscr{R}$ must exceed $11/D_u^2$, and in most practical equipment the system instabilities will prevent our resolving the Doppler ambiguity by this method when D_u is less than about 0.03.

In short-pulse radars (D_u typically less than 0.01), the use of differentiated range data offers a better approach to Doppler ambiguity resolution. The optimum accuracy of a single-pulse delay estimate is

$$\sigma_t \cong \sqrt{\frac{\tau}{2B_a\mathscr{R}_1}} \qquad \text{[see Eq. (3.41)].}$$

If an optimum differentiator is used to obtain radial velocity from a pulse train of duration t_o, the error in target velocity will be

$$\sigma_v = \frac{c\sigma_t}{2t_o}\sqrt{\frac{12}{n}} = \frac{c}{2t_o}\sqrt{\frac{6\tau}{B_a\mathscr{R}}},$$

where $\mathscr{R} = n\mathscr{R}_1$ represents the total energy ratio of the train. Converting this to a Doppler frequency error, we have

$$\sigma_{fd} = \frac{2\sigma_v}{\lambda} = \frac{f_o}{t_o}\sqrt{\frac{6\tau}{B_a\mathscr{R}}}.$$

For ambiguity resolution with 99.7 percent success,

$$\frac{f_r}{\sigma_{fd}} = \frac{n}{f_o}\sqrt{\frac{B_a\mathscr{R}}{\tau}} \geq 6.$$

The fractional bandwidth $d = B_a/f_o$ of the receiver may be introduced to obtain a clearer interpretation of this constraint:

$$nd\sqrt{\frac{\mathscr{R}}{B_a\tau}} = nd\sqrt{2n(S/N)} \geq 6.$$

Thus, for example, if a system operates at 3 GHz within a receiver bandwidth of 3 MHz, $d = 0.001$, and the constraint is

$$n\sqrt{2n(S/N)} \geq 6000.$$

Resolution of Doppler ambiguity by differentiation of range data is not an unreasonable requirement for a tracking radar with these parameters, since (S/N) will normally exceed unity, and 200 to 300 pulses will often be available for processing. Systems with greater fractional bandwidth would not need as many pulses, while those in which nonoptimum measurements are made would need more pulses, and hence longer periods before the ambiguity could be resolved. In search radar, however, it is seldom possible to resolve ambiguity by either of the means described above.

4.7 Effects of Practical Signal Processing

Detector Characteristics

Referring to the correlator configuration of Fig. 4.2, we see that the final detection process for Doppler data consists of a comparison of difference-to-sum ratios for filters of different center frequencies. For maximum sensitivity to frequency shift, and minimum cross-talk with any time-delay error, the detector should be phase-sensitive, ignoring the components of Δ which are in quadrature with Σ. The phase reference is provided by the stronger (sum) signal, and the detector loss will be set by the S/N ratio of the sum channel.

$$L_x = \frac{(S/N) + 1}{(S/N)} = \frac{S + N}{S} \qquad \text{[see Eq. (2.53)].}$$

Measurements on Noncoherent Pulse Train

When noncoherent measurements are made over a train of n pulses, the energy ratio may be high enough for reliable detection and for elimination of noise ambiguities, while the single-pulse ratio $(S/N)_1$ is still below unity. This leads to appreciable detector loss, and the effective value of $\mathscr{R}$ in the equations must be reduced by L_x.

Error for n-Pulse Noncoherent Measurement

$$\frac{\sigma_f}{B_{3a1}} = \frac{\sqrt{L_x}}{K_{f1}B_{3a1}\sqrt{\mathscr{R}}} = \frac{\sqrt{(S/N)_1 + 1}}{K_{f1}B_{3a1}(S/N)_1\sqrt{2nL_{m1}}} \qquad (4.48)$$

(The single-pulse difference slope K_{f1} and the error are both normalized to the width of the spectral envelope.)

The same form of detector loss applies to the discriminator of Fig. 4.6, and a similar loss appears in any type of discriminator at low S/N ratio.

In search radar, the detector loss will depend upon the average ratio $(S/N)_m/L_p$.

Error in Noncoherent Search Radar

$$\frac{\sigma_f}{B_{3a1}} = \frac{\sqrt{L_pL_x}}{K_{f1}B_{3a1}\sqrt{2nL_{m1}(S/N)_m}} = \frac{\sqrt{(S/N)_m + L_p}}{K_{f1}B_{3a1}(S/N)_m}\sqrt{\frac{L_p}{2nL_{m1}}} \qquad (4.49)$$

Wideband Discriminator

In Sec. 4.5, it was shown that a wideband discriminator $(B_h\tau \gg 1)$ would have a difference slope considerably less than optimum, especially when a linear output-vs-frequency curve is desired (triangular discriminator). Most of this loss can be recovered if the discriminator is followed by a lowpass (video) filter, matched to the input pulse. At high S/N ratios, the discriminator slope can be found from curves such as Fig. 4.10, or from corresponding equations, by using the two-sided bandwidth $2B_v$ of the video filter in place of B_h. The only loss caused by excessive

B_h is then the detector loss described above. Equations (4.43), (4.48), and (4.49) will apply, with K_f computed for a discriminator of width $2B_v$, and (S/N) and L_{m1} computed from actual filter bandwidth. Discriminators whose bandwidths exceed the spectrum width are often used when an unpredictable frequency shift must be accommodated, and this discussion indicates that the loss need not be large for strong signals once the signal has been brought near the center of the discriminator response. The additional error encountered in off-center operation is discussed below.

Off-Center Measurements

As with the off-axis target in angular measurement (Sec. 2.6), there will be an additional noise error if the signal lies off the null point of the discriminator. Part of this increase results simply from a reduction in signal amplitude, caused by the reduced filter transfer function applicable to the off-center signal. We describe this as a loss L_h, which is defined as a function of frequency error ϵ_f:

$$L_h \equiv \left[\frac{\psi(0, \epsilon_f)}{\psi(0, 0)} \right]^2 = \left[\frac{\Sigma(\epsilon_f)}{\Sigma(0)} \right]^2. \tag{4.50}$$

For signal spectra which are narrow relative to the filter bandwidth, we have

$$L_h \cong \left[\frac{H(0)}{H(\epsilon_f)} \right]^2.$$

A second component of off-center noise error is generated by the amplitude-normalization process, where the ratio Δ/Σ is sensitive to sum-channel noise voltage as well as to difference-channel noise and actual signal amplitude. This component is given by

$$\sigma_2 = \frac{\epsilon_f \sqrt{L_h L_{mv}}}{\sqrt{\mathscr{R}}}, \tag{4.51}$$

where L_{mv} is the matching loss associated with the bandwidth of the data filter or normalization time constant. Thus, the error is proportional to frequency offset and inversely proportional to the energy ratio evaluated over the data interval or normalization smoothing time.

The ratios of normalization error σ_2 and of total noise error $\sqrt{\sigma_f^2 + \sigma_2^2}$ to the ideal error $(\sigma_f)_{min}$ are plotted in Fig. 4.13 as functions of the offset ϵ_f for pulsed signals passed through the triangular discriminator. The error increases more sharply for moderate ratios of filter bandwidth to signal bandwidth, because of the more rapid increase in L_h. In each case, it has been assumed that the signal-to-noise ratio is large ($L_x = 1.0$), and that the video filter following the discriminator is matched to the pulse for optimum K_f. According to Fig. 4.10, this requires $B_v \cong 1/\tau$, and results in $K_f B_{3a} \cong 1.25$. Since the ideal (matched-filter) discriminator would have $K_f B_{3a} \cong 1.63$, the total noise is always somewhat greater than $(\sigma_f)_{min}$.

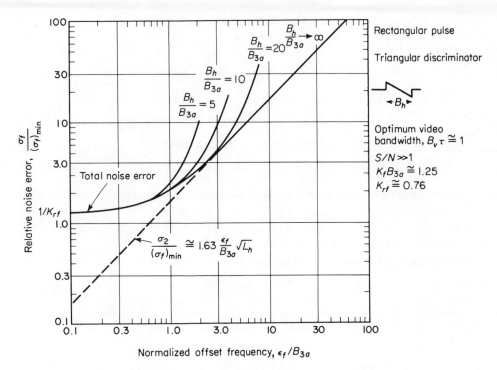

Fig. 4.13   Relative error in off-center discriminator operation.

If the S/N ratio is permitted to fall to unity or below (in the bandwidth B_h), the noise will be larger and will increase more steeply with magnitude of the offset.

Range-Doppler Cross-Talk

The preceding analysis of Doppler error has been based on the assumption that target range (time delay) is known accurately. In the case of the correlator, the multiplier function must be matched in time to the signal, within a fraction of the signal duration, if full performance is to be obtained. For small errors in delay, there will appear at the difference-channel multiplier output a voltage in quadrature with the sum-channel signal and the frequency error signal. This will normally be rejected by the phase-sensitive Δ/Σ ratio detector, and will cause only a small error in Doppler data if the phase shifts through the filters are not perfectly matched.

In the discriminator systems, there is considerable tolerance to delay error, especially when time-gating is introduced after the discriminator. The output of the sum channel may be inspected for presence of a signal peak, and the discriminator output sampled at that time to obtain an efficient estimate. The price paid for this flexibility is that the frequency must be known well enough before the measurement to heterodyne the signal into the linear response region of the

Table 4.10

ERROR EQUATIONS FOR SINGLE PULSE

Ideal

$$(\sigma_f)_{min} = \frac{1}{\alpha_a \sqrt{\mathscr{R}_1}} = \frac{1}{K_{of} \sqrt{\mathscr{R}_1}} \qquad (4.12)$$

Mismatched Filter

$$\sigma_{f1} = \frac{1}{K_f \sqrt{\mathscr{R}_1}} = \frac{\sqrt{\eta_f}}{K_f \sqrt{2(S/N)}} = \frac{1}{K_f \sqrt{2L_{m1}(S/N)}} \qquad (4.13)$$

Ideal (normalized to 3-db spectral width out of matched correlator)

$$\frac{(\sigma_f)_{min}}{B_o} = \frac{1}{\alpha_a B_o \sqrt{\mathscr{R}_1}} \simeq \frac{1}{1.63 \sqrt{\mathscr{R}_1}} \qquad (4.21)$$

Mismatched Filter (normalized to 3-db width of input signal spectrum)

$$\frac{\sigma_{f1}}{B_{3a}} = \frac{1}{K_f B_{3a} \sqrt{\mathscr{R}_1}} \simeq \frac{K_s(\alpha_h/\alpha_c)}{1.63 \sqrt{\eta_f \mathscr{R}_1}} \qquad (4.23)$$

(See Figs. 4.4, 4.5, and 4.10 for values of normalized slope $K_f B_{3a}$.)

Pulse Compression System (uniform spectrum over B_a)

$$\frac{\sigma_{f1}}{B_a} \simeq \frac{K_s(\alpha_h/\alpha_{h1})}{1.81 D \sqrt{\eta_f \mathscr{R}_1}} \qquad (4.29)$$

Off-Center Measurement [For a target displaced ϵ_f from the discriminator null, decrease the energy ratio by the loss L_h, given in Eq. (4.50), and add a second component of error, σ_2.]

$$\sigma_2 = \frac{\epsilon_f \sqrt{L_h L_{mv}}}{\sqrt{\mathscr{R}}} \qquad (4.51)$$

Brief Definitions of Symbols

B_a = width of rectangular spectrum

B_o = 3-db width of matched correlator output spectrum

B_{3a} = 3-db width of signal input spectrum

$D = B_a \tau_a$ = dispersion factor of coded waveform

K_f = relative difference slope (Table 4.3)

K_{of} = value of K_f for matched correlator

K_s = spectral broadening factor, Eq. (4.29)

L_h = detuning loss factor, Eq. (4.50)

L_{m1} = IF matching loss to single pulse

L_{mv} = video matching loss to pulse

$\mathscr{R} = 2E/N_o$ = total signal energy ratio

$\mathscr{R}_1$ = single-pulse energy ratio

S/N = single-pulse (IF) signal-to-noise power ratio

α_a = rms signal duration (Table 4.4)

α_c = rms duration of correlator output (Table 4.4)

α_h = weighting function duration (Table 4.4)

α_{h1} = weighting function (voltage) duration, Eq. (4.18)

ϵ_f = frequency displacement from center of filter

η_f = correlator efficiency, Eq. (4.9)

σ_{f1} = single-pulse rms error

$(\sigma_f)_{min}$ = minimum error (matched correlator)

discriminator, or else that a large number of well-matched filters must be used to cover the range of unknown frequencies.

The problem is much more severe when pulse-compression signals such as linear FM are used. In this case, the response function has an elongated, diagonal ridge which extends over a time interval τ_a and a frequency interval B_a, within which the range and Doppler interpolation must take place. If neither range nor Doppler is known accurately before the measurement, there will remain after measurement an uncertainty along this ridge which is large relative to the time and frequency errors of Table 3.6 and Eq. (4.29). However, as pointed out by Rihaczek (1965b), the uncertainty in range can be reduced to the value predicted by equations in Table 3.6, regardless of Doppler shift, if we consider range data to have been taken at a time displaced by a fixed amount from the actual instant of reflection.

4.8 Summary of Doppler Noise Error

The basic equations for thermal noise in Doppler measurement are repeated in Table 4.10 for single-pulse estimates. Table 4.11 gives the corresponding equations

Table 4.11

ERROR EQUATIONS FOR COHERENT PULSE TRAIN

Ideal

$$(\sigma_f)_{min} = \frac{1}{\alpha_a\sqrt{\mathscr{R}}} = \frac{1}{K_{of}\sqrt{\mathscr{R}}} \tag{4.40}$$

Mismatched Filter

$$\sigma_f = \frac{1}{K_f\sqrt{\mathscr{R}_f}} = \frac{L_{m1}}{K_f\sqrt{\mathscr{R}}} = \frac{\sqrt{\eta_f}}{K_f\sqrt{2(S/N)_f}} \tag{4.41}$$

Ideal (normalized to 3-db spectrum width at correlator output)

$$\frac{(\sigma_f)_{min}}{B_o} = \frac{1}{\alpha_a B_o\sqrt{\mathscr{R}}} \simeq \frac{1}{1.63\sqrt{\mathscr{R}}} \tag{4.40}$$
$$\tag{4.20}$$

Mismatched Filter (normalized to 3-db width of input signal spectrum)

$$\frac{\sigma_f}{B_{3a}} = \frac{1}{K_f B_{3a}\sqrt{\mathscr{R}_f}} \simeq \frac{K_s(\alpha_h/\alpha_c)}{1.63\sqrt{2(S/N)_f}} \tag{4.43}$$

Scanning Radar (with optimum fine-line filter)

$$\sigma_f = \frac{\sqrt{L_p}}{1.88 t_o\sqrt{2n(S/N)_m}} \tag{4.47}$$

Definitions of New Symbols

$\mathscr{R}_f$ = energy ratio available to coherent estimator, Eq. (4.36)

$(S/N)_f$ = signal-to-noise power ratio out of fine-line filter

t_o = time-on-target, or observation time, Eqs. (1.23), (1.24)

σ_f = n-pulse rms error

for coherent pulse trains, and Table 4.12 covers noncoherent trains. The effect of detector loss is included only in Table 4.12, because the S/N ratio in the single-pulse and coherent cases will almost always be adequate to avoid significant loss from this source. In special cases where S/N is less than about 10, as with a wide-band discriminator followed by a video filter, the appropriate loss from Eq. (2.53) may be included to reduce the effective values of $\mathcal{R}$ in Tables 4.10 and 4.11 as well.

Table 4.12

ERROR EQUATIONS FOR NONCOHERENT PULSE TRAIN

(added subscripts 1 refer to single-pulse spectrum)

General Expression with Processing Loss L

$$\frac{\sigma_f}{B_{3a1}} = \frac{\sqrt{L}}{K_{f1}B_{3a1}\sqrt{\mathcal{R}}} \approx \frac{K_s(\alpha_h/\alpha_c)\sqrt{L}}{1.63\sqrt{\mathcal{R}}}$$

With Detector Loss L_x

$$\frac{\sigma_f}{B_{3a1}} = \frac{\sqrt{L_x}}{K_{f1}B_{3a1}\sqrt{\mathcal{R}}} = \frac{\sqrt{(S/N)_1 + 1}}{K_{f1}B_{3a1}(S/N)_1\sqrt{2nL_{m1}}} \tag{4.48}$$

Search Radar with Beamshape Loss L_p

$$\frac{\sigma_f}{B_{3a1}} = \frac{\sqrt{L_p L_x}}{K_{f1}B_{3a1}\sqrt{2nL_{m1}}(S/N)_m} = \frac{\sqrt{(S/N)_m + L_p}}{K_{f1}B_{3a1}(S/N)_m}\sqrt{\frac{L_p}{2nL_{m1}}} \tag{4.49}$$

Relationship of Energy Ratio to S/N Power Ratio

$$\mathcal{R} = n\mathcal{R}_1 = 2nL_{m1}\left(\frac{S}{N}\right)_1 \tag{1.30}$$

$$\mathcal{R}_f = \frac{n\mathcal{R}_1}{L_{m1}} = 2n\left(\frac{S}{N}\right)_1 = \left(\frac{S}{N}\right)_1\left(\frac{f_r}{\beta_n}\right) \tag{4.38}$$

$$\left(\frac{S}{N}\right)_f = \frac{\eta_f \mathcal{R}_f}{2} = \frac{n\mathcal{R}_1}{2L_{m1}L_{mf}} = n\eta_f\left(\frac{S}{N}\right)_1 = \frac{(S/N)_1(f_r/\beta_n)}{2L_{mf}} \tag{4.39}$$

$$n = f_r t_o = \frac{f_r}{(2\beta_n)} \quad \text{(tracker with servo bandwidth } \beta_n)$$

$$n = \frac{f_r \theta_3}{\omega} \quad \text{(search radar scanning at } \omega \text{ rad/sec)}$$

Any received signal other than the direct reflection from the desired target can cause error in measurement, unless it is excluded from the measurement channel. Interfering signals may arrive from other targets illuminated by the radar, from the desired target via secondary paths, from extraneous electrical equipment, or from noise sources within the radar. Errors from thermal noise have been discussed in the preceding chapters, and we will now consider the other sources of interference. There is a close parallel between analysis of noise errors and analysis of errors from interfering signals, and this will be developed in the first section. Following sections will discuss special cases of interference: point targets, distributed targets, multipath reflections, and active electromagnetic interference.

5.1 Error Caused by Interfering Signals

Basic Measurement Process

The basic process of measurement will be reviewed here briefly, in order to show how the analysis for noise must be modified to apply to other interference types. This basic process is best illustrated by the monopulse tracking antenna, with sum and difference receiving channels as shown in Fig. 5.1. For angular measurement, the sum channel serves as a source of a reference signal, against which the difference-channel output is compared to determine the amplitude and direction of the angular error, measured with respect to the tracking axis. After amplification of both signals, the receiver forms a ratio of difference to sum, which is the normalized "error voltage," proportional to angular error. In range or Doppler measurement, the same basic process is carried out for interpolation of target position within the radar resolution cell, using either simultaneous or sequential sampling of adjacent subcells.

5

Multiple

Signal

Problems

in

Measurement

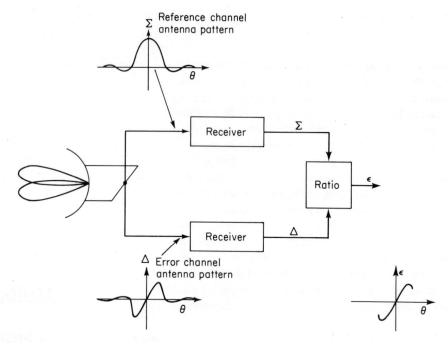

Fig. 5.1 Monopulse angle measurement.

We assume that the noise or interfering signals entering the resolution cell which contains the target are smaller than the desired target signal, so that detection and tracking are possible, but with accompanying error in measurement. Because the different sources of interference are independent, we can calculate their effects separately and add their mean-squared errors to obtain total error. The analysis fails if the sum of the errors is large enough to drive the system into a region of nonlinear response, or to prevent reliable detection.

Expression for Noise Error

The noise error in the coordinate z can be expressed in terms of four factors:

$$\sigma_z = \frac{\sqrt{\eta}}{K_z\sqrt{2(S/N)n}}. \tag{5.1}$$

The difference slope K_z expresses the sensitivity of the z-difference channel in voltage normalized to an ideal sum channel receiving the same signal, when the channel gains have been set for equal noise outputs. It depends only on the configuration of the measurement system and its match to the signal characteristics. Figure 5.2 shows the response of a typical measurement channel, with the slope near the tracking point identified as K_z. The ratio S/N is the signal-to-noise power ratio observed

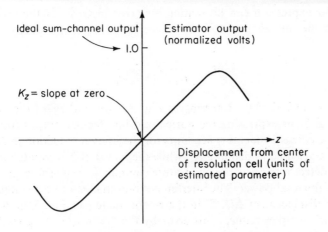

Fig. 5.2   Estimator error response.

on a single sample in the sum channel. A power ratio is used here in preference to energy ratio, because most interfering signals are better described in terms of power levels. The sum-channel efficiency factor η is needed to relate S/N to the ratio which would have been observed in an ideal sum channel, as was used to normalize K_z. Finally, the number of independent samples of noise averaged in the output reading is given by n, which depends on the relative bandwidths of the filter and the single-sample noise spectrum.

Substitution of Interference for Noise

The magnitude of the received interference can be expressed as a "signal-to-interference ratio," S/I, analogous to S/N. This will refer to the power ratio observed in the sum channel and will determine the detectability of the target when interference is present. We will show how S/I is computed from the "response functions" of the radar system in angle and in the range-Doppler plane when the intensity and location of the interference are known relative to those of the signal. Using the ratio S/I, the error caused by interference can be expressed as

$$\sigma_z = \frac{(\Delta/\Sigma)_i \sqrt{\eta}}{K_z \sqrt{2(S/I)n_e}}. \tag{5.2}$$

Here, we have replaced S/N from Eq. (5.1) with S/I, and have denoted the number of independent samples of interference by n_e to emphasize the fact that interference is not always as random as noise, and may remain correlated over part or all of the measurement interval. In order to be able to use the same difference slope K_z which we have derived earlier for noise, a term $(\Delta/\Sigma)_i$ must be included in the numerator. This is the ratio of voltage gain in the z-difference channel to that in the sum channel, for interfering signals at a given location.

A simpler expression can be written for error in terms of the ratio of signal power, S, in the sum channel to interference power, I_Δ, in the difference channel:

$$\sigma_z = \frac{\sqrt{\eta}}{K_z\sqrt{2(S/I_\Delta)n_e}}. \tag{5.3}$$

Here, $(S/I_\Delta) = (S/I)(\Sigma/\Delta)_i^2$. Referring to Fig. 5.1, we can regard the voltage ratio $\sqrt{\eta I_\Delta/2S} = \Delta/\Sigma_o$ as expressing the normalized interference output from the error detector. The efficiency factor η accounts for the known reduction in S relative to the ideal receiver used in defining K_z, while the factor of 2 accounts for the ability of the error detector to reject the quadrature component of interference, assumed to contain half the total power. The interference output cannot be distinguished from a real target displacement $\Delta/K_z\Sigma_o$ in the z-coordinate (see Fig. 5.2). Averaging of n_e interference samples reduces this error by the factor $\sqrt{n_e}$ to give the value σ_z in Eq. (5.3). In an open-loop measuring system, this error appears as an output voltage proportional to σ_z. In a tracking servo system, the interference voltage forces the servo to vary from the true target by σ_z to generate the nulling voltage.

For small displacements of the interfering signal from the target ($z_i \ll z_3$, where z_3 is the 3-db resolution width), we have

$$(\Delta/\Sigma)_i \cong K_z z_i/\sqrt{\eta}.$$

> *Error for Small Displacement,* $z_i \ll z_3$
>
> $$\sigma_z \cong \frac{z_i}{\sqrt{2(S/I)n_e}} \tag{5.4}$$

This agrees with the theory of the two-element target (Chap. 6), in that the error should be proportional to the displacement of the interfering source and inversely proportional to the signal-to-interference voltage ratio. For large displacements ($z_i > z_3$), the interference enters through a sidelobe response in the z-coordinate. In many systems, the difference sidelobe level is proportional to the sum sidelobe level over most of the response region. In particular, for a difference channel which is the derivative of the sum channel, $(\Delta/\Sigma)_i \cong 2$.

> *Error for Large Displacement, Derivative Difference Channel*
>
> $$\sigma_z \cong \frac{\sqrt{2\eta}}{K_z\sqrt{(S/I)n_e}} \tag{5.5}$$

In the general case, (S/I_Δ) must be calculated from the complete radar response functions, integrated over regions which contain interfering signals.

Although we have derived the preceding equations for a monopulse sum-and-difference measuring system, they are applicable to sequential-lobing and scanning systems when the proper error slope parameter K_z is used. In conical scan, for example, an equivalent difference-channel response may be found by subtracting the beam voltage patterns at the two extreme positions of the scan. The resulting difference pattern will exhibit a split main lobe and sidelobes which are almost identical to the ideal (derivative) difference pattern of the monopulse system (Fig. 5.3). For conical scan and search radar, the normalized error sensitivity parameters derived in Chap. 2 for thermal noise will replace the corresponding monopulse sensitivity in Eqs. (5.1)–(5.3). Thus, in equations to be derived below, we use $k_m = K\theta_3/\sqrt{\eta}$ for monopulse, and this can be replaced by $k_s/\sqrt{L_k}$ for conical scan or $k_p/\sqrt{L_p}$ for search radar (see Table 2.8). Also, for conical scan, the number of independent error samples, n, must be reduced by a factor of 2

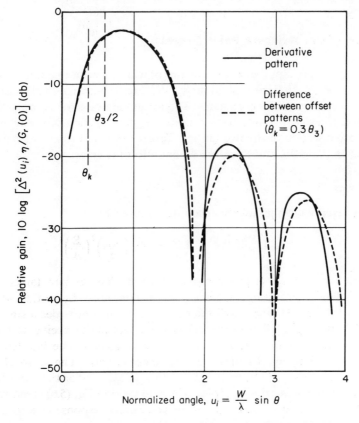

Fig. 5.3 Similarity of conical-scan and monopulse difference patterns (cosine sum illumination).

because of the time-sharing of azimuth and elevation error sensing (this consideration also applies to some types of monopulse systems, which time-share an error
channel). This reduction in number of samples does not necessarily apply to n_e
when that number is less than $n/2$.

Number of Interference Samples

The number of independent samples of interference, n_e, will often be equal to
the total number of radar pulses received during a given observation, $n = f_r t_o$
[see, for instance, Eqs. (1.25) and (1.26) in Chap. 1]. This number is increased
in certain cases by the factor $B_{nh}\tau$, when the interference extends over a receiver
bandwidth B_{nh} which is appreciably greater than $1/\tau$ and when post-detection filtering is used to recover S/N ratio. However, when the interference is an unwanted
echo from the radar transmission, its phase may change only slowly relative to
that of the target, and n_e may be smaller than n. This relative phase change (Doppler difference) will be considered in the discussion of special cases.

5.2 Errors from Adjacent Point Targets

A point target is perhaps the simplest source of tracking error to analyze. Assuming
the radar to be tracking the desired target, the interfering target will cause a tracking error in one or more coordinates. The magnitude of error is found from Eq.
(5.3), where the signal-to-interference ratio S/I_Δ is a function of the relative position and cross section of the two targets. Target range and transmitter power will
cancel out in the ratio, and therefore do not affect the error.

S/I Ratios

The difference-channel interference ratio is given by

$$\frac{S}{I_\Delta} = \left(\frac{S}{I}\right)\left(\frac{\Sigma}{\Delta}\right)_i^2 = \left(\frac{\sigma}{\sigma_i}\right)G_{st}G_{sr}\left(\frac{a_{mx}}{a_i}\right)^2\left(\frac{\Sigma}{\Delta}\right)_i^2 \tag{5.6}$$

for an interfering point target of cross section σ_i. The desired target has cross
section σ. The transmitting antenna gives a gain ratio $G_{st} = G_t(0, 0)/G_t(u_i, v_i)$
in favor of the desired target, and the receiving antenna provides a similar factor
$G_{sr} = G_r(0, 0)/G_r(u_i, v_i)$. The range-Doppler response of the receiver to the desired
target provides an amplitude gain ratio a_{mx}/a_i relative to the interfering target
when normalized to equal voltages at the receiver input. The value of a_i is the
normalized amplitude of the response function (Chap. 1) for time and frequency
displacements of the interfering signal. The final ratio in Eq. (5.6) is the sum-channel power response relative to the difference-channel response at a displacement
z_i in the measured coordinate (u, v, t_d, or f_d).

The sum-channel S/I ratio must be appreciably greater than unity to ensure reliable detection of the desired target. Since we cannot usefully discuss error unless the desired target is detected and resolved from other targets, we will assume in the remainder of the discussion that this sum-channel condition is met, and that the error remains on the linear portion of the error response curve.

Small Separation of Targets

Equation (5.6) gives S/I_Δ in the most general terms, but it can be simplified considerably in special cases. For example, if the interfering target is in the same resolution cell as the desired target we may set to unity all the factors except the cross-section ratio and the sum-to-difference response ratio in the measured coordinate:

$$\frac{S}{I_\Delta} = \left(\frac{\sigma}{\sigma_i}\right)\left(\frac{\Sigma}{\Delta}\right)_i^2 \cong \left(\frac{\sigma}{\sigma_i}\right)\frac{\eta}{K_z^2 z_i^2}, \tag{5.7}$$

$$\sigma_z \cong \frac{z_i}{\sqrt{2(\sigma/\sigma_i)n_e}}. \tag{5.8}$$

As the interfering target moves farther from the desired target in any coordinate, the decreasing response in that coordinate should be included in Eqs. (5.7) and (5.8) as a factor multiplying the cross-section ratio.

Resolution Criteria

Chapter 1 discussed resolution in terms of the radar response function, and an interfering signal was considered resolved if it fell outside the 3-db point on the response function in any one coordinate. Even though it may meet this criterion for resolution, the radar may still be subject to significant error from the interfering target. An alternate definition of resolution may be based on consideration of error: "An interfering signal may be considered resolved when the error it introduces in measurement of the desired target's position is less than a stated amount." By specifying the maximum error and minimum separations for different cross-section ratios, this definition constrains both the main lobe widths and the sidelobe levels in each coordinate.

Sidelobe Response

As an example of how Eq. (5.6) is applied to widely separated targets, let us assume two targets of equal cross section, separated in azimuth but with equal elevation, range, and Doppler. We wish to find the error in measuring azimuth of the desired target. From Eq. (5.6), S/I_Δ is completely determined by the azimuth difference pattern, the sum gain on the axis, the azimuth separation between targets, and the transmitter gain ratio. The receiving sum-channel gain at the interfering target position appears both as $[\Sigma(u_i)]^2$ and as $G_r(u_i)$, leaving the ratio $G_r(0)/[\Delta(u_i)]^2$. Figure 5.3 shows a typical difference pattern, taken from App. A. Since the difference pattern plots in that appendix are given with gain referenced to a uniformly illuminated aperture, the plotted values include the sum-channel efficiency factor η, needed in Eq. (5.3). A pattern for the actual sum-channel illumination will yield the transmitting ratio G_{st}, and the product of G_{st} and $G_r(0)/[\Delta(u_i)]^2$ gives the ratio S/I_Δ. This ratio is then used in Eq. (5.3) to find the azimuth error.

The exact position of the interfering target cannot usually be predicted, and it is more meaningful to characterize the difference pattern by an average sidelobe level over the region in which interference is expected. For estimates of maximum error, the sidelobe peaks can be used, while for rms error the average level is the most appropriate. This average level is 3 db below the peaks in the region concerned, because the lobes are approximately sinusoidal.

If the interfering target is resolved in more than one coordinate, S/I_Δ is increased by the main-lobe-to-sidelobe gain ratio in each resolved coordinate. When the relative position of an interfering target cannot be predicted, the probability of encountering interference of a given level is proportional to the volume from which such a level of interference can be received. This effect tends to equalize the contributions to rms error from targets of different response levels, since strong interference can originate in only a few resolution cells. The lower levels of interference can be expected from very broad regions of sidelobe response, and may contribute as much to the integrated rms error as the mainlobe interference.

Error Frequency

Although the Doppler separation of the two targets may be insufficient for resolution, any radial motion of the interfering target relative to the desired target will cause an oscillation of the position errors in the three spatial coordinates. This oscillation, which occurs at the relative Doppler frequency, affects the number of independent error samples which may be averaged during a given observation period, t_o. Amplitude scintillation of the two targets will cause the error to have noiselike amplitude statistics, rather than a sinusoidal shape, and this will also act to increase n_e. We may estimate an effective spectral width of the error, B_e, as the rms sum of relative Doppler frequency and target scintillation frequency. For $B_e < f_r/2$, the number of independent samples is then $n_e \cong 2B_e t_o$. Reduction in error provided by large n_e can be very desirable, but the single-sample S/I ratio must still be appreciably greater than unity to preserve tracking on the desired target. For this reason, Doppler resolution (predetection filtering) is the more effective technique of error reduction in those cases where target and system stability permit coherent operation.

5.3 Distributed Targets

Having considered how point targets cause measurement errors, we now look at distributed targets from the same point of view. The calculation is quite similar, requiring only that we revise the method for finding the difference-channel signal-to-interference ratio, S/I_Δ. In this section we describe distributed target return, outline the calculation of measurement error, and list some techniques used in radar to reduce this source of interference.

A distributed target return can be regarded as the sum of the returns from many point targets. Radar returns from the ground, sea, weather, and chaff are distributed by the nature of the reflecting surfaces. These targets each consist of many small reflectors, often spaced so that many separate scattering points lie within the radar resolution cell in each of its four dimensions. As a result, the radar sees a coherent summation of the returns from a group of these scatterers. Since they have random orientations and usually also some relative motion, the return appears to be random in both space and time. The time variation of clutter amplitude within a given resolution cell will almost always follow the Rayleigh distribution, which is also the distribution of thermal noise. The mean-square value of this amplitude (or average power) completely describes the distribution. The spatial statistics of many types of clutter can also be represented by the Rayleigh distribution, but care should be taken with ground clutter because of its larger amplitude spread.

Clutter Reflectivity

Distributed targets are generally described by a cross-section density function, which gives the concentration of radar cross section in the four radar coordinates: range, Doppler, azimuth, and elevation. Ground or sea clutter, seen by a surface-based radar, is distributed through 360 degrees in azimuth, at elevation angles near horizontal. Its range extends from zero to a few tens of kilometers (with a lower limit for sea clutter), and the range rate is narrowly distributed about zero velocity (or platform velocity, for a ship or moving vehicle). Weather and chaff occupy irregular volumes at higher elevation angles, extending to much greater ranges and having both velocity spread and a mean velocity which is determined by wind conditions (Nathanson and Reilly, 1967). The density function used to describe ground and sea clutter is the "surface reflectivity," σ^0, which indicates the radar cross section per unit area of surface (a dimensionless coefficient). Weather and chaff are described by a "volume reflectivity," η_v, which is the cross section per unit volume of clutter (usually expressed in m^2/m^3).

Thermal noise is equivalent to clutter which is spread uniformly in all radar coordinates. Thus, the receiver output produced by clutter which is spread in range, Doppler, azimuth, and elevation has the same properties as the output from noise. When a radar has been optimized for detection or measurement of targets in thermal noise, it is also optimum for targets immersed in clutter which covers the surrounding resolution cells in all four coordinates. Note also that uniformly distributed clutter will produce an error equivalent to the case in which there is no prior information about the clutter distribution, if an rms error criterion is used for system optimization.

Our objective here is to describe the effect of distributed targets, or clutter, on measurement accuracy. We give a procedure for calculating error in these cases, and present in Tables 5.1 and 5.2 a brief summary of data which may be used to estimate clutter magnitude and spectrum. Further data may be found in the references listed in the tables.

The rms velocity spread of the clutter may be used to find the spectral width for a given wavelength (Table 5.2). From this, the correlation time of the clutter and the number of independent samples may be found in cases where there is no mean velocity difference between the target and the clutter-reflected echo (see Sec. 5.4). In most cases of target tracking and measurement over a clutter surface or in volume clutter, there will be a significant mean Doppler difference, and the phase of the interfering clutter will appear almost random compared to that of

Table 5.1

CLUTTER REFLECTIVITY*

Clutter Source	*Equation*
Land clutter	$\sigma^0 = \dfrac{0.00032}{\lambda}$

[Describes reflectivity exceeded in only ten percent of resolution cells in mountainous regions, about 20 db above median. Beyond line-of-sight range R_c, reduce σ^0 by the factor $(R/R_c)^4$.]

| Sea clutter | $10 \log_{10} \sigma^0 = -64 + 6K_B + 10 \log_{10} \sin \gamma - 10 \log_{10} \lambda$ |

[Approximate average for all polarizations and wind directions. Below critical grazing angle γ_{max}, subtract the quantity $40 \log_{10} (\gamma_{max}/\gamma)$.]

| Chaff | $\eta_v = 3 \times 10^{-8}\lambda$ |

(For arbitrary weight per unit volume, encountered in typical extended corridor.)

| Rain | $\eta_v = 6 \times 10^{-14} r^{1.6} \lambda^{-4}$ |

(For matched polarization. If circular or other orthogonal polarization is used for cancellation, reduce η_v by a factor of 30 to 100.)

Definitions of Symbols and Units

σ^0 = surface reflectivity (dimensionless)
λ = wavelength (m)
K_B = Beaufort wind scale
γ = grazing angle
γ_{max} = critical angle (see Table 5.3)
η_v = volume reflectivity (m^{-1})
r = rainfall rate (mm/hr)

*from Barton, 1967.

Table 5.2

CLUTTER SPECTRUM*

Clutter Source	Wind Speed (m/sec)	Rms Velocity (m/sec)
Sparse woods	(calm)	0.02
Rocky terrain	3	$\cong 0.0$
Wooded hills	3	0.4
" "	6	0.22
" "	12	0.33
Sea echo	v_w	$v_w/8$
Chaff	0–3	0.3–0.9
"	8	1.2
Rain clouds	—	2–4

To convert rms velocity to spectral spread, multiply by $2/\lambda$:

$$\sigma_f = \frac{2\sigma_v}{\lambda} \quad \text{Hz.}$$

*Barton, 1964, p. 100. For further data, see Nathanson and Reilly, 1967.

the target. The error frequency resulting from relative motion and from target scintillation was discussed in Sec. 5.2, and the rms spread of clutter velocity is merely one more component to be added to the effective spectral width of the error.

Integrated S/I Ratios

The measurement error caused by distributed target echoes can be computed from the general expressions, Eqs. (5.2) and (5.3), when an appropriate signal-to-interference (or signal-to-clutter) ratio is inserted. The basic approach is to add the squares of the errors caused by each of the individual point reflectors which make up the distributed target. Because we have assumed that the total interfering powers are small compared with the desired target return in the sum channel, the error powers (variances) caused by the scatterers add linearly. Under these conditions, the total error can be found by inserting the total interfering power in the S/I ratios used in the general equations.

Expressions for S/I are similar to those used in the case of point targets, but must take into account the distribution of cross section of interference over the radar response functions. In place of the product of discrete values, $\sigma_i G_t(u_i, v_i)$ . . ., we must use the integral of cross-section density weighted by the radar response functions over the entire volume of interest:

$$\frac{S}{I} = \frac{\sigma G_t(0)G_r(0)\,|\psi(0, 0)|^2}{\int_v \eta_v(u, v, f_d, t_d)G_t(u, v)G_r(u, v)\,|\psi(f_d, t_d)|^2\,dv}, \tag{5.9}$$

where σ is the cross section of the desired target. The difference-channel ratio is then obtained from

$$\frac{S}{I_\Delta} = \left(\frac{S}{I}\right)\frac{\int_v \eta_v(z)\Sigma^2(z)\,dz}{\int_v \eta_v(z)\Delta^2(z)\,dz}. \tag{5.10}$$

Here, Σ and Δ are the reference and error channel response functions in z, the measured coordinate (including two-way antenna patterns in angle). Equation (5.10) presumes that the z response is separable from the other coordinates, as will often be the case. When this condition is not met, the form of Eq. (5.9) may be used to find S/I_Δ, but with the z-difference channel response replacing G_r or ψ in the denominator.

Equations (5.9) and (5.10) are written for the most general case of interfering targets distributed in all coordinates. If the cross section is concentrated at a point in any of the coordinates, this density becomes a delta function in that coordinate and the corresponding integral is removed. In particular, for the interfering point target, all integrals disappear, and Eq. (5.6) results.

Surface Clutter Case

Surface clutter may sometimes be regarded as being distributed uniformly in azimuth and range, with a Rayleigh amplitude distribution over different resolution cells of the radar. For measurements on a target lying above such a surface, the S/I ratio can be found as

$$\frac{S}{I} = \frac{\sigma G_{st}G_{sr}}{\sigma_i} = \frac{\sigma G_{st}G_{sr}}{\sigma^0} \times \frac{L_p}{R\theta_a} \times \frac{2}{\tau_n c}. \tag{5.11}$$

Assuming that the clutter is distributed uniformly over several beamwidths in azimuth, surrounding the target, the clutter power in the azimuth difference channel is the same as that in the sum channel, because the channels have been adjusted for equal noise outputs. The azimuth error is

$$\sigma_A = \frac{\sqrt{\eta_a}}{K\sqrt{2(S/I)n_e}} = \frac{\theta_a}{k_m\sqrt{2(S/I)n_e}}. \tag{5.12}$$

The range error is similarly found to be

$$\sigma_t = \frac{\sqrt{\eta_f}}{K\sqrt{2(S/I)n_e}}. \tag{5.13}$$

These expressions also apply to volume clutter which is distributed over the main lobe and major sidelobes, in which case Eq. (5.12) can be applied also to elevation error, substituting the elevation beamwidth θ_e for θ_a.

In the case of surface clutter, most of the echoes come from an elevation angle near the horizon, at an angle E_t below the target. If this is less than θ_e, we regard the clutter as a second target with small displacement. From Eq. (5.4),

$$\sigma_E \cong \frac{E_t}{\sqrt{2(S/I)n_e}} \qquad (E_t < \theta_e/2). \tag{5.14}$$

For $E_t \gg \theta_e/2$, the clutter will lie in sidelobes of the elevation pattern, characterized by a main-lobe-to-sidelobe gain ratio G_{se} for the difference channel and G_{sr} for the sum channel:

$$\sigma_E = \frac{\sqrt{\eta}}{K\sqrt{2(S/I)(G_{se}/G_{sr})n_e}} \qquad (E_t \gg \theta_e/2). \tag{5.15}$$

$$= \frac{\theta_e}{k_m\sqrt{2(S/I)(G_{se}/G_{sr})n_e}}$$

For most well-controlled illumination functions, $G_{se}/G_{sr} \cong 0.25$, and $k_m \cong 1.4$, giving

$$\sigma_E \cong \frac{\theta_e}{\sqrt{(S/I)n_e}}. \tag{5.16}$$

In ground clutter, where a non-Rayleigh distribution is often found, σ_E may be computed for several points of the σ^0 curve, and a weighted average found. Alternatively, a single σ_E value may be computed for the average (not median) value of σ^0. In all cases, (S/I) may be multiplied directly by an MTI or Doppler improvement factor if such a capability exists in the radar system.

Weather Clutter Case

Clutter from rain or snow is an example of a target distributed in all three spatial coordinates, and sometimes in Doppler as well. When the radar measures coordinates of a target immersed in uniformly distributed precipitation, Eq. (5.9) can be reduced to a simple relationship using the dimensions of the resolution cell. Assuming a Gaussian antenna pattern (two-way) and Gaussian pulse shape, we have

$$\frac{S}{I} = \frac{\sigma}{\eta_v} \times \frac{1.74}{R^2\theta_a\theta_e} \times \frac{1.88}{\tau_3 c} = \frac{3.3\sigma}{\eta_v R^2\theta_a\theta_e\tau_3 c}. \tag{5.17}$$

Assuming that the clutter is distributed uniformly over regions of significant response in both sum and difference channels, the same ratio will apply to the difference channel, and Eq. (5.3) becomes

$$\sigma_z = \frac{\sqrt{\eta}}{K_z\sqrt{2(S/I)n_e}} = \frac{R\sqrt{\eta\theta_a\theta_e\tau_3 c\eta_v}}{2.6K_z\sqrt{n_e\sigma}}. \tag{5.18}$$

The constant 1.74 in the numerator of Eq. (5.17) is simply the two-coordinate beamshape loss L_p^2 for the Gaussian beam pattern, while 1.88 is the factor of 2 applied to c for two-way transmissions, divided by the ratio τ_n/τ_3 for the Gaussian pulse. These constants do not vary appreciably for other beam and pulse shapes (see Tables 2.4 and 4.6), so Eqs. (5.17) and (5.18) may be used for any radar system.

Clutter Reduction Techniques

Because of the large measurement errors caused by clutter in many radar applications, it may be necessary to apply one of the clutter reduction techniques which have evolved over many years of radar design. For ground clutter, MTI or Doppler filtering is the most effective technique in general use, with radar fences also used in certain applications. The MTI or Doppler improvement factor represents the contribution from the ratio $[\psi(0, 0)/\psi(f_d, 0)]^2$ in Eq. (5.9), integrated over the frequency spread of the clutter. Clutter fences are merely one means of increasing the gain ratios G_{st}, G_{sr}, and G_{se}, where the fence is considered a part of the antenna system.

Against precipitation clutter, circular polarization can often provide significant improvement in S/I ratio. This is because raindrops are nearly spherical, reflecting a wave whose polarization is opposite to that which was transmitted. Complex targets tend to scatter energy in both polarizations, and only about half is rejected by the receiving antenna on the basis of polarization.

Any improvement in S/I ratio which is obtained by these techniques should be introduced in the equations before error is computed, either through modification of Eqs. (5.9) and (5.10) or as a factor multiplying (S/I) in the error equations. Furthermore, if a frequency diversity technique is used to spread the clutter spectrum, the improvement in post-detection integration can be expressed by an increase in n_e.

Finally, the importance of better resolution in reducing clutter errors can be appreciated from inspection of the various equations given above. In accordance with our alternate definitions of resolution, we can reduce the error from clutter surrounding the target by reducing the widths of the resolution cell (as given by 3-db response contours), while for more separated clutter sources we must reduce sidelobe levels in the appropriate coordinates. In the range-Doppler plane, and also in the angular coordinates for fixed aperture area, the total "volume of response" must remain constant, so that a compromise must be made between the 3-db widths and the average sidelobe levels obtained.

5.4 Multipath Error

When an echo from the desired target is received over a propagation path other than the most direct path between the target and the radar, a "multipath error" can be produced. Such errors can appear in any of the four radar coordinates. A major source of multipath error is surface reflection, as illustrated in Fig. 5.4, and we will confine the discussion to this effect. The location and form of the interfering signal is much better defined for multipath than for more general types of point or distributed interference, so the error expressions are simplified. We will describe multipath errors in elevation, azimuth, range, and Doppler.

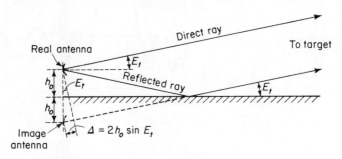

Fig. 5.4   Geometry of surface-reflected multipath interference.

Elevation Multipath Error

The effect of multipath echoes on elevation measurement may be illustrated most easily for the case of monopulse radar, although the same general forms will apply to sequential lobing and other types of antenna. Figure 5.5 shows typical patterns of an amplitude comparison monopulse antenna in the elevation coordinate. When tracking a target at elevation angle E_t, the antenna axis will be tilted upwards to this angle, and surface reflections will be received from an angle $2E_t$ below the axis. The resulting sum and difference signals are shown in the vector diagram, Fig. 5.6. The phase and amplitude of the direct ray in the sum pattern define a reference vector E_r at zero phase angle. The direct ray in the difference channel will appear in phase or 180 degrees out of phase with this reference, depending upon whether the target is slightly above or below the axis, and the servo will act to drive this error to a null. The presence of the reflected component is described by the addition of a small vector in each channel, and this vector is rotated relative to the reference phase by an angle

$$\phi = \frac{4\pi h_o}{\lambda} \sin E_t + \gamma_r, \tag{5.19}$$

where h_o is the height of the antenna and γ_r is the phase angle introduced by reflection from the surface. Since the ratio of reflected ray amplitude to direct ray is small (for targets one beamwidth or more above the surface), the effect on the reference vector is to cause small variations in both phase and amplitude, which have no appreciable effect on system operation. The reflected ray in the error channel, however, simulates an error signal and drives the servo until it is balanced by

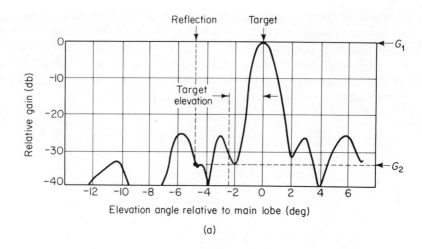

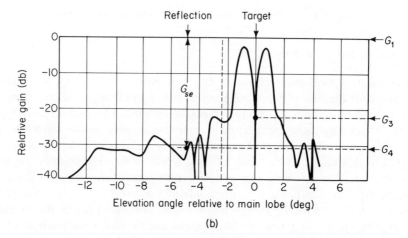

Fig. 5.5 Elevation antenna patterns for multipath error computation: (a) monopulse sum pattern; (b) monopulse difference pattern.

an equal and opposite component of target error. This error is cyclic, as the elevation angle varies.

The direct-to-multipath power ratio is

$$\frac{S}{I_\Delta} = \frac{G_1}{\rho^2 G_4} = \frac{G_{se}}{\rho^2},$$ (5.20)

and the rms error, from Eq. (5.3), is

$$\sigma_E = \frac{\rho\sqrt{\eta}}{K\sqrt{2G_{se}n_e}} = \frac{\theta_e\rho}{k_m\sqrt{2G_{se}n_e}}.$$ (5.21)

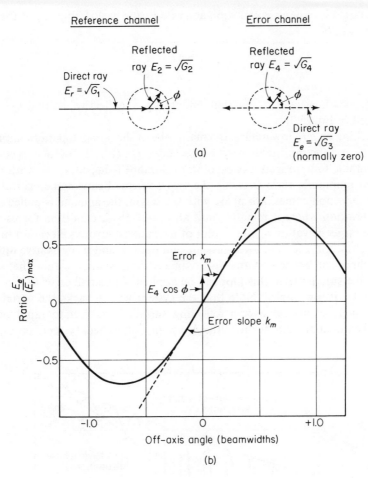

Fig. 5.6 Multipath error relationships: (a) direct and reflected rays; (b) error generated by reflected ray.

Here, ρ is the surface reflection coefficient, and $G_{se} = G_1/G_4$ is the main-lobe-to-sidelobe gain ratio for the angle $2E_t$ below the axis. An average value is taken for the interval in which σ_E is to be found. Since this ratio varies in accordance with the lobe structure of the antenna pattern, and since surface reflectivity and slope may vary at different points around the radar, it is difficult to predict the exact amplitude and phase of the multipath error, but reasonable rms estimates can usually be made. The transmitting pattern factor G_{st} does not appear in Eq. (5.21) because the target is illuminated by the sum pattern with full gain, and is assumed to scatter equal power into the direct and ground-reflected rays. Random variation in the actual power scattered in these slightly different directions will further modify the form of the multipath error, but its rms amplitude will not be changed.

If the target is moving in elevation at a rate $\dot{E}_t$, the basic frequency of the multi-path error will be

$$f = \frac{2\dot{E}_t h_o}{\lambda}.$$ (5.22)

This frequency is usually so low that smoothing is ineffective, so that $n_e = 1$ can be assumed in Eq. (5.3).

In tracking at very low angles, the main lobe of the antenna pattern illuminates the surface and the multipath error grows very large (Fig. 5.7). The target and its image combine to appear as a two-point, or dumbbell-shaped, target with varying phase and amplitude ratio between the two sources. Below some critical angle, when the reflection comes into phase with the target, the antenna is pulled sharply to zero elevation where it remains until an out-of-phase condition forces it back above the target elevation again. A plot of normalized error vs elevation in beamwidths is shown in Fig. 5.8, for Gaussian sum pattern and its derivative difference pattern. Since all beam patterns approximate these shapes within the −10-db points of the sum pattern, this plot can be applied with small error to any system which tracks at low angles. Note that the system which tracks the variations in apparent angle of arrival over a reflecting surface is subject to rapid build-up in error below about 0.8 beamwidth. A heavily smoothed system, or one con-

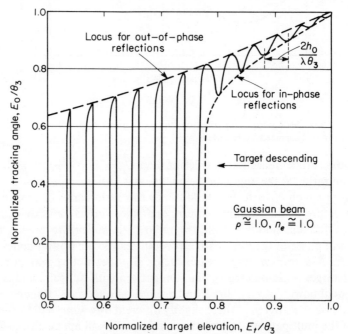

Fig. 5.7 Typical tracking angle vs target elevation.

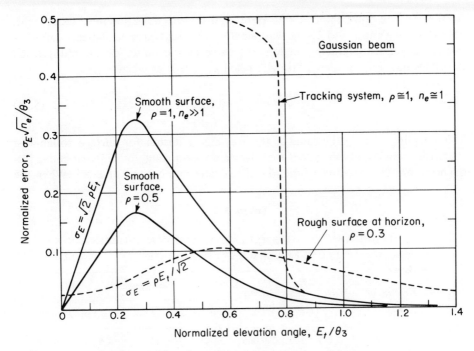

Fig. 5.8 Elevation multipath error vs target elevation.

strained to stay above $\theta_e/2$ in elevation can preserve meaningful (if not accurate) data down to the horizon, as can most trackers when $\rho < 0.5$.

Because the multipath component has approximately the same Doppler as the direct ray [the difference being given by Eq. (5.22)], neither MTI nor sophisticated Doppler processing can be expected to reduce multipath error. Hence, the target may be lost or data rendered useless even when ground clutter is brought below the signal level by a large factor.

Effect of Surface Roughness

The preceding discussion was based on specular reflection from a smooth surface, where the image target appears at an angle $-E_t$ relative to the horizon. When the surface is irregular, specular reflection can still take place at small grazing (elevation) angles, but the reflection coefficient is reduced and the image target is diffused when the target elevation rises above some critical value. If ρ_o is the reflection coefficient for a smooth surface, we may express the coefficient of specular reflection for an irregular surface of the same material as $\rho = \rho_o \rho_s$. The "specular scattering factor" ρ_s gives the rms value of the field which is reflected without being perturbed by the surface irregularities, and can be expressed as

$$\overline{\rho_s^2} = \exp\left[-\left(\frac{4\pi\sigma_h \sin\gamma}{\lambda}\right)^2\right] \tag{5.23}$$

(Beckmann and Spizzichino, 1963). Here, σ_h is the rms deviation in surface height, γ is the grazing angle, and λ is the wavelength. For cases where the flat-earth approximation is valid, the grazing angle is the same as the target elevation angle, E_t. A surface may be considered "rough" when $\overline{\rho_s^2} \leq 0.5$, or when

$$\frac{\sigma_h}{\lambda} \sin \gamma \geq 0.065. \tag{5.24}$$

This is approximately the same as the Rayleigh criterion for surface roughness.

Critical angles, above which the surface appears rough at different radar frequencies, are tabulated as a function of sea state number in Table 5.3. When the

Table 5.3

MAXIMUM ANGLES FOR SPECULAR REFLECTION AT DIFFERENT
RADAR FREQUENCIES AND SEA STATES

Sea State Number	Description of Sea	Wave Height (m)	Rms Height, σ_h (m)	Critical Angle, γ_{max} (deg)			
				$\lambda = 0.7$ m	$\lambda = 0.23$ m	$\lambda = 0.1$ m	$\lambda = 0.03$ m
1	Smooth	0–0.3	0–0.065	> 45	> 13	> 6	> 1.8
2	Slight	0.3–1	0.065–0.21	12–45	4–13	1.8–6	0.5–1.8
3	Moderate	1–1.5	0.21–0.32	8–12	2.6–4	1.2–1.8	0.3–0.5
4	Rough	1.5–2.5	0.32–0.54	5–8	1.6–2.6	0.7–1.2	0.2–0.3
5	Very rough	2.5–4	0.54–0.86	3–5	1–1.6	0.4–0.7	0.12–0.2
6	High	4–6	0.86–1.3	2–3	0.7–1	0.3–0.4	0.09–0.12
7		> 6	> 1.3	< 2	< 0.7	< 0.3	< 0.04

grazing angle exceeds about twice the critical angle, the specular component becomes insignificant, and reflection coefficient will depend upon the "diffuse scattering factor," ρ_d:

$$\rho = \rho_o \rho_d.$$

Figure 5.9 shows how ρ_s and ρ_d vary with surface roughness. Note that ρ_d rises only to about 0.25 at the critical angle where $\overline{\rho_s^2} = 0.5$. This is because both land and sea surfaces, when rough, produce appreciable masking of areas which would otherwise contribute to the energy received at the antenna. Measurements show that ρ_d seldom exceeds 0.4 for any surface, and values near 0.35 are typical.

Diffusely scattered energy arrives at the antenna from an extended area surrounding the image target (the point where the specular reflection would have taken place if the surface were smooth). This area is described by Beckmann and Spizzichino as the "glistening surface," and its size depends upon the target elevation angle and the rms slope σ_α of the rough surface (see Fig. 5.10). Because

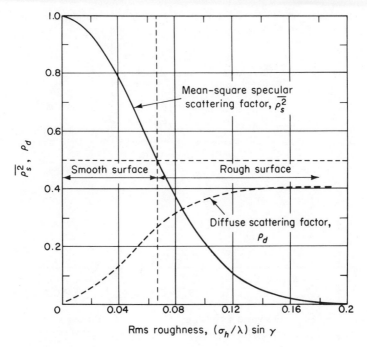

Fig. 5.9 Scattering factors vs roughness (after Beckmann and Spizzichino).

our knowledge of surface characteristics can never be accurate, we will consider the reflected energy as arriving from random-phased scatterers which are distributed uniformly over the glistening surface, but not beyond it. The powers contributed by the many scatterers add, and the total is represented by the square of ρ_d.

An important exception to the general geometry used in the previous discussions of multipath error is found when low-elevation targets are observed over a nearly level surface [Fig. 5.10(d)]. The effect of the earth's curvature here is to produce an intense concentration of scatterers at γ_{max} below the horizon, where near-specular reflection takes place. This region can be treated as a separate source of interfering echoes, independent of the diffuse glistening surface and only about half as wide in the azimuth coordinate. Antenna sidelobe levels are evaluated for this source at an angle $E_t + \gamma_{max}$ below the target, rather than at $2E_t$ as was done earlier. A rough-surface curve was included in Fig. 5.8, which illustrates the effect of this intense reflecting region on elevation multipath error.

The rms slope of the rough surface is related to the rms deviation in surface height: $\sigma_\alpha = 2\sigma_h/d_c$, where d_c is the correlation distance of the surface features. Values from 0.05 radians up to 0.25 radians are observed on both land and sea surfaces. On the sea, the rms slope is a function of wind velocity, rising fairly steeply to 0.15 radians for $v_w = 5$ m/sec (10 knots) and then more slowly to

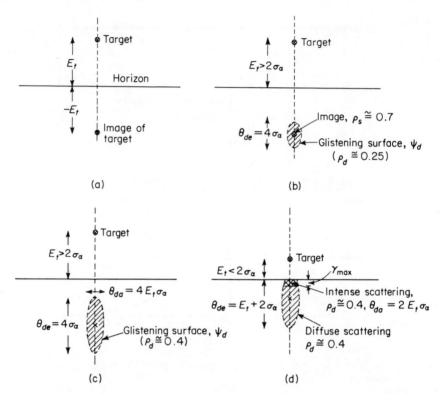

Fig. 5.10 Variation in source of reflection with elevation angle and roughness: (a) specular reflection, $\rho_s \cong 1$, $\rho \cong \rho_o$; (b) slight roughness, $(\sigma_h/\lambda) \sin E_t \cong 0.06$; (c) rough surface; (d) low elevation, curved earth.

0.25 radians at $v_w = 15$ m/sec (30 knots). Thus, values near 0.2 radians can be considered typical for most rough surfaces.

S/I Ratio for Diffuse Reflection

When evaluating the effect of diffuse scattering on radar error, we calculate the signal-to-multipath ratio by integrating the antenna gain over the glistening surface, ψ_d:

$$\frac{S}{I_\Delta} = \frac{G_1 \psi_d}{\rho_o^2 \rho_d^2 \int_{\psi_d} G_d \, d\psi} = \frac{\overline{G_{se}}}{\rho^2}. \tag{5.25}$$

This may be compared to Eq. (5.20) for the case of specular reflection, in which the gain G_4 at an angle $2E_t$ below the axis is brought to bear on the reflected ray. The integration in Eq. (5.25) is a special case of Eqs. (5.9) and (5.10) for distributed interference, expressing the fact that only the resolution of the receiving pattern is effective against multipath reflections.

For small values of surface slope σ_α, and for cases where specular reflection predominates, the integration is equivalent to smoothing the antenna pattern over an interval $\pm 2\sigma_\alpha$ in elevation. When the glistening surface spreads over a region which includes portions of the main beam as well as the sidelobes, the integration must be carried out with due consideration for the relative widths of the azimuth pattern of the antenna and the glistening surface width θ_{da}. For simplicity, the azimuth and elevation patterns can be considered separable, and integration can be carried out in elevation only:

$$\frac{S}{I_\Delta} \cong \frac{G_1 \theta_{de}}{\rho^2 \int_{\theta_{de}} G_d \, dE} \times \frac{\theta_{da}}{\theta_a} \qquad (\theta_{da} > \theta_a). \tag{5.26}$$

Azimuth Multipath Error

The reflections from a smooth, level surface cause no error in the apparent azimuth of a target. Azimuth errors do result, however, from inclined reflecting surfaces and from rough surfaces at any angle. When a smooth, inclined surface reflects energy into the antenna, the multipath error computed for elevation measurement, Eq. (5.21), is merely rotated through the angle of inclination, producing a component in azimuth proportional to the sine of that angle. When the surface is irregular, such that several separated reflecting areas lie within the azimuth beamwidth, the azimuth error is found as a special case of interference distributed in azimuth.

Referring to Fig. 5.10, we see that the width of the region containing scatterers varies from $\theta_{da} = 2E_t\sigma_\alpha$ (for low-elevation targets) to twice this value for high elevations. When this angle does not exceed the azimuth beamwidth, we may use Eq. (5.4), replacing the small displacement z_i by $\theta_{da}/2$, and S/I by the factor $\overline{G_{sr}}/\rho_d^2$:

$$\sigma_A = \frac{\theta_{da}\rho}{2\sqrt{2\overline{G_{sr}} n_e}} \qquad (\theta_{da} < \theta_a). \tag{5.27}$$

The average sidelobe ratio $\overline{G_{sr}}$ for the sum channel appears here because it represents the departure of the azimuth difference channel from the linear slope of its central region. The average is taken over the glistening surface:

$$\overline{G_{sr}} = \frac{G_1 \psi_d}{\int_{\psi_d} G_r \, d\psi}. \tag{5.28}$$

For the low-elevation target, $\theta_{da} \cong 2E_t\sigma_\alpha$, and we have

$$\sigma_A = \frac{E_t\sigma_\alpha\rho}{\sqrt{2\overline{G_{sr}} n_e}} \qquad (E_t < 2\sigma_\alpha, \; \theta_{da} < \theta_a), \tag{5.29}$$

with $\overline{G_{sr}}$ evaluated at the horizon. For higher elevations, $\theta_{da} = 4E_t\sigma_\alpha$, and

$$\sigma_A = \frac{\sqrt{2}\,\rho E_t \sigma_\alpha}{\sqrt{\overline{G_{sr}n_e}}} \qquad (E_t > 2\sigma_\alpha,\ \theta_{da} < \theta_a), \tag{5.30}$$

with $\overline{G_{sr}}$ evaluated at the image target, $-E_t$ relative to the horizon.

At still higher angles, such that the glistening surface extends beyond the azimuth beamwidth, the reflected signal assumes the appearance of noise or distributed clutter. We can assume that the S/I ratios in the sum and difference channels are approximately equal, because of the broad distribution of energy in azimuth. Equation (5.3) yields

$$\sigma_A = \frac{\rho\sqrt{\eta}}{K\sqrt{2\overline{G_{sr}n_e}}} = \frac{\theta_a\rho}{k_m\sqrt{2\overline{G_{sr}n_e}}} \qquad (\theta_{da} > \theta_a). \tag{5.31}$$

The number of independent error samples will depend on the correlation time of the reflections. Over land, this time is quite long relative to most radar observations, being related to the correlation time of clutter echoes. If we assume a Gaussian spectrum and correlation function, the correlation time is related to the standard deviation of the power spectrum, σ_f, by

$$t_c = \frac{1}{\sqrt{2\pi}\sigma_f} = \frac{\lambda}{2\sqrt{2\pi}\sigma_v},$$

where σ_v is the rms velocity spread of the scatterers. If the surface is rocky or free of dense vegetation, σ_v will be dependent upon the tropospheric fluctuations (App. D), which can contribute only a fraction of 1 cm/sec. Wooded hills have an rms velocity spread between 0.04 and 0.3 m/sec, depending on the wind speed. The rms velocity spread of the sea reflections can be expressed approximately as a function of wind speed: $\sigma_v \cong v_w/8$. This leads to the following value of correlation time:

$$t_c \cong \frac{1.6\lambda}{v_w}. \tag{5.32}$$

Target motion can also contribute to decorrelation of the multipath error, especially over land surfaces where the reflection itself is stable. Equation (5.22) gives the elevation component of variation, which is likely to be larger than the azimuth component. Decorrelation in azimuth will occur when the beam has moved through one azimuth beamwidth: $n_e = t_o\dot{A}/\theta_a$. When more than one effect is present, the number of independent samples can be taken as the rms sum of the several effects, not exceeding the number of pulses exchanged during the observation.

Example of Multipath Error Calculation

As an illustration of the varied multipath effects encountered over rough surfaces, let us compare the errors caused by land and sea reflections on the tracker whose antenna patterns were given in Fig. 5.5. The following parameters will be assumed:

$$\lambda = 0.055 \text{ m, vertically polarized transmissions,}$$

$$v_w = 10 \text{ m/sec,}$$

$$\sigma_\alpha = 0.15 \text{ rad,}$$

$$\sigma_h = 0.25 \text{ m (sea), } 1.0 \text{ m (land).}$$

The critical grazing angle over land is

$$\gamma_{max} = 0.065(\lambda/\sigma_h) = 0.0036 \text{ rad, or } 0.2 \text{ deg.}$$

Thus, the surface may be considered rough for all tracking angles, with the scattering surface approximately at the horizon. A reflection coefficient $\rho = \rho_o\rho_d = 0.3$ can be used. Figure 5.8 shows the elevation error for this case, but is based on a Gaussian beam pattern. Above 1.0 deg elevation, the actual sidelobe pattern of Fig. 5.5(b), smoothed over intervals of about one degree, may be applied to Eq. (5.21) to obtain σ_E as a function of target elevation (Fig. 5.11). Targets in level flight at reasonable ranges will have low elevation rates, leading to correlated errors over the usual tracker response times ($n_e = 1$).

Over the sea, a more complex situation prevails. As the target rises from the horizon to $E_t = 0.8$ deg, the specular reflection image drops an equal amount below the horizon (Fig. 5.12). Above this elevation, the surface begins to appear roughened, and the source of intense scattering settles near $E = -1$ deg for any target. Simultaneously, the value of ρ_o begins to drop towards its minimum of 0.12 at a grazing angle of 7 deg. The grazing angle which applies to the reflecting facets, however, is only half the elevation angle. For example, at $E_t = 14$ deg it is facets of the sea surface inclined 7 deg to the horizontal which scatter energy to the radar. Thus, the reflection coefficient will vary with target elevation as shown in Fig. 5.12. The two separate components of multipath error are shown in Fig. 5.11. At very low angles, the curve reproduces the smooth-surface plot of Fig. 5.8, while above $E_t = 1$ deg it follows the rough-surface curve (but with decreasing ρ).

The correlation time for low-elevation errors is long, being dependent on target elevation rate. When the target enters the rough-surface region, we have, from Eq. (5.32),

$$t_c = \frac{1.6\lambda}{v_w} = \frac{1.6 \times 0.055}{10} = 0.009 \text{ sec.}$$

For a servo bandwidth $\beta_n = 1$ Hz, $t_o = 0.5$ sec, and $n_e = t_o/t_c = 50$. This will give a 7:1 reduction in error, shown in the "smoothed" plot of Fig. 5.11.

Azimuth multipath error is calculated from Eq. (5.29) for target elevations below 4 deg, and from Eq. (5.31) for higher elevations. The errors are less than 0.1 mr at all elevations, because of the narrow azimuth beamwidth.

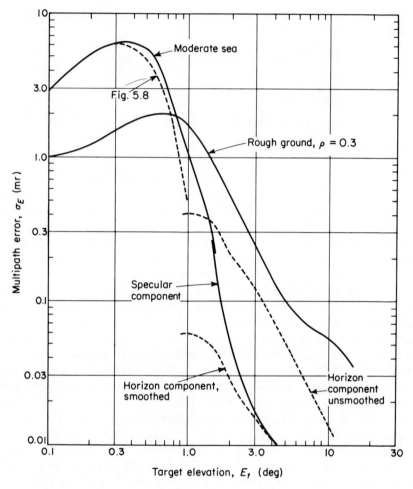

Fig. 5.11   Elevation multipath error ($\theta_3 = 20$ mr, $\sigma_\alpha = 0.15$ rad).

Range Multipath Error

The ray which is reflected from the surface of the earth (see Fig. 5.4), arriving at the receiver through sidelobe response of the antenna, is delayed by an amount equivalent to a range delay

$$\Delta R = h_o \sin E_t. \tag{5.33}$$

Since the phase of this signal varies relative to that of the direct ray, the reflected component may add to or subtract from the direct component in the region of overlap. The center of energy for the combined signal will tend to oscillate about the range of the direct component, with a peak deviation equal to the delay ΔR mul-

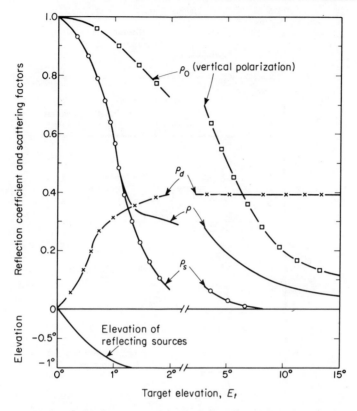

Fig. 5.12 Reflection over rough sea surface ($\lambda = 0.055$ m; sea state 3).

tiplied by the amplitude ratio of the reflected to the direct component. This amplitude ratio, from Fig. 5.5, is $\rho/\sqrt{\overline{G_{sr}}}$, where ρ is the surface reflectivity and $\overline{G_{sr}} = G_1/G_2$ is the power ratio of main (reference) lobe response to sidelobe response in the angle of the reflected ray. The resulting range error has an rms value given by

$$\sigma_{rm} = \frac{\rho h_o \sin E_t}{\sqrt{2G_{sr}}}. \tag{5.34}$$

This oscillating error may be accompanied by a bias error approaching $\Delta R/\sqrt{\overline{G_{sr}}}$ if the reflected signal lies entirely within the range gate ($\tau_g \geq \tau + 4\,\Delta R/c$). If the gate is only slightly longer than the echo pulse, the multipath error will be symmetrical about zero, as shown in Fig. 5.13.

Since the gate will exclude those portions of the reflected signal which are received more than $(\tau_g - \tau)/2$ after the end of the direct signal, the maximum bias is limited to the delay

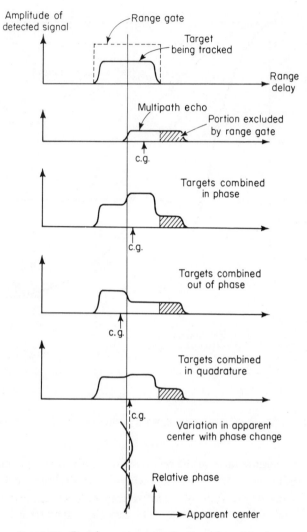

Fig. 5.13   Ranging system waveforms with multipath.

$$(\Delta t)_{max} = (\tau_g - \tau)\frac{\rho}{2\sqrt{G_{sr}}}. \tag{5.35}$$

Also, the maximum cyclic component will occur when the reflected signal is delayed by exactly one-half the pulse width:

$$(\sigma_t)_{max} = \frac{\rho\tau}{\sqrt{8G_{sr}}}. \tag{5.36}$$

There will, of course, be no error at all when the delay ΔR exceeds the gate width [more exactly, when it exceeds the quantity $(\tau_g + \tau)/2$]. In such cases, the first reflected signal energy will arrive after the end of the late gate, and will be excluded from the measurement. This limit, plus the relatively short pulse lengths normally found in the precision tracking radar and the 20 db or more of sidelobe rejection, will normally reduce the range multipath error to values of a meter or less. For example, if the pulse length is 1 μsec, the surface reflectivity 0.3, and the sidelobe ratio 20 db, the maximum cyclic error would be about 2 m rms, and the maximum bias error about 1 m. These errors would be encountered only when the radar was sited high above the reflecting surface, such that low-angle reflections could arrive with delays $h_o \sin E_t$ approaching 100 m.

Doppler Multipath Error

The multipath error in Doppler appears as a second signal component, whose amplitude, relative to that of the direct component, may be written

$$\frac{E_2}{E_r} = \rho \sqrt{\frac{G_2}{G_1}} = \frac{\rho}{\sqrt{\overline{G_{sr}}}}$$

(see Fig. 5.5). The sidelobe attenuation in the sum pattern is represented by $\overline{G_{sr}}$, averaged over the angles from which reflections arrive. The response of the Doppler discriminator to the combined signal is best described in terms of the vector diagram of Fig. 5.6(a), which shows the reflected component as a small vector rotating about the end of the direct component. The direct signal itself rotates about the origin at a rate which corresponds to the Doppler shift relative to the transmitted signal. The presence of the reflected ray causes the combined signal alternately to lead and lag the direct signal by a phase shift

$$\Delta\phi = \frac{E_2}{E_r} \sin \phi = \frac{\rho}{\sqrt{\overline{G_{sr}}}} \sin \left[\frac{4\pi h_o}{\lambda} \sin E_t + \gamma_r\right]. \tag{5.37}$$

This phase will change as the elevation angle changes, and for low elevation angles we may write the frequency error of the combined signal, in Hz, as $1/2\pi$ times the derivative of Eq. (5.37).

$$f = \frac{2h_o \dot{E}_t \rho}{\lambda \sqrt{\overline{G_{sr}}}} \cos \phi. \tag{5.38}$$

The rms value of the multipath error in frequency is

$$\sigma_f = \frac{\sqrt{2}\, h_o \dot{E}\rho}{\lambda \sqrt{\overline{G_{sr}}}} = \frac{\sqrt{2}\, h_o \rho v_t \sin \alpha_t}{\lambda R \sqrt{\overline{G_{sr}}}}. \tag{5.39}$$

The angle α_t represents the angle in the vertical plane between the radar beam and the target velocity vector.

Assume, for example, a low-elevation target at a range of 50 km, with a vertical velocity of 1600 m/sec. The elevation angle rate will be 0.032 rad/sec. If the antenna height above the surface is 16 m, the wavelength 10 cm, the ground reflectivity $\rho = 0.3$, and the sidelobe ratio 25 db, the rms error in Doppler will be

$$\sigma_f = \frac{\sqrt{2} \times 16 \times 0.032 \times 0.3}{0.1 \times 17.8} = 0.12 \text{ Hz.}$$

This error corresponds to a velocity of 0.6 cm/sec, and indicates that the multi-path error should not be very troublesome in a narrow-beam system with good sidelobe rejection. As the antenna height is increased, the FM sidebands representing the reflected component may fall outside the passband of the discriminator. These sidebands are separated from the direct signal by the multipath cyclic frequency $f = 2h_o\dot{E}/\lambda$ [see Eq. (5.22)]. In the above example, $f = 10$ Hz, placing the sidebands within the width of the usual tracking filter. If the separation increases beyond this, or if very narrow bandwidths are used, the error may be reduced still further.

5.5 Electromagnetic Interference

Measurement errors caused by electromagnetic interference are difficult to describe quantitatively, because the types of interference are so varied. Equations (5.2) and (5.3) provide the basis for calculating the error for a particular source, but in each case the appropriate S/I ratio is determined differently. We will not attempt here to give detailed procedures for this. However, the general procedure can be described.

The errors from various types of reflecting targets have been found by analogy to thermal noise, inserting additional terms in the error equations, when necessary, to describe departures of the reflected interference from a uniform distribution in the radar coordinates. The same procedure can be followed with electromagnetic interference. In most cases, this type of interference will originate from one or a few discrete locations, so that it will be subject to the antenna gain ratios G_{sr} and G_{se}. It can have any distribution in apparent range and Doppler, with random noise as one extreme and coherent repeater pulses as the other. These two extreme cases can be analyzed using the thermal noise expressions with an appropriate S/N, or the two-target expressions with appropriate S/I and n_e (see Table 5.4).

Because the interfering signals may be extremely large (especially in the case of random pulse interference), the details of receiver response must be inspected carefully for nonlinear effects. Most of our discussions have assumed that operation of the measurement system is linear, so that any noise error is additive. This theory can often be applied to interference, but we must always consider the possibility of signal suppression by strong interference. In a properly designed receiver, this should have the effect of eliminating some signal pulses, possibly replacing them with noise impulses of limited amplitude. Such impulses can be regarded

Table 5.4

EFFECTIVE NUMBER OF SIGNAL AND NOISE SAMPLES FOR LINEAR SYSTEM

(a) Random impulses entering system on m out of n signal samples, output filter matched to n:

$$n_e = \frac{n^2}{m}. \tag{5.40}$$

(b) Single impulse, output filter matched to n signal pulses:

$$n_e = n^2. \tag{5.41}$$

(c) Coherent interference, offset by frequency $\Delta f < \beta_n$ from signal:

$$n_e \cong 1. \tag{5.42}$$

as having a uniform spectrum over the region from zero to $f_r/2$ in the output of the error detector, and thus they can be smoothed effectively in the data filter at the output. This filter, especially if instrumented by digital devices, can be designed in such a way as to reject entirely any input which falls too far from the normal distribution of signals and tracking noise.

Where the interference originates at the desired target, the possibilities of angular errors are greatly reduced. Most intentional jamming from the target is intended to confuse the range tracker of the radar, because of the difficulty of angle jamming. Once the range gate has been forced off the desired echo, discontinuance of the jamming leaves the radar with no angular information either, until reacquisition can take place on the echo. However, during the jamming transmission, there may be an actual improvement in radar angular tracking, at least in monopulse radar where the jamming constitutes a strong point source.

5.6 Summary of Multiple-Signal Errors

The basic expressions for measurement error in any coordinate z are summarized in Table 5.5. In general, it is necessary first to compute the sum-channel ratio (S/I), to ensure that this ratio is well above unity. Nonlinear operation and suppression of the desired signal will take place when this ratio is near or below unity, and target detection is also uncertain. The ratio (S/I_Δ) for the sum-channel signal to the difference-channel interference is then found by multiplication of (S/I) by the ratio $(\Sigma/\Delta)_i^2$ for the response function in the z-coordinate. As noted, this latter ratio is approximately unity when the interference is distributed broadly over the main lobe and the major sidelobes (as with thermal noise).

Several special cases are summarized in Table 5.6, with notes on extension to systems having Doppler resolution, and to conical-scan and search systems. The material given earlier in the text provides guidance in applying the basic theory to a variety of different cases.

Table 5.5

BASIC ERROR EXPRESSIONS FOR INTERFERING SIGNALS

In terms of signal-to-interference ratios

$$\sigma_z = \frac{(\Delta/\Sigma)_i \sqrt{\eta}}{K_z \sqrt{2(S/I)n_e}} \tag{5.2}$$

$$= \frac{\sqrt{\eta}}{K_z \sqrt{2(S/I_\Delta)n_e}} \tag{5.3}$$

For small displacement, $z_i \ll z_3$

$$\sigma_z \cong \frac{z_i}{\sqrt{2(S/I)n_e}} \tag{5.4}$$

Large displacement, derivative difference channel

$$\sigma_z \cong \frac{\sqrt{2\eta}}{K_z \sqrt{(S/I)n_e}} \tag{5.5}$$

Interference from discrete reflecting object

$$\frac{S}{I_\Delta} = \underbrace{\frac{\sigma}{\sigma_i}}_{\substack{\text{Target} \\ \text{term}}} \times \underbrace{G_{st}G_{sr}}_{\substack{\text{Antenna} \\ \text{term}}} \times \underbrace{\left[\frac{a_{mx}}{a_i}\right]^2}_{\substack{\text{Receiver} \\ \text{term}}} \times \underbrace{\left[\frac{\Sigma}{\Delta}\right]_i^2}_{\substack{z\text{-coordinate} \\ \text{difference} \\ \text{channel}}} \tag{5.6}$$

$$\underbrace{\hspace{6cm}}_{\text{Sum-channel ratio, }S/I}$$

Error for unresolvable target echoes

$$\sigma_z \cong \frac{z_i}{\sqrt{2(\sigma/\sigma_i)n_e}} \tag{5.8}$$

Interference from distributed sources

Integrate interference power over volume of response functions, as in Eqs. (5.9) and (5.10). For interference broadly distributed over main lobe and major sidelobes, $(S/I_\Delta) \cong (S/I)$.

Brief definitions of symbols

σ_z = rms error in z-coordinate

$(\Delta/\Sigma)_i$ = z-difference-channel to sum-channel interference voltage ratio

η = efficiency factor (η_a for angle, η_f for range and Doppler measurement)

K_z = relative difference slope in z-coordinate

(S/I) = sum-channel signal-to-interference ratio (power)

(S/I_Δ) = sum-channel-signal to difference-channel-interference ratio (power)

n_e = effective number of signal and interference samples integrated at output

z_i = location of interference in z-coordinate

σ = cross section of desired target

σ_i = cross section of interfering target

G_{st} = target-to-interference gain ratio of transmitting pattern

G_{sr} = target-to-interference gain ratio of receiving pattern

a_{mx} = peak filter output for signal

a_i = peak filter output for interference

Table 5.6

ERROR EXPRESSIONS FOR SPECIAL CASES

Broadly distributed clutter

$$\text{Azimuth} \quad \sigma_A = \frac{\sqrt{\eta_a}}{K\sqrt{2(S/I)n_e}} = \frac{\theta_a}{k_m\sqrt{2(S/I)n_e}} \tag{5.12}$$

$$\text{Elevation} \; \sigma_E = \frac{\sqrt{\eta_a}}{K\sqrt{2(S/I)n_e}} = \frac{\theta_e}{k_m\sqrt{2(S/I)n_e}}$$

$$\text{Range} \quad \sigma_t = \frac{\sqrt{\eta_f}}{K\sqrt{2(S/I)n_e}} \tag{5.13}$$

Elevation measured over surface clutter

$$\sigma_E \cong \frac{E_t}{\sqrt{2(S/I)n_e}} \quad (E_t < \theta_e/2) \tag{5.14}$$

$$\sigma_E = \frac{\theta_e}{k_m\sqrt{2(S/I)(G_{se}/G_{sr})n_e}} \quad (E_t \gg \theta_e/2) \tag{5.15}$$

$$\cong \frac{\theta_e}{\sqrt{(S/I)n_e}} \quad \text{(most common illuminations)} \tag{5.16}$$

(The effect of Doppler resolution is to increase S/I, often by a factor much greater than n_e.)

Multipath error in elevation

$$\sigma_E = \frac{\theta_e \rho}{k_m\sqrt{2\overline{G_{se}}n_e}} \tag{5.21}$$

(Evaluate $\overline{G_{se}}$ at or near $E = -E_t$, or $2E_t$ below target, for specular reflection. Integrate over glistening surface for diffuse reflection, as in Eqs. (5.25) and (5.26). In most cases of multipath, $n_e \cong 1$ and no Doppler resolution is available.)

Multipath error in azimuth (*diffuse reflection*)

$$\sigma_A = \frac{\theta_{da}\rho}{2\sqrt{2\overline{G_{sr}}n_e}} = \frac{E_t\sigma_\alpha\rho}{\sqrt{2\overline{G_{sr}}n_e}} \quad (E_t < 2\sigma_\alpha, \theta_{da} < \theta_a) \qquad \left\{ \begin{array}{l} (5.27) \\ (5.28) \end{array} \right.$$

$$\sigma_A = \frac{\sqrt{2}\,E_t\sigma_\alpha\rho}{\sqrt{\overline{G_{sr}}n_e}} \quad (E_t > 2\sigma_\alpha, \theta_{da} < \theta_a) \tag{5.30}$$

$$\sigma_A = \frac{\theta_a\rho}{k_m\sqrt{2\overline{G_{sr}}n_e}} \quad (\theta_{da} > \theta_a) \tag{5.31}$$

Multipath error in range and Doppler

$$\sigma_{rm} = \frac{\rho h_o \sin E_t}{\sqrt{2\overline{G_{sr}}}} \tag{5.34}$$

$$\sigma_f = \frac{\sqrt{2}\,h_o\dot{E_t}\rho}{\lambda\sqrt{\overline{G_{sr}}}} \tag{5.39}$$

Additional definitions of symbols

σ_A = rms error in azimuth

σ_E = rms error in elevation

σ_t = rms error in time delay

θ_a = half-power azimuth beamwidth

Table 5.6—*Cont.*

θ_e = half-power elevation beamwidth

k_m = normalized monopulse slope

E_t = target elevation angle

G_{se} = difference-channel sidelobe ratio

ρ = surface reflection coefficient

θ_{da} = azimuth width of glistening surface

σ_α = rms surface slope

h_o = height of antenna above surface

σ_{rm} = multipath error in range

σ_f = multipath error in Doppler

(For conical-scan radar, substitute $k_s/\sqrt{L_k} \cong 1.2$ in place of k_m, and restrict n_e to be $\leq n/2$. For search radar, substitute $k_p/\sqrt{L_p} = 1.45$ in place of k_m.)

The characteristics of the radar target itself may lead to measurement errors of several types. In angular measurement at short range and in range and Doppler measurements at all ranges, the wandering of the center of reflection relative to the geometric center can be an important source of error. This will be a significant effect when the target's physical dimensions exceed about one percent of the radar resolution cell in any coordinate. For example, in a radar with a beamwidth of one degree, the angular glint of a target ten meters wide will be important at ranges less than fifty kilometers.

The apparent radar center will wander on most targets which are a number of wavelengths long. Such a target can be described as a group of separated scatterers. When the target rotates relative to the line of sight, the relative phases of the scattered signals will change, causing the apparent position, Doppler shift, and cross section to vary. When the target is smaller than the radar resolution cell, it will appear as a single point whose location depends on radar frequency, target aspect angle, and internal target motions (such as aircraft vibration or engine motion).

This chapter summarizes the general characteristics of radar glint and scintillation in three sections. The first considers the properties of the wavefront as it is received at the radar antenna, and the movement of the apparent target center (glint). The second section describes the fluctuation in target amplitude or cross section (scintillation), in terms of amplitude distributions, time correlation functions, and frequency spectra. The correlation functions and spectra are also useful in describing the properties of glint in regard to smoothing and differentiation. The third section summarizes the effects of scintillation on various practical measuring systems in which sequential samples of cross section are used. No attempt is made to catalog the properties of particular targets. For this data,

6

Target

Induced

Errors

the books of Skolnik (1962) and Barton (1964), and the references listed in those volumes, should be consulted.

6.1 Target Glint

The ideal radar can do no more than measure the properties of the reflected signal as it is received at the radar antenna. This constitutes the "available" information, and any imperfections in the measurement technique can only increase the measurement error. For this reason, we consider first the properties of the reflected signal itself.

When the target is smaller than the radar resolution cell, the angle of arrival may be expressed as the rate of change of carrier phase with distance across the aperture. Similarly, Doppler shift is the rate of change of signal phase (relative to a fixed carrier) with respect to time, and range (or time delay) is the rate of change of phase with carrier frequency. The utility of these expressions will become clear in the following paragraphs, in which two-point and multiple-point targets are discussed. These target models are admittedly artificial, but they lead to results which are as accurate as the prior knowledge of target structure.

Two-Point Target

The two-point target is the simplest example of a target which exhibits glint as its aspect angle changes. Meade (1955) and Howard (1959) have discussed the angular glint of the two-point target. Their results are summarized here and extended to describe the variations in apparent range and Doppler.

The two-point target consists of two scatterers at the ends of a nonreflecting rod. The target is assumed to rotate about its center, as shown in Fig. 6.1. For rotation in a plane containing the radar, the echo signal amplitude at the radar antenna will vary, giving an instantaneous voltage

$$E_s(t) = \cos\left[\frac{2\pi ct}{\lambda} + \phi\right] + k \cos\left[\frac{2\pi ct}{\lambda} - \phi\right],$$

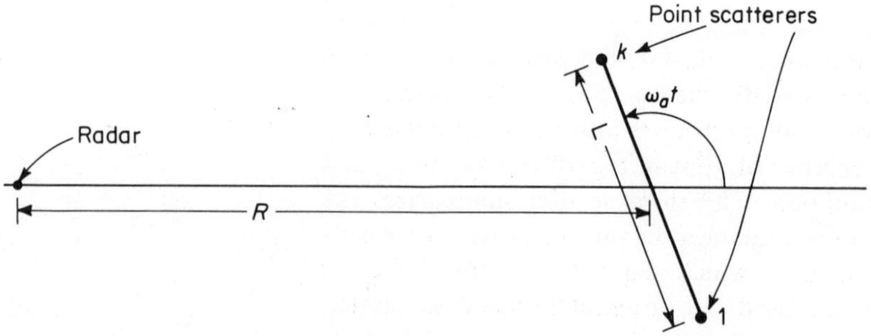

Fig. 6.1   Two-point target.

where $\phi = (2\pi L/\lambda)\cos\omega_a t$, L is the distance between the two scatterers, λ is the radar wavelength, c the velocity of light, ω_a the rotation rate of the target in rad/sec, and k is the amplitude of the signal from the second scatterer relative to the first. The following analysis assumes that the distance R between the radar and the target is very much larger than L, and that L is larger than λ, so that the relative amplitudes remain constant as the relative phase changes with aspect angle $\omega_a t$.

The variation in observed cross section with phase angle ϕ can be expressed as

$$\frac{\sigma}{\sigma_{max}} = \frac{1 + k^2}{(1 + k)^2}\left[1 + \left(\frac{2k}{1 + k^2}\right)\cos 2\phi\right]. \tag{6.1}$$

This ratio is plotted in Fig. 6.2(a). The expressions for target glint are simplified if we consider a normalized cross section or signal power S', such that

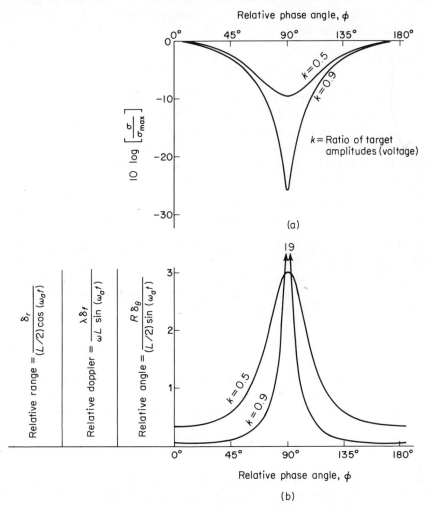

Fig. 6.2 Two-point target amplitude and apparent position.

$$S'(\phi) = \frac{1 + k^2}{1 - k^2}\left[1 + \left(\frac{2k}{1 + k^2}\right)\cos 2\phi\right].$$

(6.2)

The glint errors in position and Doppler will be proportional to the reciprocal of this quantity, or $1/S'$, and curves for this are plotted in Fig. 6.2(b). Both the cross section and the error may be expressed as functions of time if we substitute $\phi = (2\pi L/\lambda)\cos \omega_a t$.

Apparent angle of arrival at the radar antenna varies with target aspect angle, and can be calculated as the slope of the phase front across the aperture, in the plane of rotation. Since the far-field pattern of the target, consisting of amplitude and phase terms, rotates in space at twice the rotation rate of the target, there is a direct proportionality between the phase slope across the aperture and the time rate of change of phase observed at the center of the aperture. The results of Meade and of Howard lead to the following expression for angular error, measured relative to the center of the target:

$$\delta_\theta = \frac{L\sin \omega_a t}{2RS'} \qquad \text{(rad)}.$$

(6.3)

The relative angle scale in Fig. 6.2(b) shows δ_θ vs ϕ for $k = 0.5$ and $k = 0.9$. We note that the quantity $(L/R)\sin \omega_a t$ is the angle subtended at the radar by the target, and that this angle is modulated by the term $1/2S'$.

The error in apparent Doppler, δ_f, is proportional to the time rate of change of phase observed at the radar, the same function used in derivation of δ_θ. A simple renormalization gives

$$\delta_f = \frac{\omega_a L\sin \omega_a t}{\lambda S'} \qquad \text{(Hz)}.$$

(6.4)

The relative Doppler scale on Fig. 6.2(b) permits us to use the same curves for this error, emphasizing the functional similarity between δ_f and δ_θ. This interrelationship would permit an estimate of ω_a to be made from measurements of δ_f and δ_θ without having to observe an entire rotation of the target. If we wish to express the error in apparent radial velocity, we write

$$\delta_v = \frac{\lambda\delta_f}{2} = \frac{\omega_a L\sin \omega_a t}{2S'}.$$

(6.5)

Finally, for the case when L is less than the radar's range resolution, we find that the variation in apparent range has the same functional form. Taking the rate of change of signal phase with carrier frequency to equal the delay time, we find

$$\delta_r = \frac{L\cos \omega_a t}{2S'}.$$

(6.6)

The peak of the range variation occurs when the two scatterers lie along the line of sight, and for other aspect angles the quantity $L\cos \omega_a t$ expresses the radial separation between the two scatterers. The relative range scale on Fig. 6.2(b) permits us to use the basic curves to find range glint error vs ϕ for this target.

The apparent position and Doppler shift of the two-point target will fluctuate as the scatterers change phase. At times when the amplitude is smallest, the peak error is largest, and this apparent position can easily fall outside the physical limits of the target. The worst case occurs when the scatterers are of equal size ($k = 1$), permitting the amplitude to fall to zero and the error to become infinite at $\phi = 90$ deg. For all other values of ϕ, the apparent position for $k = 1$ lies midway between the two scatterers.

Although it is a rather artificial case, the two-point target illustrates the gross properties of any complex target. It shows that the deviations from the geometric center in range, angle, and Doppler have the same functional form when the target dimensions are smaller than the size of the radar resolution cell. The reason for this will be made clearer below, when we consider the properties of a target composed of many point scatterers.

Multiple-Point Target

Radar targets such as ships and aircraft extend over many wavelengths at the frequencies used for precision radar measurement, causing the radar echo to resemble that from a group of isolated point scatterers. The group can often be considered rigid, but rotating relative to the radar line of sight. This causes the target's apparent position to deviate from the geometric center. Both measurements and analysis have shown that the apparent position can lie outside the physical extent of the target ten to twenty percent of the time. This section summarizes these results, and describes the analogy between a complex target and bandlimited noise.

Measurements of the apparent angle to an SNB aircraft have been reported by Hastings, et al. (1952). These data, shown in Fig. 6.3, illustrate the case where

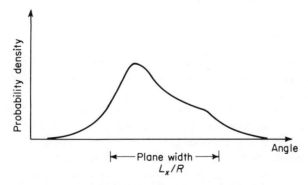

Fig. 6.3 Distribution of angle noise for nose-on view of SNB aircraft (from Hastings, et al.).

the apparent center of reflection deviates beyond the physical limits of the target. The fluctuation of apparent range for this same target at the same general aspect angle was reported by Howard and Lewis (1955). These data are shown in Fig. 6.4. It is interesting to note that the apparent range deviation exceeds the physical limits of the target a smaller fraction of the time than does the apparent angle. This is attributable to shadowing of the tail surfaces by the forward surfaces of the aircraft, which shortens the apparent length. The data do not indicate the absolute position of the geometric center of the aircraft, but we would expect the range data to be biased forward from the geometric center.

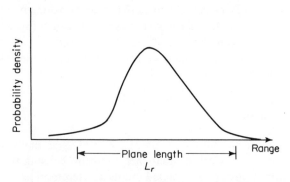

Fig. 6.4 Distribution of range noise for nose-on view of SNB aircraft (from Howard and Lewis).

Recent data on Doppler spectra of aircraft and rotating satellites show a similar bias caused by shadowing. The scattering surfaces which face the radar tend to dominate the spectrum. If these surfaces have a radial velocity differing from that of the center of gravity, there may be a Doppler bias, causing the integrated Doppler data to drift linearly from the actual (and measured) range. This effect is especially pronounced when rotating surfaces such as propellers make large contributions to the total radar cross section.

Analogy to Noise

The results obtained above can be reproduced by making an analogy between target scatterers distributed over a part of the radar resolution cell and bandlimited noise distributed over a section of the frequency spectrum. Table 6.1 lists the analogies between bandlimited noise and apparent angle, Doppler, and range. For the angle case, the rectangular noise spectrum of width B (Hz) is analogous to a uniform scatterer distribution L_x/λ wide in the cross-range direction. The noise voltage waveform (or Fourier transform of the spectrum) is analogous to the target echo amplitude and phase observed as functions of aspect angle α. Apparent target angle as observed by the radar is found from the rate of change of phase in the α-coordinate. Curves of the probability density and cumulative probability for rate of change of phase angle have been given by Rice (1948), and these are replotted in Fig. 6.5, on a normalized scale to describe the distribution of apparent angle of arrival.

The same analogy can be used to find the variation in apparent Doppler. As with the two-point target analyzed earlier, the variation in phase across the aperture is proportional to the variation in phase with time at the center of the aperture, so that $\delta_f = \delta_\theta(2R\omega_a/\lambda)$. This will apply when the scattering surfaces are rigid, rotating together at ω_a rad/sec. Additional scales on the curves of Fig. 6.5 show how these can be applied to apparent Doppler.

Variation in apparent range can be treated in a similar way, and scales for this

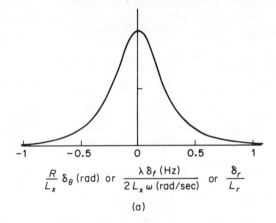

$$\frac{R}{L_x}\,\delta_\theta\,(\text{rad}) \quad \text{or} \quad \frac{\lambda\,\delta_f\,(\text{Hz})}{2\,L_x\,\omega\,(\text{rad/sec})} \quad \text{or} \quad \frac{\delta_r}{L_r}$$

(a)

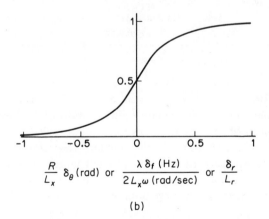

$$\frac{R}{L_x}\,\delta_\theta\,(\text{rad}) \quad \text{or} \quad \frac{\lambda\,\delta_f\,(\text{Hz})}{2\,L_x\,\omega\,(\text{rad/sec})} \quad \text{or} \quad \frac{\delta_r}{L_r}$$

(b)

Fig. 6.5 (a) Probability density and (b) cumulative probability vs apparent angle, δ_θ, or apparent Doppler, δ_f, or apparent range, δ_r, for a target composed of a large number of random scatterers spread over a distance L_r in range and L_x in crossrange.

case also appear in Fig. 6.5. In this case the noise spectrum is analogous to the scatterer distribution in range. The Fourier transform of the scatterer response is the frequency response of the target, and the apparent range (delay) is found from the rate of change of phase with frequency (see Table 6.1). Shadowing is not accounted for here, so the results are not as realistic as in the case of angle and Doppler. We could make a correction for this effect by using an equivalent value of L_r somewhat less than the actual extent of the target in range.

Table 6.1

TARGET-NOISE ANALOGIES

Narrowband Noise	Angle and Doppler Measurement	Range Measurement
Noise spectrum (a function of frequency, f)	Scatterer distribution in crossrange coordinate	Scatterer distribution in range coordinate
Noise voltage waveform (a function of time, t)	Target reflectivity pattern (a function of aspect angle, α)	Coherent target response (a function of carrier frequency, f_o)
Phase of noise voltage, $\phi(t)$	Phase of reflected signal, $\phi(\alpha)$	Phase of reflected signal, $\phi(f_o)$
Rate of change of phase, $d\phi/dt$ (statistics given by Rice, 1948)	Apparent angle of arrival, δ_θ (rad) $= (\lambda/2\pi R)\, d\phi/d\alpha$ and apparent Doppler, δ_f (Hz) $= (\omega_a/\pi)\, d\phi/d\alpha$	Apparent range, $\delta_r = (c/4\pi)\, d\phi/df_o$

Equivalent rms Error

Although the probability density shown in Fig. 6.5(a) has an infinite standard deviation, the radar output will remain finite because of the restricted dynamic range and bandwidth of the signal processing circuits. A Gaussian curve may be fitted to the curve of Fig. 6.5(a), and the standard deviation of this Gaussian fit will be an excellent representation of the apparent target. By fitting the Gaussian curve in such a way that it has the same probability of exceeding the physical limits of the target (fifteen percent), we obtain

$$\left. \begin{aligned} \sigma_\theta &= 0.35 \frac{L_x}{R} \qquad \text{(rad)}, \\[6pt] \sigma_f &= 0.35 \left(\frac{2L_x \omega_a}{\lambda} \right) \qquad \text{(Hz)}, \\[6pt] \sigma_r &= 0.35\, L_r \qquad \text{(units of } L\text{)}. \end{aligned} \right\} \tag{6.7}$$

The results given above were derived for a target consisting of scatterers distributed uniformly over L_x and L_r. For other distributions, the Gaussian curve will give an even closer representation of the apparent target, with the value of the constant varying from about 0.2 to the 0.35 given in Eq. (6.7). Alternatively, we may describe an actual target scatterer distribution in terms of an effective span L_{nx} or L_{nr}, defined in the same way as the noise bandwidth of a filter, and apply the constant 0.35 to find the rms glint error in each coordinate.

6.2 Target Scintillation

Amplitude Distribution

The amplitude of the multiple-point target echo, like the apparent position, is a random variable. The amplitude distribution differs from the position distribution, but the two variables are not statistically independent. Each scatterer will contribute an echo component with variable phase, and when the spacing between components is several wavelengths the relative phases tend to be random. Thus, the coherent sum of the scattered components tends to be a two-dimensional Gaussian variable. The probability density of signal voltage is then a Rayleigh distribution, and the signal power distribution is exponential. This corresponds to the "Case 1" and "Case 2" models of Swerling (1954). Figure 6.6 gives this probability density and cumulative distribution, plotted against cross section, σ,

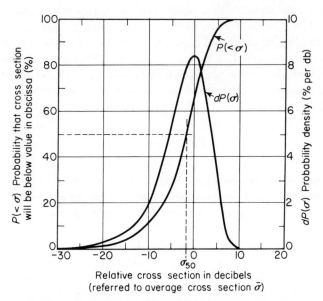

Fig. 6.6 Rayleigh target amplitude distribution, plotted against cross section in db.

on a db scale with the average $\bar{\sigma}$ as a reference. The fluctuation is seen to extend about 10 db above and 20 db below the average, for practical purposes.

Correlation in Time and Frequency

Thus far we have discussed the distribution of possible values of position, Doppler shift, and amplitude of the echo from a multiple-point target. The degree of correlation of the error or the amplitude from one observation to the next is also important, because this will determine the response of some measurement systems as well as the effectiveness of smoothing which may be applied to the output data. Decorrelation may be caused by target rotation relative to the line of sight, by change in the radar frequency or by internal motion of the target. We will consider the first two effects quantitatively and will note the type of changes to be expected when internal motion is encountered.

When the target scatterers are uniformly distributed, the correlation time interval, t_c, and frequency interval, f_c, are given by

$$t_c = \frac{\lambda}{2\omega_a L_x} \quad \text{(sec)}, \tag{6.8}$$

$$f_c = \frac{c}{2L_r} \quad \text{(Hz)}, \tag{6.9}$$

where λ is the radar wavelength, c is the speed of light, and ω_a is the rotation rate of the target. For other scatterer distributions, equivalent target spans L_{nx} and L_{nr} can be used in place of the actual cross-range and radial spans in these equations.

These correlation intervals describe time or frequency differences between measurements which may be considered uncorrelated. If measurements are made continuously or at very small intervals in time, the effective number of independent samples will be only

$$n_e = 1 + \frac{t_o}{t_c} \tag{6.10}$$

when both end points are considered and the total observation period is t_o. Similarly, measurements made continuously or at closely-spaced intervals over a frequency band Δf will yield only n_e independent samples of error, where

$$n_e = 1 + \frac{\Delta f}{f_c}. \tag{6.11}$$

Equations (6.8) and (6.9) apply to the coherent echo signal, and with small error can be used also to describe the correlation intervals in measurements derived coherently from this signal (i.e., angle, Doppler, and range). The relationships follow directly from the analogy summarized in Table 6.1. The target is composed of a group of scatterers with random phase, spread uniformly over an area of length L_r and width L_x. The echo signal observed by the radar is the two-dimensional Fourier transform of the target scattering function, where the transform coordinates are frequency $f_o = c/\lambda$ and aspect angle α. The impulse response of the target extends over an interval $2L_r/c$ in time, and the corresponding Nyquist interval in radar frequency is $f_c = c/2L_r$. In aspect angle, α, the Nyquist interval for a target of width L_x is $\lambda/2L_x$. As the target aspect angle changes at rate ω_a, the corresponding time interval is $t_c = \lambda/2\omega_a L_x$.

As an example, consider an aircraft of length $L_r = 10$ m and width $L_x = 10$ m, rotating at $\omega_a = 0.01$ rad/sec relative to the radar line of sight. If the radar wavelength is $\lambda = 0.03$ m, the correlation time $t_c = 0.15$ sec, and the correlation frequency $f_c = 15$ MHz. As far as angle and Doppler glint errors are concerned, the radar will observe about six independent error samples per second. Range glint error will change very slowly if the scatterers are distributed radially along a line (like a fuselage at nose or tail aspect), but will change at the same rate as angle glint error if the scatterers are distributed in cross-range over the entire length of the target. If more samples are required in a short time period, the radar may use frequency diversity, obtaining independent samples in each 15-MHz interval of the spectrum. Tuning over a five percent bandwidth at 10 GHz, it would be able to observe some thirty-four independent samples of range glint error within any short interval of time. Angle and Doppler glint errors would change slowly with frequency if the predominant source of glint were a linear set of scatterers like a wing, oriented normal to the line of sight. However, if scatterers are distributed in width and depth simultaneously, the use of frequency diversity could decorrelate angle and Doppler glint errors as well as range glint.

The correlation intervals for amplitude of the echo signal, as observed after envelope detection, will be smaller than those for coherent data. The reduction in the intervals (and broadening of the corresponding spectra—see below) results from the nonlinear process of envelope detection, which forms spectral components at video at frequencies up to the maximum spread of the coherent spectrum.

Glint and Scintillation Spectra

The time and frequency characteristics of target glint and scintillation may also be described in terms of spectral density functions, the Fourier transforms of correlation functions. In the case of angle and Doppler glint, the frequency spectrum is directly analogous to the distribution of scatterers in the cross-range coordinate, represented by x in Fig. 6.7. Conversion from cross-range to frequency is made by the relationship $f = 2\omega_a x/\lambda$, on the assumption that the target remains rigid while it rotates at ω_a rad/sec. The bandwidth of the rectangular spectrum corresponds to the width of the target, for uniform scatterer distribution. For other distributions, as shown in the figure, the noise bandwidth of the spectrum corresponds to the equivalent target width L_{nx}, defined under Eq. (6.7).

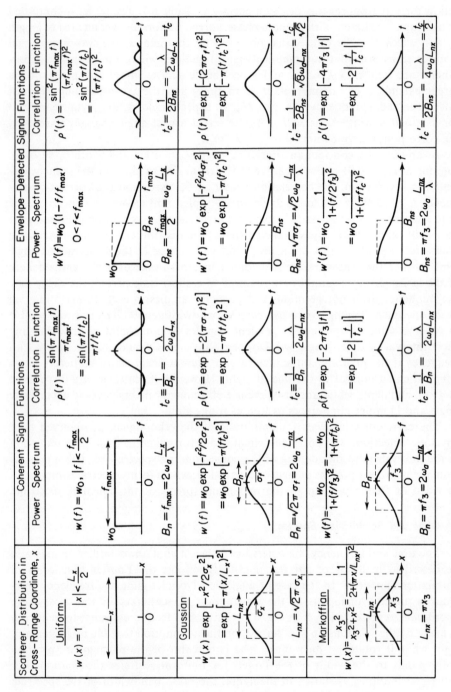

Fig. 6.7   Frequency spectra and correlation functions for different scatterer distributions in the cross-range coordinate.

Gaussian and Markoffian spectra represent the scatterer distributions which lead, respectively, to Gaussian and to exponential correlation functions in time. Although these distributions extend to infinity, they can be truncated at levels below the ten percent point without altering significantly the results derived below. In particular, the Markoffian spectrum has proven useful in describing targets which are subject to internal motion and vibration. These time-varying components can produce modulations at frequencies well beyond the upper limits set by rotation of the target as a whole, and the contributions to modulation power often tend to fall off slowly with frequency.

In the case of scintillation, the effect of the envelope detector can be described with acceptable accuracy by squaring the correlation function, or convolving the spectrum with itself. The results are shown in the second part of Fig. 6.7. The noise bandwidth of the spectrum is increased by a factor between 1 and 2, and the correlation time decreased by this same factor. In the aircraft example given earlier, where the scatterer distribution was assumed to be uniform, the noise bandwidth remains $2\omega_a L_x/\lambda = 6$ Hz, and the correlation time remains 0.15 sec. A Gaussian distribution would have an increase in spectral width by 1.4, and a Markoffian distribution would have twice the spectral width, after envelope detection. The spectral width may be increased further by such factors as propeller modulation and skin vibration. Many aircraft whose physical span places a limit $f_{max} = 2\omega_a L_x/\lambda$ on their rotation-induced spectra can be described satisfactorily by a Markoffian spectrum whose half-power width $2f_3 = 2\omega_a L_x/\lambda$. For the ten-meter target in the earlier example, the resulting correlation time for amplitude scintillation would be only 0.05 sec, instead of 0.15 sec.

The radial distribution of scatterers is equivalent to a time-delay spectrum, the transform of the frequency correlation function, shown in Fig. 6.8. This concept may be difficult to grasp, owing to the reversal of the normal terminology between time and frequency. However, in considering measurement errors, it is important to account for the correlation of errors between measurements made at adjacent frequencies. If, as in many systems, the radar tunes linearly over a band of frequencies, then the two variables may be interchanged again, and correlation time determined as the interval required for the system to tune over the correlation frequency f_c of the target.

When output data are smoothed to reduced glint and scintillation errors, we are concerned with the number of independent error samples included in the smoothing time. The effective number of samples for either time or frequency variation has been given in Eqs. (6.10) and (6.11). These may be used in conjunction with Figs. 6.7 and 6.8 to find the number of samples for any specific case. If both time and frequency change are involved, a two-dimensional correlation function must be used. While this is relatively simple for the Gaussian functions, it can present rather formidable difficulties in the other cases unless one or the other coordinate can be found to represent the dominant source of decorrelation. A general expression for the effective number of independent samples observed can be written as

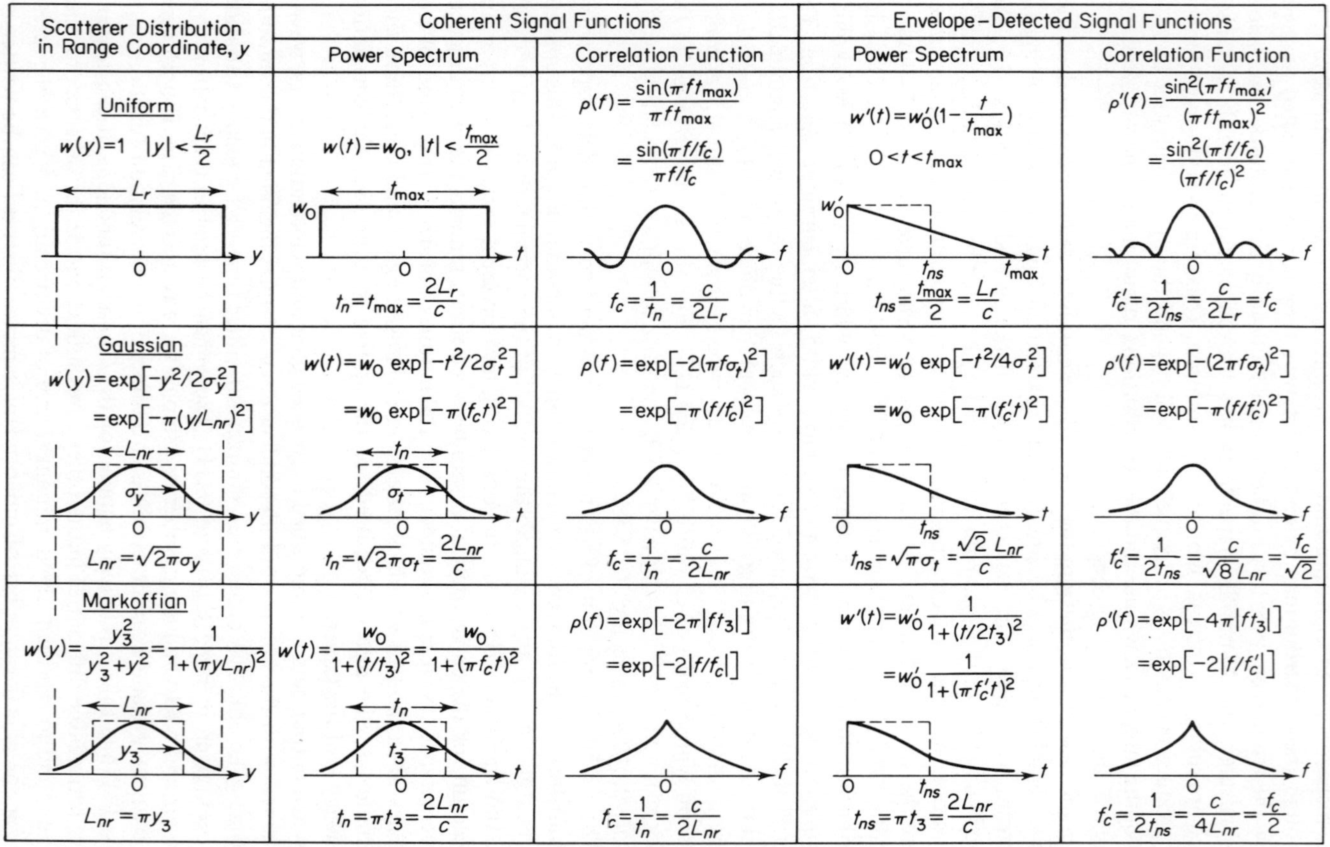

Scatterer Distribution in Range Coordinate, y	Coherent Signal Functions		Envelope-Detected Signal Functions	
	Power Spectrum	Correlation Function	Power Spectrum	Correlation Function
Uniform $w(y)=1$ $\;|y|<\frac{L_r}{2}$; L_r	$w(t)=w_0,\ |t|<\frac{t_{max}}{2}$; $t_n=t_{max}=\frac{2L_r}{c}$	$\rho(f)=\frac{\sin(\pi f t_{max})}{\pi f t_{max}}=\frac{\sin(\pi f/f_c)}{\pi f/f_c}$; $f_c=\frac{1}{t_n}=\frac{c}{2L_r}$	$w'(t)=w_0'(1-\frac{t}{t_{max}})$, $0<t<t_{max}$; $t_{ns}=\frac{t_{max}}{2}=\frac{L_r}{c}$	$\rho'(f)=\frac{\sin^2(\pi f t_{max})}{(\pi f t_{max})^2}=\frac{\sin^2(\pi f/f_c)}{(\pi f/f_c)^2}$; $f_c'=\frac{1}{2t_{ns}}=\frac{c}{2L_r}=f_c$
Gaussian $w(y)=\exp[-y^2/2\sigma_y^2]=\exp[-\pi(y/L_{nr})^2]$; $L_{nr}=\sqrt{2\pi}\,\sigma_y$	$w(t)=w_0\exp[-t^2/2\sigma_t^2]=w_0\exp[-\pi(f_c t)^2]$; $t_n=\sqrt{2\pi}\,\sigma_t=\frac{2L_{nr}}{c}$	$\rho(f)=\exp[-2(\pi f\sigma_t)^2]=\exp[-\pi(f/f_c)^2]$; $f_c=\frac{1}{t_n}=\frac{c}{2L_{nr}}$	$w'(t)=w_0'\exp[-t^2/4\sigma_t^2]=w_0\exp[-\pi(f_c't)^2]$; $t_{ns}=\sqrt{\pi}\,\sigma_t=\frac{\sqrt{2}\,L_{nr}}{c}$	$\rho'(f)=\exp[-(2\pi f\sigma_t)^2]=\exp[-\pi(f/f_c')^2]$; $f_c'=\frac{1}{2t_{ns}}=\frac{c}{\sqrt{8}\,L_{nr}}=\frac{f_c}{\sqrt{2}}$
Markoffian $w(y)=\frac{y_3^2}{y_3^2+y^2}=\frac{1}{1+(\pi y/L_{nr})^2}$; $L_{nr}=\pi y_3$	$w(t)=\frac{w_0}{1+(t/t_3)^2}=\frac{w_0}{1+(\pi f_c t)^2}$; $t_n=\pi t_3=\frac{2L_{nr}}{c}$	$\rho(f)=\exp[-2\pi|f t_3|]=\exp[-2|f/f_c|]$; $f_c=\frac{1}{t_n}=\frac{c}{2L_{nr}}$	$w'(t)=w_0'\frac{1}{1+(t/2t_3)^2}=w_0'\frac{1}{1+(\pi f_c't)^2}$; $t_{ns}=\pi t_3=\frac{2L_{nr}}{c}$	$\rho'(f)=\exp[-4\pi|f t_3|]=\exp[-2|f/f_c'|]$; $f_c'=\frac{1}{2t_{ns}}=\frac{c}{4L_{nr}}=\frac{f_c}{2}$

Fig. 6.8   Time spectra and frequency correlation functions for different scatterer distributions in the range coordinate.

$$n_e = \frac{n}{1 + 2\sum_{k=1}^{n-1}(1 - k/n)p(k)},$$

where $p(k)$ is the correlation between amplitudes of the 0th and kth pulses.

6.3 Measuring-System Error

The previous sections have described the errors in apparent position and Doppler, inherent in the reflected signal as it reaches the radar, and the characteristics of the amplitude scintillation. Monopulse angle-measuring systems and most range and Doppler error detectors, such as time and frequency discriminators or cycle counters, make their measurements rapidly enough to approach the accuracy available from the received signal, at least at high S/N ratios. Equation (6.7) gives the single-sample value of this error, and the smoothed value at the system output can be estimated from the correlation intervals described above. Many practical systems, however, take samples sequentially, in such a way as to increase the error when echo amplitude scintillation interacts with the measurements process. This section summarizes the effects of sequential lobing, automatic gain control, and uniform scanning in time and frequency on accuracy of measurement. In all cases, the additional error can be attributed to the fact that the target amplitude changes during the period of measurement, introducing a form of cross-talk between the amplitude and the estimation process.

Conical Scan and Sequential Lobing

Radars which scan a receiving antenna beam in a circular pattern around the target, or which switch between beam positions on each side of the target, will be sensitive to changes in target cross section occurring during the measurement cycle. Even a small change in echo amplitude during this interval will be interpreted as an angle error, equivalent to displacement of the apparent target center away from the scan axis.

The scintillating component of echo power can be considered as noise entering the error detector, and that portion of the scintillation power which lies near the scan frequency f_s will produce an error at the output. The magnitude of the angle error can be found from the thermal noise equations of Chap. 2 if the servo-channel signal-to-noise ratio $n(S/N) = (f_r/2\beta_n)$ is replaced by the ratio of steady signal power to scintillation power within the servo passband, or $1/W(f_s)\beta_n$. The resulting error expression for the scintillation component (Barton, 1964, p. 290) is

$$\sigma_s = \frac{\theta_3\sqrt{W(f_s)\beta_n}}{k_s}, \tag{6.12}$$

where $W(f_s)$ is the spectral density of scintillation power near f_s, expressed as (fractional modulation)^2/Hz. The error slope k_s has been plotted for both one-way and two-way operation in Fig. 2.9. In the case of scintillation error, the sensitivity is not changed by crossover loss, which has an equal effect on the steady signal and the scintillating component.

For a multiple-point target, the total fractional modulation at all frequencies is typically around 0.5 rms, a value of 0.52 characterizing the Rayleigh distribution. Using the Markoffian spectrum of Fig. 6.7, we set the total fractional modulation equal to the square root of the area under this spectrum:

$$0.5 = \sqrt{W_o \beta_n} = \sqrt{W_o/2t_c'},$$

$$W_o = 0.5 \, t_c'.$$

If the scan frequency is well above $1/t_c'$, we may write

$$W(f_s) \cong \frac{W_o}{\pi^2 t_c'^2 f_s^2} = \frac{1}{2\pi^2 t_c' f_s^2}.$$

The resulting tracking error component can be written

$$\sigma_s \cong \frac{\theta_3}{k_s} \sqrt{\frac{\beta_n}{2\pi^2 t_c' f_s^2}} = \frac{0.225\theta_3}{f_s k_s} \sqrt{\frac{\beta_n}{t_c'}}. \tag{6.13}$$

Caution should be used, however, in attempting to apply the Markoffian spectrum or any other mathematical model to actual scintillation spectra. Aircraft targets have peaks and asymmetries in their scintillation spectra which can cause the error to be appreciably larger than that predicted by Eq. (6.13).

Automatic Gain Control

Automatic gain control, or AGC, is often used as a means of normalizing the error signal, so that it represents a measure of off-axis angle regardless of target size and range. Scintillation which is not suppressed by AGC can modulate the error signal, causing a corresponding variation in any servo lag error which might be present. To minimize this effect, a fast (short time-constant) AGC would be used, both in monopulse and conical-scan systems. One of the advantages of monopulse is that it permits operation with faster AGC than can normally be used in conical scan (Barton, 1964, pp. 291–293).

A second effect of AGC involves the interaction of scintillation and glint. Fluctuations in echo amplitude and apparent position are not independent. The deep amplitude fades tend to coincide with the largest glint errors, an effect illustrated in Fig. 6.2 for the two-point target. Because of this, a slow AGC loop will reduce the tracking servo gain during amplitude fades, when the glint error is greatest, reducing the rms error caused by glint. Dunn and Howard (1959) have discussed this effect, and its relationship to the modulation of lag error. Their results are summarized in Fig. 6.9, where the rms sum of lag variation and glint is plotted as a function of average lag. The minimum glint error is observed with no AGC action, for the multiple-point target used in their measurements. The two-to-one reduction in glint without AGC clearly establishes that correlation exists

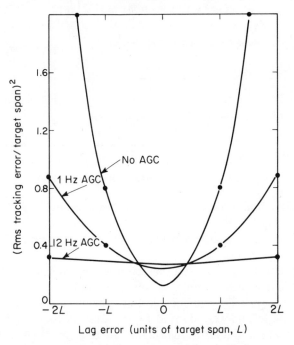

Fig. 6.9 Radar tracking noise power as a function of tracking error for different AGC bandwidths (from Dunn and Howard, 1959).

between scintillation and glint. However, tracking radars cannot take advantage of this glint reduction in practice, because of the presence of lag error. When target maneuvers induce a tracking lag greater than about one-third the target span, scintillation will modulate this lag sufficiently to increase the total noise component of tracking error. Because of this, fast AGC is normally used in precise trackers, in spite of the increased glint which results.

Uniform Scanning

Search radars measure angle by processing sequentially the signals received as the beam scans past the target. The measurement may be restricted to azimuth angle, in a fan-beam search radar, or may extend to elevation as well in a raster-scanning system with a pencil beam. When the target echo is of constant amplitude,

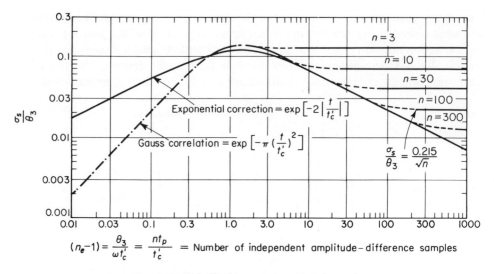

$$(n_e-1) = \frac{\theta_3}{\omega t_c'} = \frac{n t_p}{t_c'} = \text{Number of independent amplitude-difference samples}$$

Fig. 6.10   Scintillation error in a scanning radar.

receiver noise sets a limit on angle accuracy, as discussed in Chap. 2. With the more common fluctuating target, amplitude variations introduce appreciable errors in measurement, even at very large S/N ratios, setting a more severe limit on achievable accuracy.

Figure 6.10 plots the normalized rms scintillation error, σ_s/θ_3, as a function of the number of independent samples of amplitude difference observed during the time-on-target, $t_o = n t_p$. The symbols have been defined in Sec. 6.1, and illustrated in Fig. 6.7. These results were derived for a Gaussian beam, a Rayleigh target, and a processor which minimizes the thermal noise errors (matched to the steady component of target echo). The indicated performance is not quite as good as that which could be achieved with a maximum-likelihood estimator, optimized for a scintillating target at high S/N ratio (Swerling, 1956; Bernstein, 1955). However, the data given in Fig. 6.10 are applicable to most of the practical measurement processors used in search radar, which cannot adapt in advance to unknown signal levels.

In search radar, when the target correlation time is less than the time between pulses ($t_c' < t_p$), the error depends only on the number of pulses per beamwidth, $n = f_r t_o$, and is given by

$$\sigma_s = \frac{0.215 \theta_3}{\sqrt{n}}. \tag{6.14}$$

This situation corresponds to the Swerling Case 2 target model. The Swerling Case 1 model assumes complete correlation of amplitudes during the time-on-target ($t_c'/t_o \rightarrow \infty$), for which case there is no scintillation error at all. In practical search radar situations, however, the correlation time seldom exceeds about $10 t_o$, even for targets whose detection probabilities can be calculated accurately from the Case 1 statistics.

Some processors form an estimate of angle by taking the midpoint of the interval in which the signal lies above a threshold level. Neglecting any possible effects from integration and imperfect compensation for the resulting time lag, we can analyze such a system as though it were based on the use of two pulses, one at each edge of the beam. The error in this measurement, owing to scintillation, is given by

$$\sigma_s = 0.19\theta_3\sqrt{1 - \rho(nt_p)}, \tag{6.15}$$

where $nt_p = t_o$ is the time-on-target and $\rho(t)$ is the target correlation function. It is assumed here that the threshold is set at the 3-db point of the pattern. If $\rho(t)$ is taken to be Gaussian, with correlation time $t'_c \ll t_p$, we have

$$\rho(nt_p) = \exp[-\pi(nt_p/t'_c)^2] \cong 1 - \pi(nt_p/t'_c)^2,$$
$$\sigma_s \cong 0.33(nt_p/t'_c)\theta_3. \tag{6.16}$$

Scintillation errors are independent of range and tend to limit the accuracy at intermediate ranges where both thermal noise and angle glint are small (Dunn and Howard, 1959). The errors can be reduced to some extent by data smoothing, either digitally or in a tracking servo, over periods of several scans. In search radar, the error is almost always independent from scan to scan.

Frequency Scan

In radars with frequency-scanned antennas, there will be amplitude decorrelation caused by the change in frequency as the beam is moved over the target. When the beam motion is rapid, the time-decorrelation effects may be neglected, and data of Fig. 6.10 may be used with a change in scale of the abscissa to express the number of independent frequency samples: $(n_e - 1) = \Delta f/f'_c = nf_p/f'_c$. It is assumed here that the frequency change is in uniform steps and that Δf is the change necessary to move the beam over its 3-db width (analogous to t_o in time).

When two samples at the edges of the beam are used as the basis for measurement, the two-dimensional correlation function may be used in Eq. (6.16):

$$\rho(nt_p, nf_p) = \rho(nt_p)\,\rho(nf_p).$$

For Gaussian correlation functions, we have

$$\rho(nt_p, nf_p) = \exp\left[-\pi\left(\frac{nt_p}{t'_c}\right)^2\right]\exp\left[-\pi\left(\frac{nf_p}{f'_c}\right)^2\right].$$

Then, for targets which are well correlated over the scan ($nt_p \ll t'_c$, $nf_p \ll f'_c$), we can write

$$\rho(nt_p, nf_p) \cong 1 - \pi\left(\frac{nt_p}{t'_c}\right)^2 - \pi\left(\frac{nf_p}{f'_c}\right)^2,$$

$$\sigma_s \cong 0.33\theta_3\sqrt{(nt_p/t'_c)^2 + (nf_p/f'_c)^2}. \tag{6.17}$$

In most frequency-scan systems, it is possible to separate the time and frequency decorrelation, considering only the effects of frequency in the "rapid-scan" coordinate, and only those of time in the "slow-scan" coordinate. This permits the data of Fig. 6.10 to be used in estimating errors in both coordinates of a raster-scanning system.

Several aspects of radar measurement involve discrete processes. These fall into two categories: sampling, and quantization. Sampling is the process of selecting portions of the available information, while quantization is the process of encoding a sample into a digital word. The basic properties of sampling and quantization are summarized in this chapter, including the effects of the various errors introduced by these processes.

7.1 Sampling

Target information available to the radar is generally sampled in both time and space. Pulsed radar samples the reflecting properties of the target at regular time intervals. A tracker, while following a target, takes these samples uniformly during the period of tracking, while a search radar typically takes its samples in bursts as it scans past the target. The received echoes represent samples of the target's position and reflecting properties. If the time between pulses or samples is shorter than the time required for the target to change position or aspect, then the received data contain essentially all the information which could be obtained about the target during the period of observation.

Information is also available in space. A large receiving antenna aperture will gather more information about the angular position of the target than would be obtained over a small aperture. Information in space can be sampled, just as information in time, by combining the signals received in elemental antennas. Arrays and interferometers are examples of antennas that sample spatial information. The discussion of time sampling, contained in the next section, can be applied to space sampling by using the analogies between time and frequency, on the one hand, and aperture and far-field pattern, on the other.

7

Discrete

Processes

in

Measurement

Time Sampling

Any measured variable can be represented by a series of regularly spaced samples, as shown in Fig. 7.1. Values of the function $a(t)$ are recorded at intervals of T seconds. Shannon (1949) has established that these samples will contain all the information in $a(t)$ if the two-sided bandwidth B_a which describes the spectrum of $a(t)$ is less than $1/T$ Hz. In other words, if Shannon's condition is met, $a(t)$ can be reconstructed without error from the series of samples.

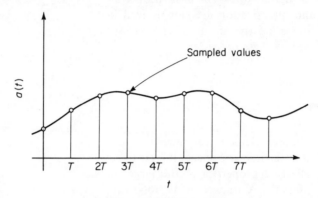

Fig. 7.1 Sampling in time.

The reconstruction of $a(t)$ is carried out by convolving the sampled function $a_s(t) = a(nT)$, shown in Fig. 7.2(a), with the function $(\sin \pi t/T)/(\pi t/T)$. This process, illustrated in Fig. 7.2(b), can be expressed mathematically as

$$a(t) = a_s(t) \otimes \text{sinc}\left(\frac{t}{T}\right)$$

$$= \sum_{n=0}^{\infty} a(nT) \, \text{sinc}\left(\frac{t - nT}{T}\right). \tag{7.1}$$

Here, the sampled function can be written

$$a_s(t) = \sum_{n=0}^{\infty} a(nT) \, \delta(t - nT),$$

and Woodward's sinc function is defined by

$$\text{sinc}\left(\frac{t}{T}\right) = \frac{\sin(\pi t/T)}{\pi t/T}.$$

We must remember the requirement that the spectrum of $a(t)$ lie entirely in the frequency band between $\pm 1/(2T)$. For a concise derivation of this sampling theorem, see Woodward (1953), p. 31.

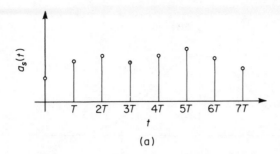

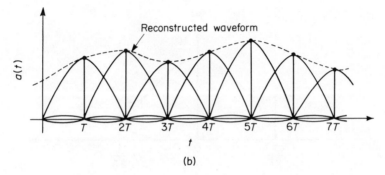

Fig. 7.2   (a) Sampled function $a_s(t)$. (b) $a_s(t)$ convolved with sinc (t/T).

In cases where the spectrum is wider than $1/T$, an effect known as "aliasing" will arise. The power (or energy) which lies beyond $\pm 1/(2T)$ in the spectrum is folded back into the region $\pm 1/(2T)$, when $a(t)$ is reconstructed, as shown in Fig. 7.3. The original spectrum in Fig. 7.3(a) has tails which extend beyond $\pm 1/(2T)$. After sampling, this energy is added to the portion of the spectrum which would have been reproduced accurately by sampling, causing a false indication of spectral density in that region. Distortion in the reconstructed function will result, but its form is not easy to predict because it depends on the phase and amplitude of the spectrum $A(f)$ beyond $\pm 1/(2T)$. For waveforms with finite average power, the distortion is describable as an additive waveform with mean square value equal to the area which lies under $|A(f)|^2$ and outside $\pm 1/(2T)$. For waveforms with finite energy, the distortion tends to increase the rise time and to smooth sudden changes in amplitude or phase of the function.

Aliasing errors can be reduced by filtering the waveform before sampling, so as to restrict the spectrum to the region $\pm 1/(2T)$. This filtering sacrifices the information in the higher frequencies, but avoids contamination of the low-frequency information by folded-over energy from the part of the spectrum which cannot be reproduced properly. Prefiltering is not always possible, e.g., in the pulsed radar case where the target reflection is the sample of target characteristics. Here, the continuous target data are not available at the radar receiver for filtering, and some aliasing effect is inevitable whenever the target spectrum width exceeds the radar repetition frequency.

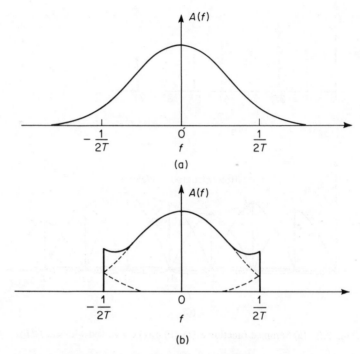

(a)

(b)

Fig. 7.3   (a) Spectrum wider than $1/T$. (b) Spectrum after sampling and reconstruction showing the effects of aliasing.

The discussion thus far has been concerned with real, low-pass waveforms. It can be applied equally well to bandpass waveforms by regarding $a(t)$ as a complex modulation of a carrier. The real bandpass waveform is then

$$\text{Re}\,[a(t)e^{j2\pi f_o t}],$$

where f_o is the carrier frequency and Re indicates "the real part of." The real and imaginary parts of $a(t)$ are low-pass waveforms, and a sample of each is necessary to describe one point in time on the bandpass waveform. Therefore, each sampled point of a bandpass waveform is a two-dimensional sample, containing twice as much information as the corresponding sample of a low-pass waveform. The two dimensions of the bandpass waveform can also be regarded as amplitude and phase.

Errors in the reconstructed waveform can also be caused by errors in timing of the samples, introduced at any stage in the sampling or reconstruction process. The effect on Eq. (7.1) can be approximated by multiplying the derivative of the function $a(t)$ by the time error ϵ_t:

$$a(t) = \sum_{n=0}^{\infty} a(nT)\,\text{sinc}\left(\frac{t-nT}{T}\right) + \epsilon_t \sum_{n=0}^{\infty} \frac{da(nT)}{dt}\,\text{sinc}\left(\frac{t-nT}{T}\right). \tag{7.2}$$

The mean square error in the function is

$$\sigma_a^2 = \sigma_t^2 \int_{-\infty}^{\infty} (2\pi f)^2 \, |A(f)|^2 \, df, \tag{7.3}$$

where σ_t^2 is the mean square value of ϵ_t in seconds. The integral in Eq. (7.3) is $E\beta_a^2$, the mean square bandwidth of $a(t)$ times the total signal energy (see Chap. 3 and App. B). Thus, Eq. (7.3) can be rewritten as

$$\frac{\sigma_a^2}{E} = (\sigma_t \beta_a)^2 = \left(\frac{\sigma_t}{T}\right)^2 (\beta_a T)^2. \tag{7.4}$$

By analogy to the antenna illumination functions of App. A, we see that the value of β_a for a rectangular spectrum $1/\tau$ wide is $1.83/\tau$. When the sample frequency, $1/T$, is much larger than the bandwidth of the waveform sampled, the factor $1.83T/\tau$ is small and the effect of timing errors is reduced.

Space Sampling

A linear array antenna takes samples of a signal in space, which are analogous to voltage samples taken of a waveform in time. In either case, increasing the number of independent samples will improve signal detectability by increasing the collected energy. Resolution is also improved in proportion to the number of independent samples. More time samples give better frequency resolution, and more space samples give better angular resolution.

The output of an antenna represents a sum of space samples. The antenna has a pattern which can be steered over a visible angle (-90 to $+90$ deg from the array normal) by properly phasing the space samples before summation. The beam is analogous to a filter response formed by phasing and summing the time samples of a waveform. Figure 7.4 illustrates this analogy, and shows that the presence of grating lobes in an array is caused by the same problem that leads to aliasing in the frequency domain: when the sample spacing is too large, signals originating outside the visible angle (reproduced frequency band) will be folded back into that region, leading to ambiguity and erroneous indications.

The design problems of array antennas are discussed in considerable detail in the literature (see, for instance, Hansen, 1966). The brief discussion here is intended only to call attention to the existence of the problem, to the extent that it may affect the accuracy and reliability of angular measurements.

7.2 Quantization

Each data sample taken by the radar may be processed in either analog or digital form. If it is to be digital, an analog voltage or shaft position must be quantized

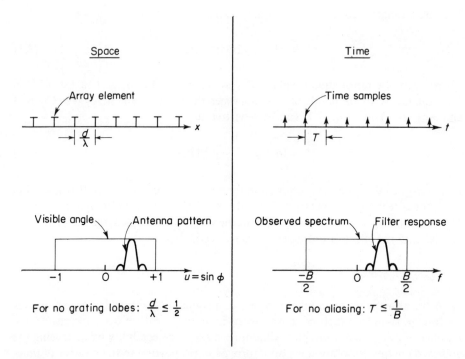

Fig. 7.4   Space-time analogies.

and encoded into a digital word. Here, we describe the quantization process and discuss the errors, systematic and random, which are introduced.

Digital phase shifters are used in some array antennas to control the phase of each element signal, so that a collimated, steerable beam can be formed by summing the element signals. The effects of phase quantization on array performance are also discussed here.

Amplitude Quantization

An ideal quantizer is a zero-memory, nonlinear device with a transfer characteristic as shown in Fig. 7.5(a). Input intervals Δ wide are uniquely related to encoded digital outputs. The transition points are equally spaced and permanently fixed. Figure 7.5(b) shows the systematic error introduced by the ideal quantizer. It has a peak error $\Delta/2$, and average error of zero, and an rms error $\Delta/\sqrt{12}$. The probability density of systematic quantization error is shown in Fig. 7.5(c).

Practical quantizers have additional errors caused by electrical or mechanical limitations. The two most important are noise and calibration errors. Noise error is jitter in the transition levels, and is usually independent from one sample to the next. Calibration error is a fixed nonlinearity in the transfer characteristic of the quantizer. It is usually set by manufacturing tolerances or by the limiting errors

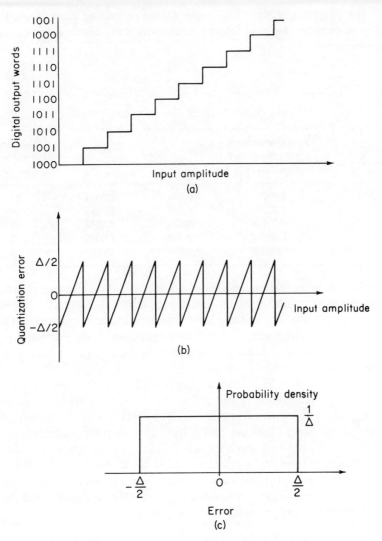

Fig. 7.5 Description of ideal quantizer: (a) quantizer transfer characteristic; (b) systematic quantization error; (c) probability density of quantization error.

in the calibration adjustment, and is often affected by such environmental factors as temperature and humidity.

For a particular encoder design, these errors limit the number of levels which can be meaningfully resolved. This, in turn, sets the number of binary digits (bits) of the encoded sample. Table 7.1 shows how the rms error is reduced as the number of bits is increased in a typical encoder with fixed noise and calibration error. When the number of levels is such that the ideal quantization error equals the fixed errors in the quantizer, there is little to be gained by adding more levels. For this reason, most encoders have additional errors which are about equal to the ideal quantization error.

Table 7.1

TYPICAL QUANTIZATION AND ENCODER ERROR

Number of Bits, N	Rms Quantization Error (for a range from 0 to 1)	Fixed Error rms	Total rms	Reduction in Error with Last Bit, %
3	0.0375	0.008	0.0384	50
4	0.0175	0.008	0.0193	50
5	0.00845	0.008	0.0116	40
6	0.00416	0.008	0.0090	22
7	0.00206	0.008	0.0083	5

The effect of quantization on the signal spectrum is quite complicated, and there is no analysis available which describes it in simple form. To illustrate the effect, we consider two examples which represent extreme situations. One is a voltage ramp, illustrating what happens to a spectrum much narrower than half the sample rate. The other is a white spectrum in the band from zero to half the sample rate.

The ramp is shown in Fig. 7.6(a) as it appears both before and after quantization. The voltage quantum is Δ and the voltage rate of change is $\dot{a}$. Sampling is so rapid that the points merge into a line. The spectrum of the quantized ramp is the sum of the ramp spectrum and a quantization error spectrum, as shown in Fig. 7.6(b). The combined spectrum is widened by the quantization lines, which fall off as $1/f$ rather than $1/f^2$ as does the ramp spectrum.

A white spectrum representing a fluctuation which is large compared to Δ will still be white after quantization. The average power will be increased, because of the added quantization error, so that the total power is

$$\overline{a_{sq}^2} = \overline{a_s^2} + \frac{\Delta^2}{12},$$ (7.5)

where a_{sq} is the quantized data sample and the overbar denotes averaging.

Voltage Encoders

Voltage encoders are used to quantize radar receiver outputs such as signal amplitude and angle error. They accept a video input, and produce a digital output representing the input voltage at the time a trigger is received. The trigger may be applied at a uniform rate, as when sampling successive range bins in a search

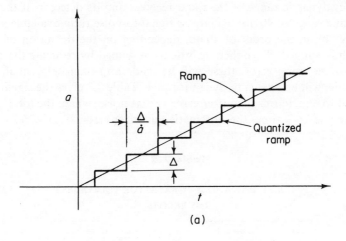

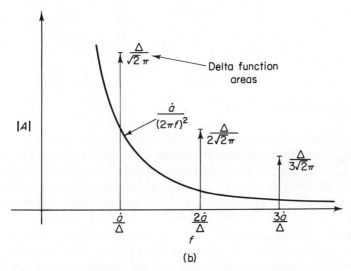

Fig. 7.6 Effect of quantizing on ramp signal: (a) quantized ramp;
(b) spectrum of quantized ramp.

radar, or at preselected points in time, as in range-gated outputs of a tracker.
Speed limitations in the voltage encoder will establish the minimum separation
between samples. Circuit bandwidth limits the rate of recovery of the encoder,
after a sample has been taken. Another sample cannot be encoded until the residue
of the previous sample has decayed to a value comparable to the systematic quantization error. Voltage encoders now available can quantize to seven bits every
200 nanoseconds. Where fewer levels are needed, the encoder can be made faster
because there will be less circuitry to limit the bandwidth.

The number of bits needed in a voltage encoder is usually determined by the simultaneous dynamic range of the radar receiver and its detector. If the receiver has a dynamic range of 40 db, the largest signal-to-noise ratio available at its output will also be in the order of 40 db, depending on the definition of dynamic range which is applied. Since there is little to be gained by reducing the quantizer errors below the noise error, these will be made approximately equal, and the number of bits will be related to dynamic range. Table 7.2 gives the signal-to-noise ratio needed to equate rms quantizer error to rms noise, where the total quantizer error is taken as $\sqrt{2}$ times the systematic quantizer error, or $\Delta/\sqrt{6}$.

Table 7.2

ENCODER BITS AND S/N RATIO FOR EQUIVALENT
RMS ERRORS

Number of Bits	Signal-to-Noise Ratio (db)
4	32
5	38
6	44
7	50
8	56

Shaft-Angle Encoders

Shaft-angle encoders are used to quantize radar pointing angles, through a mechanical connection to the antenna mount. The output indicates the orientation of the boresight axis of the antenna relative to a fixed angular coordinate system. Target angle is found by adding boresight corrections, which account for the difference between the mechanical and electrical axes, and angle error measurements, to locate the target relative to the electrical axis, or beam center.

Since shaft-angle encoders make an absolute, rather than a relative measurement, their outputs contain ten to twenty bits of information. A variety of techniques is used, the complexity increasing with the accuracy required. Atmospheric errors ultimately limit the number of useful bits to about twenty. Long (1963) describes a twenty-bit shaft-angle encoder with 1.4 arc-sec (0.007 mil) repeatability, and 6 arc-sec (0.03 mil) absolute accuracy. The unit weighs about four pounds and takes ten readings per second.

Range Encoders

Range encoders are used in radars where accurate range tracking is required or where range data will be processed or transmitted in digital form. Range is usually quantized by counting cycles of a clock, starting with the transmitted pulse and stopping with the received echo. Where high precision is required, the fine range bits are obtained by interpolating between cycles of the clock, using a tapped delay line with coincidence indicators on the taps.

Irreducible atmospheric errors limit range accuracy to about one foot, in most cases. To illustrate what this requires of a range encoder, consider a rather extreme problem where one-foot accuracy is needed at 1000 miles. This requires a 32-bit range word, where the first 29 bits can be obtained from a counter operating on a 6-MHz clock. The last three bits are obtained from a delay-line interpolator. The absolute clock frequency must be known to one part in 10^7, to prevent a one-foot range error at this range. A crystal oscillator with temperature control can provide this stability.

Phase Quantization in Array Antennas

Digital phase shifters are used in certain array antennas to control the phase of the element signals, for beam forming and steering. A phase shifter is associated with each element, introducing discrete phase steps of $2\pi/2^m$ rad, where m is the number of binary bits in the control word. Because phase shift is quantized, a phase error of the type shown in Fig. 7.7 appears across the aperture when the beam is steered away from the line normal to the aperture plane. The degradation in antenna gain, sidelobe level, and pointing accuracy, caused by phase-shift quantization, is described below.

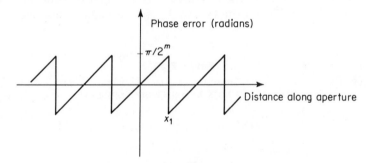

Fig. 7.7 Phase quantization error.

Gain Reduction

All the element signals are added in phase when a wavefront arrives from the direction of the antenna pattern peak. Phase-shift quantization error causes the peak antenna power gain to be reduced by the factor

$$L_q \cong 1 + \frac{\pi^2}{3 \cdot 2^{2m}}.$$

Since this loss is quite small, in most cases, it can be expressed in decibels by the approximation

$$(L_q)_{\mathrm{db}} \cong \frac{4.343\pi^2}{3 \cdot 2^{2m}} \cong \frac{14.3}{2^{2m}}, \tag{7.6}$$

where m is the number of bits controlling the phase shifter (see Table 7.3). This result applies both to linear and to planar arrays, because it indicates the portion of the power, transmitted or received by each element, which is diverted from the main beam to some other point in the pattern. For this reason, L_q is referred to as a "nondissipative loss."

Rms Sidelobes

The power lost from the main beam must reappear in sidelobes. Although this power is not generally distributed uniformly in angle, its net effect can be expressed as a relative rms sidelobe voltage level, σ_{sl}. This is measured with respect to the main-lobe gain, and can be written as

$$\frac{\sigma_{sl}^2}{G_m} = \frac{L_q - 1}{T\eta_a} = \frac{\pi^2}{3 \cdot 2^{2m} T \eta_a}, \tag{7.7}$$

where T is the total number of elements in the array, η_a is the aperture illumination efficiency, and m is the number of bits of phase quantization. The element gain does not enter into the expression because it is common to both the main-lobe and sidelobe gains. Table 7.3 shows how L_q and normalized sidelobe level vary with m.

Table 7.3

QUANTIZING LOSS AND RMS SIDELOBE LEVEL

Number of Bits, m	$(L_q)_{\mathrm{db}}$	$(T\eta_a\sigma_{sl}^2/G_m)_{\mathrm{db}}$
3	0.228	−12.9
4	0.057	−18.9
5	0.0142	−24.9
6	0.00356	−30.9
7	0.0089	−36.9
8	0.0022	−42.9

Peak Sidelobes

The phase error illustrated in Fig. 7.7 is a known, systematic error for any particular scan angle. Its effect on the sidelobe pattern can be computed by expanding the phase error in a Fourier series. It follows directly that the antenna pattern is approximately the sum of the error-free pattern and the convolution of this pat-

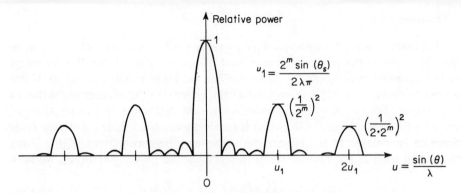

Fig. 7.8 Antenna pattern with phase quantization error.

tern with the terms of the Fourier expansion, as shown in Fig. 7.8. This figure applies to a linear array or to a planar array which is steered in one dimension. As the scan angle, θ_s (measured from the axis normal to the array), is increased from zero, the error sidelobes move farther away from the main lobe. The offset angle of each sidelobe is a multiple of

$$\frac{2^m \sin \theta_s}{2\pi},$$

where θ_s is the scan angle in radians. The peak sidelobe power, relative to the main lobe, is

$$\frac{G_i}{G_m} = \left[\frac{1}{i2^m}\right]^2,$$

where i is the number of the sidelobe being considered. An easy rule of thumb for the largest quantization sidelobe ($i = 1$) is

$$(G_1/G_m)_{\text{db}} \cong -6\,m. \tag{7.8}$$

Finally, the situation in which the phase error drops to zero, as seen from the direction of the beam, repeats whenever there is an integral number of phase quanta between the phases of successive elements. This is when

$$\left(\frac{2^m}{2\pi}\right)\left(\frac{d}{\lambda}\right)\sin \theta_s = 0, 1, 2, \ldots; \tag{7.9}$$

where d/λ is the distance in wavelengths between adjacent antenna elements. At these points in scan angle, a new set of sidelobes appears and sweeps away from the beam as θ_s is varied in either direction. There are about 2^m such points in the 180-deg sector.

Pointing Error

The same method of analysis which gave sidelobe level and location can be used to compute the pointing error caused by quantization. Here, the derivative of the error-free pattern (or the pattern of the monopulse difference beam, if this is used to measure angle) is convolved with the terms in the Fourier expansion of phase error. The largest pointing errors occur near the points of zero phase error, given by Eq. (7.9) above. As the beam is scanned away from these points, the error drops in proportion to the pattern derivative or difference pattern, $F_d(\theta)$. Using the largest term in the Fourier series to represent the pointing error, ϵ_θ, we have

$$2^m(w/\lambda)\epsilon_\theta \cong 1.2\frac{F_d(u_s')}{F_o} = 1.2\sqrt{\frac{G_d(u_s')}{G_o}}. \tag{7.10}$$

Here, G_d/G_o is the normalized pattern derivative, or monopulse difference pattern (plotted in App. A), $u_s' = (w/\lambda)\sin\theta_s$ is normalized scan angle, and w/λ is the aperture width in wavelengths. Figure 7.9 is a plot of this normalized error vs scan angle for the case of a Taylor pattern having -30-db sidelobes (App. A, Fig. A.15). This relatively large error, which only occurs at 2^m points in the total scan, can be reduced by introducing a systematic defocus into the array, offsetting

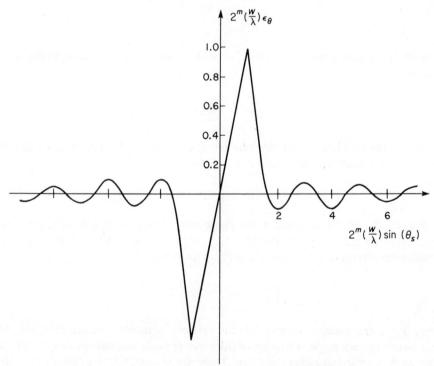

Fig. 7.9 Pointing error with phaseshift quantization.

the transition points of the different phase shifters. Such a defocus causes a smoothing of the error pattern, and reduces the large peaks. The rms pointing error remaining can be calculated by assuming a random noise error such that the signal-to-noise energy ratio is

$$\mathcal{R} = \frac{3 \cdot 2^m T \eta_a}{\pi^2} = \sigma_{sl}^2 / G_m \qquad \text{(see Table 7.3)}. \tag{7.11}$$

Noise errors were discussed in Chap. 2, and the equations summarized in Table 2.8 may be applied to find the minimum rms quantizing error by substituting Eq. (7.11). It should be kept in mind that this substitution can only be made in those cases where steps have been taken to eliminate the systematic variation of pointing error illustrated in Fig. 7.9. For the monopulse system described by Eq. (2.27), the random quantizing error can be expressed in this way as

$$(\sigma_{\theta q})_{min} = \frac{\pi}{K\sqrt{3 \cdot 2^m T}}. \tag{7.12}$$

When normalized to the 3-db beamwidth, this error is

$$\frac{(\sigma_{\theta q})_{min}}{\theta_3} = \frac{\pi\sqrt{\eta_a}}{k_m 2^m \sqrt{3T}} \simeq \frac{1.12}{2^m \sqrt{T}} = \frac{\sigma_\psi}{1.61\sqrt{T}}, \tag{7.13}$$

where the rms quantizing error in each phase shifter is $\sigma_\psi = 2\pi/(2^m\sqrt{12})$ rad.

The purpose of this handbook is to present the theoretical background of radar error analysis, and to relate the results to design parameters of practical equipment. To this end, we have discussed in considerable detail the effects of noise, interference, target characteristics, and sampling processes on the accuracy of coordinate data. In this chapter, we discuss briefly the other sources of error in radar measurement and give procedures for combining the individual error estimates into an overall description of radar performance.

The first section defines fundamental terms used in describing errors, and shows how the various components of error in any radar coordinate may be combined. Succeeding sections give a basic procedure for analysis of radar errors, and apply this procedure to angle, range, and Doppler coordinates. The important sources of error for each case are listed, and data are given to help in estimating those components which have not been analyzed elsewhere in this book. Two final sections cover the analysis of velocity errors in differentiated radar data, and the combination of radar coordinate errors to find the system error in Cartesian or other coordinate systems.

8

Radar

Error

Analysis

8.1 Mathematical Models of Error

The error in a given measurement may be defined as the difference between the value indicated by the measuring instrument and the "true" value of the measured quantity:

$$x = U_{\text{measured}} - U_{\text{true}}. \qquad (8.1)$$

The purpose of error analysis is to provide a description of the error which will permit its magnitude to be estimated for any set of operating conditions, without the necessity of running calibrations or tests for all possible combinations of conditions which may be encountered. In general, the error will vary with the time at which the measurement is made, the value of the quantity to be measured, and with the environmental conditions.

It is common practice to divide errors into systematic (or bias) and random (accidental or noise) components. The former are characterized by their predictability, and may be corrected, at least partially, by a process of calibration applied to the instrument before and after the measurement. In an extreme case, if the error is constant for all conditions, a single number will describe its magnitude,

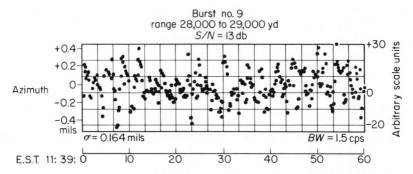

Fig. 8.1 Typical radar tracking error plot derived from boresight telescope.

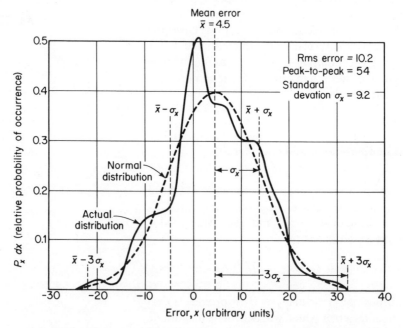

Fig. 8.2 Distribution of amplitude of error in typical radar track, compared with normal distribution.

and subtraction of this number from all values measured by the instrument will provide error-free data. One calibration will determine the value of initial bias error and will permit the reduction of residual bias error to zero. More generally, the error will assume some value within a limited range of values which are centered about a mean (or true bias) error, indicated in Figs. 8.1 and 8.2. For any measured set of n error values, we may determine this mean error from the expression

$$\bar{x} = \frac{1}{n} \sum_{i=1}^{i=n} x_i. \tag{8.2}$$

Other terms used to describe the magnitude of the error include the maximum (or peak) error x_m, the peak-to-peak error x_{p-p}, the rms error x_{rms}, and the probable error x_{50}. The rms error is defined for any distribution by the expression

$$(x_{\text{rms}})^2 = \frac{1}{n} \sum_{i=1}^{i=n} x_i^2. \tag{8.3}$$

Thus, x_{rms} is exactly what its title states: the square root of the mean value of the square of the individual error values. The probable error is the value which is exceeded in 50 percent of the readings, and can have any value from zero to the value of the peak error.

In those cases where the mean error is not zero, the rms error may be written as the sum of the bias and the variable components, added in an "rms fashion":

$$(x_{\text{rms}})^2 = \bar{x}^2 + \frac{1}{n} \sum_{i=1}^{i=n} (x_i - \bar{x})^2. \tag{8.4}$$

The above relationships are not limited to any particular types of error, and may be applied to any arbitrary or measured distribution of error values. Figure 8.2, for example, was obtained from the experimental data of Fig. 8.1 by simply counting the errors occurring within each increment of three scale divisions (0–2, 3–5, etc.), and plotting a smooth curve through the resulting points. The rms values shown were computed by using Eq. (8.3), and the equivalent normal distribution curve (with equivalent values of $\bar{x}$ and σ_x; see below) was then drawn to show how well the data could be approximated by a simple mathematical distribution function.

Amplitude Distribution

In many practical cases, the error values are distributed according to the "normal distribution" or Gaussian curve, centered about the mean value $\bar{x}$. This curve is defined by the probability distribution function

$$dP_x = \frac{1}{\sqrt{2\pi\sigma_x^2}} \exp\left[-\frac{(x - \bar{x})^2}{2\sigma_x^2}\right] dx. \tag{8.5}$$

This is the same distribution which was used to describe the IF noise voltage amplitudes in the radar receiver. The standard deviation of x for this distribution, designated σ_x, is simply the rms value of the variable component of error:

$$\sigma_x^2 = \frac{1}{n} \sum_{i=1}^{i=n} (x_i - \bar{x})^2. \tag{8.6}$$

The square of the standard deviation is the "variance" of the error. Although the term *standard deviation* should be limited to cases where the normal distribution is found, the symbol σ_x, is often applied to designate the rms value of any type of variable error, and will be so used here. The probable error x_{50}, for the special case of the normal distribution, is given by

$$x_{50} = 0.6745\sigma_x. \tag{8.7}$$

The normal distribution is often assumed to represent errors of unknown characteristics, and it closely approximates the distribution of many actual errors. In most cases, the peak error observed will correspond to a deviation of $3\sigma_x$ from the mean value, and the peak-to-peak error will be about $6\sigma_x$, when the error appears "noisy" in character. The probability of exceeding the 3σ deviation for the normal distribution is 0.3 percent, which implies that one out of 300 independent observations will deviate from the mean by an amount greater than 3σ. When a set of data points is taken over a period of one minute, using a system with a band-width of several cps, we would expect only one or two excursions beyond this value. A deviation of 4σ would be expected only once in 20 or 30 such sets of data, if it followed the normal distribution, and hence we would be likely to measure our peak error near the 3σ point.

The error contributed by a single element in the system may follow a regular pattern, such as those shown in Fig. 8.3, rather than the random noise pattern. It will be noted that the waveforms are given in order of increasing ratio of peak to rms error, and that these ratios vary from unity for the square wave to three for random noise. True Gaussian noise, of course, would have an infinite "peak" error, but we would have to wait a long time, as mentioned above, to see the 3σ level exceeded.

Time Functions and Frequency Spectra

The variation of error with time, and the resulting frequency spectrum of the error are important factors in classifying and describing the performance of a tracking system, and must be known if the effect of smoothing or differentiation is to be determined. The bias error defined above, if it is truly constant for periods of time which are long compared to the calibration and operation times of the system, can be removed by the calibration procedure. Hence, it is of little importance in error analysis. The residual bias error observed in most tests and evaluations of equipment is taken as that portion of error which does not change during the period

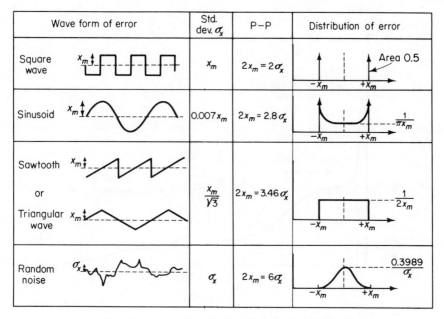

Wave form of error	Std. dev. σ_x	P–P	Distribution of error
Square wave x_m	x_m	$2x_m = 2\sigma_x$	Area 0.5, $-x_m$, $+x_m$
Sinusoid x_m	$0.007\,x_m$	$2x_m = 2.8\,\sigma_x$	$\frac{1}{\pi x_m}$, $-x_m$, $+x_m$
Sawtooth x_m or Triangular wave x_m	$\dfrac{x_m}{\sqrt{3}}$	$2x_m = 3.46\,\sigma_x$	$\frac{1}{2x_m}$, $-x_m$, $+x_m$
Random noise σ_x	σ_x	$2x_m = 6\sigma_x$	$\frac{0.3989}{\sigma_x}$, $-x_m$, $+x_m$

Fig. 8.3 Relationships between rms, peak, and peak-to-peak errors.

of an individual test or operation, which may be as short as a few seconds or one minute. When errors are introduced as functions of the measured quantity, the speed at which this quantity varies will determine how much of the error appears as "bias" and how much as "noise." For this reason, no sharp line can be drawn separating bias from noise components in error analysis. An arbitrary time period must be chosen, and those errors which do not change appreciably during this time may be classed as "bias."

The frequency spectrum of the error may be obtained from the observed time function by using one of the harmonic analysis techniques based on the use of the Fourier integral or transform. A typical error spectrum for an angle-tracking system is shown in Fig. 8.4. True bias is represented by an "impulse function" at zero frequency, with infinitesimal width and an area σ_o^2 equal to $\bar{x}^2$. The apparent bias observed over time intervals shorter than t_o is represented by the area σ_b^2 beneath the spectral density curve between zero frequency and the frequency $f_1 = 1/(2t_o)$. Above f_1 there appears a low-frequency error component, which can be approximated by a Markoffian spectrum:

$$W_a(f) = W_o \frac{f_a^2}{f_a^2 + f^2}. \tag{8.8}$$

The variance of this component is given by the value of the integral of W_a between zero and infinity:

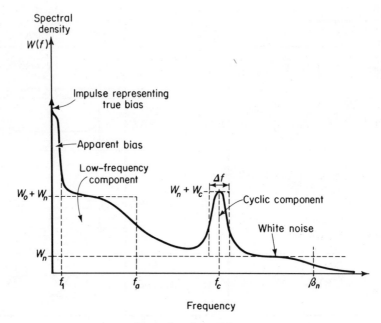

Fig. 8.4 Typical angle-tracking error spectrum.

$$\sigma_a^2 = \frac{\pi}{2} W_o f_a. \tag{8.9}$$

It represents the spectrum which results when broadband (white) noise is passed through a single-section low-pass filter consisting of a series resistor and parallel capacitor, where $RC = t_a = 1/(2\pi f_a)$.

Above the frequency f_a in Fig. 8.4 we see two more error components. The random or white-noise component extends with uniform spectral density W_n to the limit of the observed spectrum, as set by the bandwidth of the measuring device. The variance of the random noise error is

$$\sigma_n^2 = W_n \beta_n. \tag{8.10}$$

Superimposed upon the white noise is a cyclic component, occupying a narrow band of frequencies centered at f_c. If this component is a pure sinusoid, given by the wave form $x_c = X_c \sin(2\pi f_c t)$, it should appear as an impulse function with area $\sigma_c^2 = X_c^2/2$ at the frequency f_c. In a spectrum obtained from experimental data over a finite period, the same area will be distributed over a narrow band of frequencies, and the error may be approximated by a rectangular spectrum of amplitude W_c and width Δf, chosen so that $\sigma_c^2 = W_c \Delta f$.

The variance of the total error is given by the area under the entire spectrum, or

$$\sigma_x^2 = \int_0^\infty W(f)\, df = \sigma_o^2 + \sigma_b^2 + \sigma_a^2 + \sigma_n^2 + \sigma_c^2. \tag{8.11}$$

In the case shown, no error is correlated with any other, and the process of rms addition yields the total variance. It might be expected that two bias components would add directly, rather than in an rms fashion. However, the area σ_b^2 represents a slowly varying error which will sometimes have the same polarity as the true bias and sometimes oppose it. When evaluated over a large number of intervals, each of duration t_o, the sum $(\sigma_o^2 + \sigma_b^2)^{1/2}$ represents the rms value of the bias observed, and the probable bias will be approximately 67 percent of this value, in accordance with the normal distribution.

8.2 General Procedure for Error Analysis

Radar error analysis is a complex problem, because some of the major components of error are dependent upon range, elevation angle, signal strength, target dynamics, and environmental factors. If we are given the radar and target parameters and a description of the environment, the error level for a specific point in space can be computed as the rms sum of all the individual components evaluated at that point. This procedure can be quite cumbersome, however, when we are interested in a description of error over the entire coverage volume of the radar. To simplify such an analysis, we outline here a general procedure which can be applied to each measured coordinate, and which is based on a division of the coverage volume into four regions:

(a) The region of optimum accuracy, where the full instrumental capabilities of the radar are realized;

(b) The long-range region, where thermal noise is the predominant source of error;

(c) The short-range region, where glint and dynamic lag predominate; and

(d) The low-elevation region, where multipath, surface clutter, and refraction effects predominate.

These regions are not mutually exclusive, nor are they sharply defined. However, boundaries may be drawn where the variable components reach the level set by instrumental error. The outer limits of the coverage volume may then be found by considering the range or elevation at which errors are large enough to preclude continuation of target tracking, and a few intermediate contours may be drawn in to describe regions in which tracking is still possible at degraded accuracy.

When the rms error in any radar coordinate reaches about one-sixth the width of the resolution cell, continued tracking becomes uncertain. This sets the minimum value of "beamsplitting ratio" (or interpolation ratio, in range and Doppler) at about six to one. The dynamic range of errors is the ratio of the maximum error to the rms instrumental error, a level set by one or more error components which are independent of target location and motion. The examples discussed below show that beamsplitting or interpolation ratios greater than one hundred to one are very difficult to realize, and that many systems are restricted to ratios near ten to one. In systems with a small dynamic range of error (less than two to one, in many search radar systems), detailed error analysis is often neglected. It becomes important only when attempts are made to refine the data by smoothing or curve-fitting over long periods of time, and in such cases the information on spectral distribution of error power is of critical importance.

The following general steps in error analysis will be illustrated in detail by examples of tracker and scanning-radar measurement.

1. *Optimum Accuracy*. The radar is characterized by a "minimum error level" or instrumental accuracy, σ_e, in each measured coordinate. This error is the sum of those components which are essentially independent of signal strength, target dynamics, range, and elevation angle, for the class of targets being considered. In most cases, σ_e can be calculated or estimated from design and test data on the radar. The region of optimum accuracy is then defined as that portion of the coverage volume in which the variable components do not exceed σ_e, and in which the total error will not exceed $\sqrt{2}\,\sigma_e$.

2. *Required S/N*. Preparatory to calculating the transition to the long-range region, we find the energy ratio $\mathscr{R}$ and the S/N ratio required to make the thermal noise error equal to σ_e, when the radar is operated at the minimum available servo bandwidth β_n or with maximum observation interval t_o.

3. *Transition to Long-Range Region*. From the radar equation, we find the transition range R_2 which provides the required $\mathscr{R}$ and S/N. If the required S/N is less than unity, the reference range R_0 for $S/N = 1$ should be used instead, because the servo gain and bandwidth will show a sharp deterioration beyond this point.

4. *Linear Error at Target*. For the type of target under consideration, we calculate the glint and lag errors as linear displacements of the tracking point from the target, when the radar is operated with the maximum servo bandwidth available for short-range operation. The lag errors will be equal to the target rates (velocity in m/sec, acceleration in m/sec²) divided by the corresponding servo error coefficients (Barton, 1964, p. 295). We obtain the total linear error σ_L as the rms sum of glint and lag errors.

5. *Transition to Short-Range Region*. We now find the range R_1 at which the instrumental error σ_e is equivalent to the linear error (in angle, this is where $R_1\sigma_e = \sigma_L$). For tracking systems in which no compensation is made for the "geometrical acceleration" of straight-line target paths (Barton, 1964, pp. 301–306), we must check to ensure that this acceleration term does not govern the lag error at R_1. If it does, the transition range must be increased to bring total lag and glint within σ_e.

6. *Surface Clutter Error vs. Elevation*. Using the surface clutter reflectivity σ^0 and Eq. (5.11), we express S/I as a function of the sum-channel sidelobe ratio, for the range R_1 or the maximum range R_c at which clutter is within the line of sight. A smooth curve at the rms level of the antenna patterns (as in Fig. 5.5) is used to account for surface roughness, and from this a plot is made giving S/I as a function of elevation angle. This will give directly the lower limit of reliable tracking, E_{min},

where $S/I = 1$ at R_1 or R_c, and E_{min} for other ranges may be estimated from knowledge of σ^0 and S/I variation with range. Now, the clutter-induced error is calculated from equations in Chap. 5, and this is plotted as a function of elevation. The transition to the low-elevation region, E_1, is found where clutter error equals σ_e.

Where weather or chaff clutter is a problem, similar calculations are made for these terms to find both range and elevation boundaries.

7. *Multipath Error vs. Elevation.* Again using the smoothed elevation pattern data, with an appropriate surface reflection coefficient ρ, we calculate and plot the multipath error as a function of target elevation E_t, according to equations in Chap. 5. Pattern values are read either at E_t or $2E_t$ below the axis, depending on the type of reflection. From this plot, we may read the low-elevation transition E_2 for error equal to σ_e, applicable to targets at all ranges. In addition, if the multipath error reaches one-sixth the width of the resolution cell at any angle above the horizon, this may establish a tracking limit E_{min} for targets at any range.

8. *Tropospheric Fluctuation.* The tropospheric fluctuation error is now found from data in App. D, using path lengths L corresponding to targets at R_2, E_2 and at R_1, E_1 (or R_1, E_2 if this is a higher limit). This is a check to ensure that fluctuation is well below σ_e in these regions. If the fluctuation exceeds $\sigma_e/2$, an upwards adjustment is made to E_1 or E_2, to keep total error below $\sqrt{2}\,\sigma_e$.

9. *Residual Refraction Bias.* In elevation and range, we must verify that the residual error after correction of average bias is well within σ_e. The total bias errors are read from Figs. D.2 and D.3, for a target at R_2, E_2, and the percentage of residue is estimated from knowledge of the correction procedure. The possible ionospheric errors are read from Figs. D.19 and D.20. If either the tropospheric residue or the ionospheric error exceeds $\sigma_e/2$, the low-elevation transition E_2 should be adjusted upwards to keep the total error below $\sqrt{2}\,\sigma_e$.

10. *Coverage Summary.* Maximum and minimum tracking ranges are set by the values R_o and σ_L already calculated, since the continuity of tracking cannot be assured much beyond R_o, and the target will also be lost when σ_L reaches about one-sixth the width of the resolution cell. These limits are plotted on a vertical coverage chart, along with E_{min}, to describe the tracking coverage of the radar. Contours corresponding to R_1, E_1, R_2 and E_2 are also plotted to describe the region of optimum accuracy and the regions where other types of error are dominant. After inspection of the resulting contours, we may wish to compute ranges and elevation angles corresponding to other levels of error, intermediate between σ_e and one-sixth the width of the resolution cell.

11. *Discrete Interference.* Having described the regions in which noise, clutter, multipath, and target dynamics are likely to impose limitations on tracking accuracy, we must still consider the errors caused by interference from discrete sources: sidelobe response to large fixed targets, main-lobe response to objects near the desired target, and response to strong, active electromagnetic sources anywhere in the coverage region. The interference from discrete reflecting objects is confined to a few resolution cells, where it may cause large errors for brief intervals during a track. Active interference is more difficult to characterize in the absence of

a specific problem. Because errors from either source are transient and may fall outside the normal distribution of other error components, they are not generally included as additional rms terms in the calculation of total error. Instead, it proves convenient to describe those combinations of circumstances which can lead to significant errors from discrete objects, and also those which can cause loss of track. In many cases, provisions can be made in system design to "edit" errors from the output data when they exceed some critical value, and to reacquire a track which may have been lost as a result of discrete interference.

8.3 Angle Error Analysis

Components of angle error which are discussed elsewhere in this book are as follows:
- (a) Thermal noise (Chap. 2);
- (b) Clutter and interference (Chap. 5);
- (c) Multipath reflections (Chap. 5);
- (d) Target glint and scintillation (Chap. 6);
- (e) Quantization and array error (Chap. 7);
- (f) Dynamic lag (App. C; also Barton, 1964, pp. 293–310);
- (g) Atmospheric propagation (App. D).

Before combining this material into a description of total angular error, we will consider briefly the other components of error which enter into most practical systems:
- (h) Monopulse network error; and
- (i) Servo and mechanical error.

The material will initially cover the problems of a precision mechanical tracker, but will then be generalized to include scanning or array antenna systems.

Monopulse Network Error

The basic elements of an amplitude comparison monopulse system are shown in Fig. 8.5, with two types of error in the electrical elements:

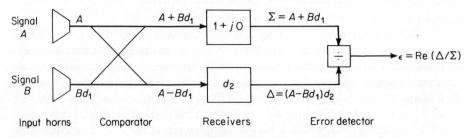

Fig. 8.5 Monopulse network error.

Precomparator error, $d_1 = 1 + a_1 + j\phi_1$,

Postcomparator error, $d_2 = 1 + a_2 + j\phi_2$.

The unbalance errors for amplitude, a_1 and a_2, are assumed small with respect to unity, and phase unbalance errors ϕ_1 and ϕ_2 are small with respect to one radian. Precomparator error is associated with RF components, and is often stated as a null depth G_n in the difference pattern, relative to the sum-pattern maximum (see Fig. 5.5):

$$G_n = \frac{G_1}{G_3} = \frac{4}{\phi_1^2}. \tag{8.12}$$

Restricting our concern to the first- and second-order terms, we may write the shift in null position, relative to an ideal (error-free) system in terms of an error voltage,

$$\epsilon(0) = \mathrm{Re}\left(\frac{\Delta}{\Sigma}\right)_0 = \frac{-a_1}{2} - \frac{a_1 a_2}{2} + \frac{a_1^2}{4} + \frac{\phi_1 \phi_2}{2} + \frac{\phi_1^2}{4}. \tag{8.13}$$

The normalized difference slope which relates this voltage to angle is

$$k_m \equiv \frac{\theta_3}{\Sigma} \frac{\partial \Delta}{\partial \theta}. \qquad \text{[see Eq. (2.9)]}$$

Thus an error voltage ϵ is equivalent to an angle error $\Delta\theta = \epsilon\theta_3/k_m$, and Eq. (8.13) may be converted to angle with this scale factor.

The monopulse network error will also affect the difference slope, changing the error-free value k_m to a value k'_m such that

$$\frac{k'_m}{k_m} = 1 + a_2 + \frac{a_1^2}{4} - \phi_1\phi_2 - \frac{3\phi_1^2}{4} \tag{8.14}$$

(retaining first- and second-order terms only).

In practice, a major portion of precomparator error is usually eliminated by offsetting the feed or data system to center the "RF null" of the system (as measured, for instance, with receiving equipment carefully adjusted to ideal performance, $a_2 = \phi_2 = 0$). This displaces the zero point by the amount

$$\epsilon_o = \frac{a_1}{2} - \frac{a_1^2}{4} - \frac{\phi_1^2}{4}. \tag{8.15}$$

Assuming that the precomparator errors remain fixed, the introduction of postcomparator errors (usually associated with gain and phase variations in active receiver elements) will now cause an error signal

$$\epsilon_2 = \epsilon(0) - \epsilon_o = \frac{\phi_1\phi_2}{2}. \tag{8.16}$$

The corresponding angular error for this case is

$$(\Delta\theta)_2 = \frac{\theta_3\phi_1\phi_2}{2k_m} = \frac{\theta_3\phi_2}{k_m\sqrt{G_n}}. \tag{8.17}$$

The rms error σ_θ caused by postcomparator error after zeroing to the RF null is found from Eq. (8.17) by inserting the rms value of ϕ_2. This postcomparator phase variation can result from any of the following factors:

1. Tuning of the receiver such that the signal is not centered in the IF passband;
2. Change in signal power level at the receiver input;
3. Temperature effects on receiver components. In addition, of course, there may be changes in precomparator errors a_1 and ϕ_1, taking place after calibration and zeroing. These will cause the null point to shift in accordance with Eq. (8.13), and can result from any of the following:
4. Change of operating frequency within the band;
5. Ambient temperature or solar heating effects on the RF structure; or
6. Polarization changes in the received signal.

In a phase monopulse system, the error-sensing circuits operate in such a way as to interchange a_1 and ϕ_1 in all of the error equations given above. The null depth becomes $G_n = 4/a_1^2$, so that the final expression in Eq. (8.17) remains the same.

Servo and Mechanical Error

The significant sources and magnitudes of servo and mechanical error are dependent almost entirely upon the details of radar component design, and we will make no attempt here to generalize on these factors. A design analysis for a precise mechanical tracker must consider at least the following problems in arriving at the error estimate:

(a) Torque applied to the antenna by wind and gravity, and resulting in an offset of the servo null;

(b) Mechanical deflection of the antenna and pedestal caused by wind and gravity torques, and by accelerations in tracking;

(c) Electrical noise and drift in the servo amplifiers;

(d) Ability to align the radar axes precisely to a given coordinate system, and stability of this alignment with age, temperature, and operating stresses;

(e) Precision of the bearings which support each axis, and of the couplings or gears which drive the data output devices.

The bias and noise error from each of these sources may be combined on an rms basis, along with quantization (or analog) errors in the data output devices and monopulse network error, to arrive at the instrumental accuracy, σ_e, used in error analysis. In sequential lobing or scanning systems, a target scintillation error is substituted for monopulse network error.

Monopulse Tracker Example

To illustrate the analysis procedure for a precise monopulse tracker, the published data on the AN/FPS-16 instrumentation radar will be used (Barton, 1964, pp. 343–346), and a medium-sized aircraft will be assumed as the target. The steps follow the general procedure outlined in Sec. 8.2 above.

Step 1: Optimum Accuracy. From the reference data, the instrumental accuracy in both elevation and azimuth may be taken as

$$\sigma_e = 0.1 \text{ mr.} \quad \text{(Barton, 1964, p. 344)}$$

Step 2: Required S/N. The following radar parameters are given in the reference:

$$B = 1.6 \text{ MHz}, \quad \beta_n = 0.5 \text{ Hz (min)},$$
$$\tau = 1.0 \text{ } \mu\text{sec}, \quad \theta_3 = 20 \text{ mr},$$
$$f_r = 160 \text{ pps}, \quad k_m = 1.57.$$

From these, we calculate

$$L_m = 1.0 \text{ db} \quad \text{(from Fig. 3.17)},$$

$$n = \frac{f_r}{2\beta_n} = 160 \quad \text{[(from Eq. (1.25)]},$$

$$\mathscr{R} = \left(\frac{\theta_3}{\sigma_e k_m}\right)^2 = 16200 \quad \text{[from Eq. (2.34)]},$$

$$\frac{S}{N} = \frac{\mathscr{R}}{2nL_m} = 40 \quad \text{[from Eq. (1.30)]}$$

(assuming negligible collapsing loss or video matching loss).

Step 3: Long-Range Transition. The other needed radar and target parameters for the calculation are

$$P_t = 1.0 \text{ MW}, \quad\quad \overline{NF_o} = 10 \text{ db},$$
$$G_t = G_r = 44.5 \text{ db}, \quad\quad L = 4 \text{ db},$$
$$\lambda = 5.6 \text{ cm}, \quad\quad \bar{\sigma} = 5 \text{ m}^2.$$

The radar equation gives

$$R_o = \left[\frac{P_t G^2 \lambda^2 \bar{\sigma}}{(4\pi)^3 kT_o \overline{NF_o} BL}\right]^{1/4} = 440 \text{ km, or } 240 \text{ naut mi},$$

$$R_2 = R_o\left(\frac{S}{N}\right)^{-1/4} = 175 \text{ km, or } 95 \text{ naut mi}.$$

Step 4: Linear Error at Target. We assume target and servo characteristics as follows for short-range operation:

$$L_x = 20 \text{ m}, L_{nx} = 14 \text{ m}, \quad K_v = 500 \text{ sec}^{-1},$$
$$v_t = 300 \text{ m/sec}, \quad\quad\quad \beta_n = 6 \text{ Hz (max)},$$
$$a_t = 5 \text{ m/sec}^2.$$

From these, we compute

$$K_a \cong 2.5\,\beta_n^2 = 90 \text{ sec}^{-2} \qquad [\text{from App. C}]$$

$$\text{Glint error} \cong 0.35 L_{nx} = 5 \text{ m} \qquad [\text{from Eq. (6.7)}]$$

$$\text{Velocity lag} = \frac{v_t}{K_v} = 0.6 \text{ m},$$

$$\text{Acceleration lag} = \frac{a_t}{K_a} = 0.055 \text{ m},$$

$$\text{Linear error}, \ \sigma_L = (5^2 + 0.6^2 + 0.055^2)^{1/2} \cong 5 \text{ m}.$$

Step 5: Short-Range Transition. Neglecting geometrical acceleration, we find

$$R_1 = \frac{\sigma_L}{\sigma_e} = 50 \text{ km, or 27 naut mi.}$$

For a straight-line passing course at this range, the maximum geometrical acceleration in azimuth will be

$$(\ddot{A})_{max} \cong 0.65\left(\frac{v_t}{R_1}\right)^2 = 0.02\,\frac{\text{mr}}{\text{sec}^2}.$$

The resulting lag is negligible at this range, and will remain below the glint error for all ranges at which tracking is possible. Lag in elevation will be still smaller.

Step 6: Extended Clutter Error. We may assume a surface clutter reflectivity $\sigma^0 = 0.005$ out to a line-of-sight range of $R_c = 20$ km. At this range, we calculate

$$\frac{S}{I} = \frac{\bar{\sigma} G_{st} G_{sr}}{\sigma^0} \times \frac{L_p}{R\theta_a} \times \frac{2}{\tau c} = 0.022\,G_{sr}^2, \qquad [\text{from Eq. (5.11)}].$$

The smoothed pattern plots from Fig. 5.5 are shown in Fig. 8.6, along with S/I data for range R_c. The lower elevation limit at this range is $E_{min} \cong 0.9$ deg for unity S/I. For an aircraft moving at high velocity over the clutter, rms error may be computed by assuming the clutter power to be spread uniformly between the prf lines in the spectrum, so that

$$n_e = f_r t_o = \frac{f_r}{2\beta_n} = 13.3,$$

$$\sigma_A = \frac{\theta_a}{k_m\sqrt{2(S/I)n_e}} = \frac{17}{G_{sr}}\,\text{mr} \qquad [\text{from Eq. (5.12)}],$$

$$\sigma_E = \frac{\theta_e}{k_m\sqrt{2(S/I)(G_{sr}/G_{se})n_e}} = \frac{17}{\sqrt{G_{sr}G_{se}}}\,\text{mr} \qquad \begin{matrix}(E > 0.6 \text{ deg}) \\ [\text{from Eq. (5.15)}].\end{matrix}$$

These are plotted in Fig. 8.7, from which we can read $E_1 \cong 2.9$ deg.

Further calculation for light rain ($r = 1$ mm/hr) gives $S/I = 2$ at $R = 100$ km,

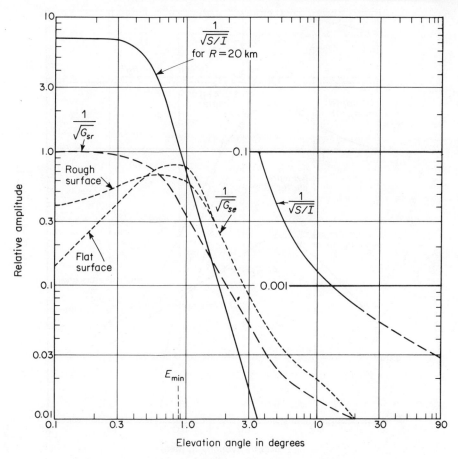

Fig. 8.6   Typical smoothed antenna patterns and S/I ratio.

at which range $\sigma_A = \sigma_E \cong 0.5$ mr for $\beta_n = 0.5$ Hz. Thus, any areas in which rain is present must be excluded from the region of optimum accuracy.

Step 7: Multipath Error. We have computed the multipath error for rough ground in Chap. 5 for the antenna patterns of this radar. The elevation error, shown in Fig. 8.7, is given by the following equations:

$$\sigma_E = \frac{\rho\theta_e}{k_m\sqrt{2G_{se}n_e}} = \frac{2.7 \text{ mr}}{\sqrt{G_{sr}}} \quad (E_t > 0.8 \text{ deg}); \quad \text{[see Eq. (5.21)],}$$

$$\sigma_E \cong \frac{\rho E_t}{\sqrt{2}} = 0.21E_t \quad (E_t < 0.8 \text{ deg}); \quad \text{[see Fig. 5.8].}$$

We find that tracking is possible down to the horizon when the target is beyond the range of clutter, and that $E_2 = 5$ deg for precise elevation data.

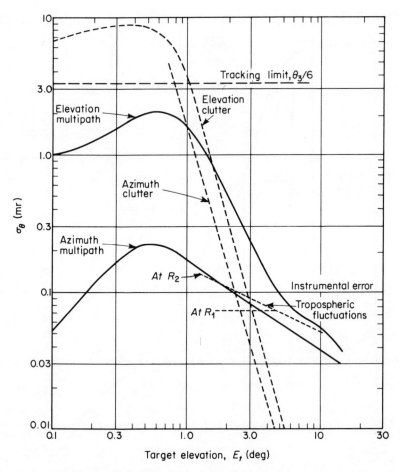

Fig. 8.7   Typical angle errors vs. elevation. ($\theta_3 = 20\,\text{mr}$; $\rho = \rho_o \rho_d = 0.3$; $\sigma_\alpha = 0.15\,\text{rad}$).

Azimuth multipath is given by

$$\sigma_A = \frac{\rho \theta_a}{k_m \sqrt{2 G_{sr} n_e}} = \frac{2.7\,\text{mr}}{\sqrt{G_{sr}}} \quad \left(E_t > \frac{\theta_a}{2\sigma_\alpha} = 4\,\text{deg} \right); \quad \text{[see Eq. (5.31)],}$$

$$\sigma_A = \frac{\rho E_t \sigma_\alpha}{\sqrt{2 G_{sr} n_e}} = \frac{0.032 E_t}{\sqrt{G_{sr}}} \quad (E_t < 4\,\text{deg}); \quad \text{[see Eq. (5.29)],}$$

$$\sigma_A \cong 0.032 E_t \quad \left(E_t < \frac{\theta_e}{2} = 0.6\,\text{deg} \right).$$

This curve is also plotted on Fig. 8.7, showing that the azimuth multipath error is not significant above $E_t = 3$ deg.

Step 8: Tropospheric Fluctuation. Path lengths in the troposphere are found from Fig. D.8.

$$\text{At } R_2 = 175 \text{ km}, E_2 = 3 \text{ deg}, L = 85 \text{ km}.$$

$$\text{At } R_1 = 50 \text{ km}, E_2 = 3 \text{ deg}, L = 50 \text{ km}.$$

$$\overline{\sigma}_\theta = 0.44 \sqrt{L/\sqrt{w}} \times 10^{-3} \text{ mr} \quad \text{[from Eq. (D. 23)]}$$

$$\cong 0.09 \text{ mr at } R_2, \text{ or } 0.071 \text{ mr at } R_1.$$

The fluctuation error is just significant at the low-elevation limit of coverage. Since the slope of the multipath vs elevation curve is quite steep, the upwards adjustment needed to hold total error below $\sqrt{2}\,\sigma_e$ is only a fraction of one degree, and is within the uncertainty caused by varying ground conditions.

Step 9: Residual Bias. From Fig. D.3, we read

$$\Delta E_o \cong 2 \text{ mr at } R_2 = 175 \text{ km}, E_2 = 5 \text{ deg}.$$

Correction to less than five percent residue is required to avoid significant error in elevation at this point. There will be no significant ionospheric error at the high microwave frequency used in this radar.

Step 10: Coverage Summary. The maximum and minimum range limits are

$$R_{max} = R_o = 440 \text{ km, or } 240 \text{ naut mi,}$$

$$R_{min} = R_1 \left(\frac{6\sigma_e}{\theta_3}\right) = 1.5 \text{ km, or } 0.8 \text{ naut mi} \quad \text{(from Fig. 8.8).}$$

A low-elevation limit of about $E_{min} = 1$ deg exists in the clutter region ($R < R_c = 20$ km), but beyond this range the horizon is the only limit. The four regions which have different error characteristics are shown in Fig. 8.9, along with a typical aircraft altitude limit and approximate contours for an accuracy of 1 mr. For azimuth data, the low-elevation limit is set by tropospheric fluctuation near 2.5 deg at $R_2 = 175$ km, and by multipath at this same angle for all target ranges. The only limit for an azimuth accuracy of 1 mr is that set by thermal noise at maximum range.

Step 11: Discrete Interference. Significant angular error (greater then 0.1 mr) can arise from either of the following situations: (a) randomly phased interference entering the range gate on m out of n pulses during the interval t_o, such that the quantity $(S/I_\Lambda)n^2/m$ is reduced below $\mathscr{R}/2 = 8100$; (b) coherent or partially coherent interference (relative to target phase) such that $(S/I_\Lambda)n_e$ is reduced below the same value. Probable loss of target will occur when the rms error within any interval t_o reaches $\theta_3/6$. In a linear system with a phase reference locked to the signal, this would imply an increase in interference power of 1100 relative to the values causing 0.1 mr error. In most practical cases, however, the interference also affects the sum-channel phase enough to cause detector loss or signal suppression

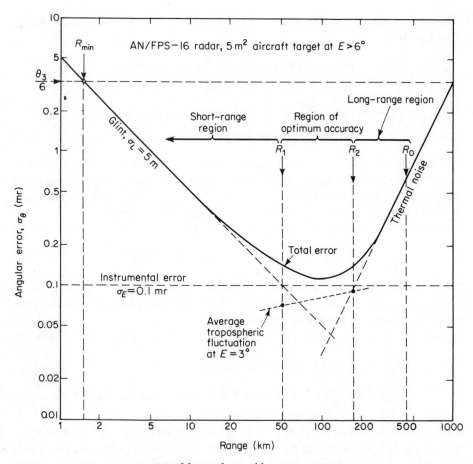

Fig. 8.8 Monopulse tracking error vs. range.

before the track is lost, and this must be taken into account. Receiver blanking, "constant false-alarm rate" techniques, and "coast" operation of the tracking circuits must also be considered.

The required difference-channel ratios S/I_Δ may be computed as follows, including detector loss but excluding any saturation or similar nonlinearities.

Type of Interference (see Table 5.4)	Required S/I_Δ to avoid:			
	Significant Error		Probable Loss of Track	
	$\beta_n = 0.5$	$\beta_n = 6$	$\beta_n = 0.5$	$\beta_n = 6$
Single impulse ($m = 1$)	0.31	3.7	0.0003	0.003
Random-phase pulses ($m = n$)	50	600	0.13	0.74
Coherent pulses ($n_e = 1$)	8100	8100	7.3	7.3

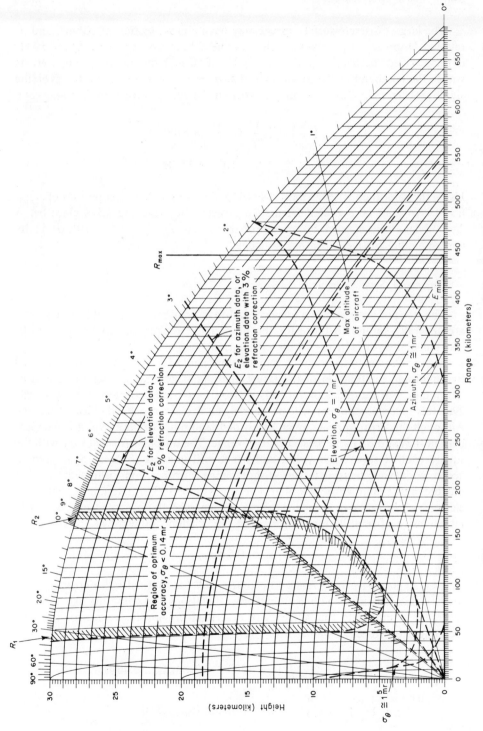

Fig. 8.9 Monopulse tracker coverage summary (drawn on Naval Res. Lab. chart; see Blake, 1968).

The largest discrete surface target may have a cross section of 10^4 m^2, and its relative Doppler frequency may lie anywhere between zero and $f_r/2 = 80$ Hz in the received spectrum. Approximately $1/n$ of all such targets will produce errors within the passband of the servo, such that $n_e = 1$. The requirement for avoiding significant errors in these cases may be written, for surface targets in the range gate,

$$\left(\frac{S}{I_\Delta}\right) = \left(\frac{\sigma}{\sigma_i}\right) G_{sr} G_{se} > 8100,$$

$$G_{sr} G_{se} > 1.6 \times 10^7, \text{ or } 72 \text{ db.}$$

Using the smoothed antenna patterns of Fig. 8.6, we see that brief periods of significant error can be expected from such targets when tracking takes place below $E = 12$ deg. Loss of track becomes probable when $G_{sr} G_{se} < 1.5 \times 10^4$, or 42 db, corresponding to tracks below $E = 2$ deg.

Sequential Scan Example

The steps in error analysis for a conical-scan or other sequential-lobing system are generally similar to those for the monopulse tracker, but differ because the instrumental error is increased by scintillation.

Step 1: Optimum Accuracy. A major component of instrumental error is caused by amplitude scintillation of the target at frequencies near the scan rate and its harmonics. For a typical aircraft target (see Chap. 6; also Barton, 1964, pp. 289–290) this error component will be near $\theta_3/60$ for a servo bandwidth of 0.5 Hz. Although this error appears much like broadband noise to the servo, it is independent of the ratio S/N and hence of target range. The output error will vary with the square root of servo bandwidth, and hence will increase if larger β_n is used to accommodate accelerating targets at short range. For the 20-mr beamwidth used in our previous example, the scintillation error will be 0.33 mr. Assuming that other elements of the radar are designed to be consistent with this accuracy (e.g., $\sigma = 0.37$ mr exclusive of scintillation), we may set the instrumental error at

$$\sigma_e = 0.5 \text{ mr.}$$

Step 2: Required S/N. Energy ratio and S/N for thermal noise error $\sigma_\theta = \sigma_e$ are now computed, using conical-scan error equations and constants. Assume that the radar is similar to the AN/FPS-16 described previously, except for the scan technique. Beam-center values are

$$\mathscr{R}_m = 2L_{k2}\left(\frac{\theta_3}{\sigma_\theta k_s}\right)^2 = 1900 \qquad \text{[from Eq. (2.46)],}$$

$$\left(\frac{S}{N}\right)_m = \frac{\mathscr{R}_m}{2nL_m} = 4.7 \qquad \text{[from Eq. (1.37)].}$$

Step 3: Long-Range Transition. The value of R_o found in the previous example also applies to the beam-center value $(S/N)_m$, giving

$$R_o = 440 \text{ km, or } 240 \text{ naut mi,}$$

$$R_2 = R_o(S/N)_m^{-1/4} = 298 \text{ km, or } 160 \text{ naut mi.}$$

The larger instrumental error masks the thermal noise error out to longer ranges than calculated in the previous example.

Step 4: Linear Error at Target. The short-range error calculation will be similar to that in the monopulse case, except that the reduced bandwidth β_n will affect K_a and acceleration lag:

$$K_a \cong 2.5\,\beta_n^2 = 0.63 \text{ sec}^{-2},$$

$$\text{Acceleration lag} = \frac{a_t}{K_a} = 8 \text{ m,}$$

$$\text{Linear error at target } \sigma_L = 9.4 \text{ m.}$$

Step 5: Short-Range Transition. Neglecting geometrical accelerations, we have

$$R_1 = \frac{\sigma_L}{\sigma_e} = \frac{9.4}{5 \times 10^{-4}} = 19 \text{ km, or } 10 \text{ naut mi.}$$

This range could be decreased slightly with a small increase in servo bandwidth (e.g., $R_1 = 14$ km at $\beta_n = 0.7$ Hz). Geometrical acceleration will be found to lie below σ_e with the target velocity assumed here.

Step 6: Extended Clutter Error. The data derived for the monopulse example may be used, but with reduced error sensitivity $k_{s2}/\sqrt{L_{k2}} = 1.3$ in place of $k_m = 1.57$. The maximum number of error samples integrated is also cut in half, so that for $\beta_n = 0.5$ Hz we have $n = f_r/(4\beta_n) = 80$. The result, for clutter power distributed uniformly over intervals f_r in the spectrum, is an error 1.7 times as great as calculated for the monopulse case. Referring to Fig. 8.7, we can find a lower limit for accurate tracking in surface clutter at $R = 20$ km by reading $E_1 = 2.1$ deg for $\sigma = 0.5/1.7$ mr. Similarly, for a target immersed in rain at 1 mm/hr rate, $\sigma = 0.5$ mr at $R = 58$ km.

Step 7: Multipath Error. The conical-scan system senses the wavefront distortion caused by the image target in the same way as does the monopulse system, with approximately equal errors (Fig. 8.7). As long as the error is small, the target illumination is reduced by the one-way crossover loss L_{k1}, and this reduces the amplitudes of both the direct and the reflected rays. The direct ray suffers this loss for a second time upon reception, while the reflected ray enters a sidelobe or the edge of the main lobe. The effective error sensitivity is then $k_{s2}/\sqrt{L_{k1}} = 1.47$, leading to errors larger than the monopulse errors by the factor $1.57/1.47 = 1.07$. In elevation, assuming sidelobe levels equal to those of the monopulse system, multipath will reach the 0.5-mr threshold at $E_2 = 2.0$ deg, while in azimuth it will remain below this level at all elevation angles.

Step 8: Tropospheric Fluctuation. This error is seldom of significance in sequential-scan systems, as a result of the larger scintillation error.

Step 9: Residual Tropospheric Bias. Only the most rudimentary correction of elevation error is needed to bring the residual error below the threshold of significance:

$$\left.\begin{array}{l} E_2 = 1.9 \text{ deg} = 33 \text{ mr} \\ R_2 = 290 \text{ km} \end{array}\right\} \quad \Delta E_o \cong 4 \text{ mr} \quad \text{(from Fig. D.3).}$$

A correction based on the world-wide average of N_o, adjusted for radar height above sea level, will bring the residue to about ten percent of ΔE_o, or 0.4 mr, which is adequate to avoid significant increase in total error.

Step 10: Coverage Summary. Limits of reliable tracking, corresponding to S/N and S/I of unity and $\sigma_\theta = \theta_3/6$ are found as in the monopulse case, but considering crossover loss.

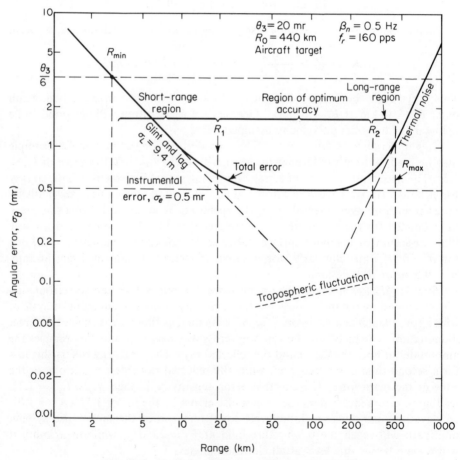

Fig. 8.10 Typical conical-scan system error vs. range.

$$R_{max} = R_o(L_{k2})^{-1/4} = 390 \text{ km, or } 211 \text{ naut mi,}$$

$$R_{min} = R_1(6\sigma_e/\theta_3) = 2.9 \text{ km, or } 1.6 \text{ naut mi} \qquad \text{(Fig. 8.10).}$$

These limits and the 0.5-mr error contour (for elevation data) are shown in Fig. 8.11. Contours for 1-mr error are also plotted for comparison with those on the monopulse coverage diagram (Fig. 8.9). The monopulse system has better accuracy over all regions, but its region of optimum accuracy is smaller because of the tighter tolerances needed to achieve the 0.1-mr error level.

Step 11: Discrete Interference. The instrumental error is exceeded when the quantity $(S/I_A)n^2/m$ drops below $\mathscr{R}/2 = 600$, considering the crossover loss L_{k2}. For different types of interference, the required S/I_A values for the conical-scan system are as follows.

Type of Interference (see Table 5.4)	Required S/I_A to avoid:	
	Significant Error	Probable Loss of Track
Single impulse ($m = 1$)	0.019	0.004
Random-phase pulses ($m = n$)	3.7	0.16
Coherent pulses ($n_e = 1$)	600.	14.

For ground targets up to 10^4 m^2 in cross section, the sidelobe ratio G_{sr} must exceed 2200, for negligible error, and this is achieved at elevations above 5 deg (see Fig. 8.6). To avoid probable loss of track, G_{sr} must exceed 350, and E_t must remain above 2.9 deg. These calculations are based on conical-scan difference sidelobes 6 db greater than those observed in the beam pattern itself (see Fig. 5.3).

Search Radar Example

In a conventional, azimuth-scanning (2D) search radar there is no elevation angle measurement. Steps in analysis of azimuth errors are as follows.

Step 1: Optimum Accuracy. Scintillation error is usually the controlling component in instrumental error. The data presented in Fig. 6.10 are used after calculating the radar time-on-target t_o and the expected correlation time t_c for targets of interest. For this example, let us assume an S-band search radar ($\lambda = 0.1$ m) with a beamwidth $\theta_a = 2$ deg (35 mr), scanning at $\omega = 10$ rpm (1.05 rad/sec). Targets will range from jet fighters ($L = 10$ m) to large transports ($L = 40$ m), with expected aspect angle rates near $\omega_a = 0.01$ rad/sec. The following calculations are made:

$$t_o = nt_p = \frac{\theta_a}{\omega} = 0.033 \text{ sec} \qquad \text{[from Eq. (1.24)],}$$

$$t_c = \frac{\lambda}{2\pi\omega_a L_x} = 0.04 \text{ to } 0.16 \text{ sec} \qquad \text{[from Eq. (6.8)],}$$

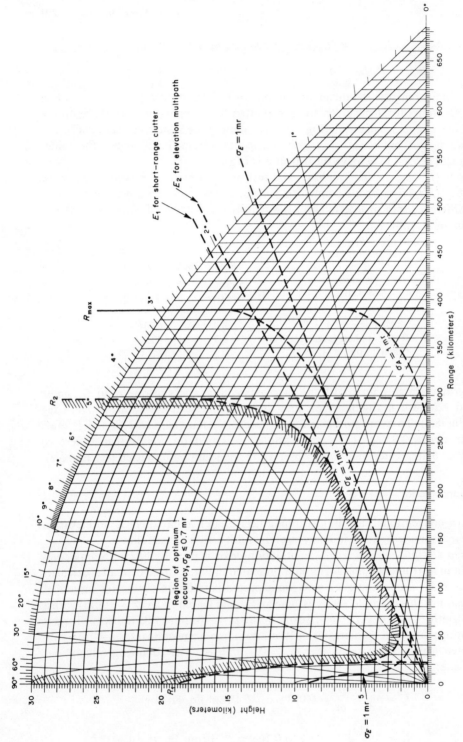

Fig. 8.11 Typical conical-scan tracking limits and accuracy (drawn on Naval Res. Lab. chart; see Blake, 1968).

$$\frac{t_o}{t_c} = \frac{nt_p}{t_c} = 0.21 \text{ to } 0.83,$$

$$\sigma_A = 0.075 \text{ to } 0.11 \, \theta_a \qquad \text{(Fig. 6.10).}$$

Pulse-to-pulse frequency agility may be used to enhance detectability and reduce scintillation error. If the frequency steps are large enough to achieve pulse-to-pulse decorrelation of signal amplitude ($f_c = 3.75$ to 15 MHz for the targets assumed here), we have, for a repetition rate $f_r = 400$ pps,

$$n_e = n = \frac{f_r \theta_a}{\omega} = 13.3 \qquad \text{[from Eq. (1.26)],}$$

$$\sigma_A = 0.06 \, \theta_a \qquad \text{(from Fig. 6.10).}$$

Thus, in the absence of monopulse processing, the search radar error due to scintillation will approach one-tenth the azimuth beamwidth, and this can be taken as the instrumental error level for a system with reasonable provisions for signal processing and beam-splitting. For our example,

$$\sigma_e = 3.5 \text{ mr or } 0.2 \text{ deg.}$$

Step 2: Required S/N. The required on-axis ratio $(S/N)_m$ is calculated with allowances for collapsing loss and other nonoptimum processes. If we assume $L_c L_x = 2$, we find

$$n\left(\frac{S}{N}\right)_m = \frac{L_x L_{p2} L_c}{2}\left[\frac{\theta_A}{k_{p2}\sigma_e}\right]^2 = 48 \qquad \text{(from Table 2.9),}$$

$$\left(\frac{S}{N}\right)_m = \frac{48}{13.3} = 3.6.$$

This value is consistent with the assumed detector loss L_x, and leaves a margin for collapsing loss. The S/N requirement for measurement is almost the same as for target detection with blip-scan ratios near fifty percent.

Step 3: Long-Range Transition. Because the signal level needed for detection will also ensure optimum accuracy, there will be no "long-range region" for this radar. The R_2 contour is identical to the coverage contour R_{50} for fifty percent blip-scan ratio.

Step 4: Linear Error. A search-radar track-while-scan system can be characterized by the data interval $t_s = 2\pi/\omega$ sec between revolutions of the antenna. The corresponding acceleration error constant, in a loop which meets reasonable stability criteria, is given by

$$K_a \cong 1/2t_s^2.$$

The lag error for an accelerating target is

$$\Delta_a = \frac{a_t}{K_a} \cong 2a_t t_s^2.$$

In our example, with $t_s = 6$ sec and a typical, slowly-maneuvering aircraft ($a_t = 1$ m/sec^2), we have

$$\Delta_a = \sigma_L = 2 \times 1 \times 36 = 72 \text{ m.}$$

This is the dominant linear error for a track-while-scan system operating in rectangular coordinates, since it greatly exceeds the glint error on aircraft targets.

Step 5: Short-Range Transition. Setting $\sigma_L = R_1\sigma_e$, we find

$$R_1 = \frac{72}{0.0035} = 20{,}500 \text{ m} = 20.5 \text{ km.}$$

Within this range, acceleration lag will exceed instrumental error.

Step 6: Extended Clutter Error. As with thermal noise at long range, the clutter limitation on accuracy is associated with detectability in the clutter (whether located on the surface or in the atmosphere). As long as the blip-scan ratio exceeds fifty percent, the rms error should remain below $0.1\theta_a$. A more serious problem caused by clutter at short range is the loss of data needed by the track-while-scan system to maintain K_a. The loop constants used in obtaining the performance of Step 4 above require that the blip-scan ratio should be near ninety percent, and this is normally assured at short range by high S/N ratio. Presence of excessive clutter requires an increase in the detection threshold, reducing the blip-scan ratio and leading to larger lag errors.

Step 7: Multipath Error. Elevation multipath error is not a problem in 2D search radar, but the presence of an azimuth error component caused by reflection from rough surfaces should be considered. Calculations for both high-elevation and low-elevation targets are carried out. In the high-angle case, we consider the diffuse reflections to reach the antenna from a region surrounding the image target (which lies $2E$ below the actual target and E below the horizon). The region extends over an elevation angle interval $\pm 2\sigma_\alpha$ from this image, and over an azimuth interval $\pm 2E\sigma_\alpha$, and the total power reflected from this region to the antenna is equal to that which would have been reflected specularly from a smooth surface. As an example, let us take a moderate sea surface, $\sigma_\alpha = 0.1$ rad, $\rho_o = 1.0$ for horizontally-polarized transmissions. Assume that our search radar uses an elevation beamwidth of 0.2 rad (12 deg), with its lower 3-db point on the horizon, and that the target is at the upper 3-db point ($E = 0.2$ rad). The reflecting region will extend from the horizon to -0.4 rad in elevation, and 0.04 rad each side of the target in azimuth, so that most of the power will lie outside the antenna response region (see Fig. 8.12). Integration of the sidelobe ratio over the glistening surface gives the average value for use in estimating azimuth error:

$$\overline{G_{sr}} = \frac{G_r(E_t)\psi_d}{\displaystyle\int_{\psi_d} G_r d\psi} = 18 \qquad \text{[see Eq. (5.28)].}$$

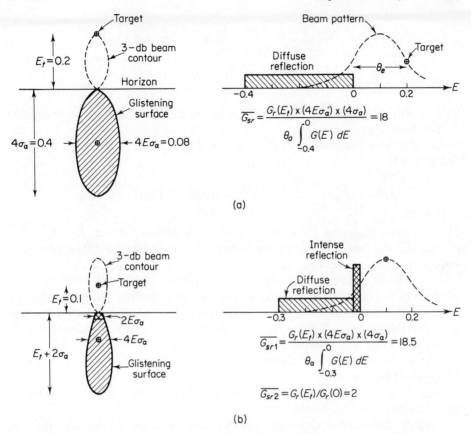

Fig. 8.12   Surface reflection calculations: (a) high-elevation case ($E \geq 2\sigma_\alpha$); (b) low-elevation case ($E < 2\sigma_\alpha$).

To find the effective number of independent error samples introduced during the time-on-target, we can assume a wind speed $v_w = 3$ m/sec, and calculate multipath signal correlation time.

$$t_c = \frac{1.6\lambda}{v_w} = 0.05 \text{ sec} \qquad [\text{see Eq. (5.32)}].$$

The number of amplitude-difference samples, which determines the performance of the search-radar azimuth estimator, is

$$(n_e - 1) = \frac{t_o}{t_c} = \frac{0.033}{0.05} = 0.66 \qquad [\text{see Fig. 6.10}].$$

Accordingly, we may set the number of effective samples in the error equations to unity, to obtain an adequate error estimate.

$$\sigma_A = \frac{\rho\sqrt{\eta}}{K\sqrt{2\overline{G_{sr}}n_e}} = \frac{\rho\theta_a\sqrt{L_{p2}}}{k_{p2}\sqrt{2\overline{G_{sr}}n_e}} \qquad \text{[see Eq. (5.31)]}$$

$$= \frac{0.4 \times 0.035}{2.03\sqrt{18 \times 1}} = 0.0016 \text{ rad.}$$

Because this error is less than half the instrumental error, it may be neglected.

For lower-elevation targets, most of the reflected power will come from the region of intense reflection near the horizon (Fig. 8.12). For this case, $\overline{G_{sr}} = 2$, and $\rho = 0.4$:

$$\sigma_A = \frac{\rho E \sigma_\alpha}{\sqrt{2\overline{G_{sr}}n_e}} = \frac{0.4 \times 0.1E}{\sqrt{2 \times 2}} = 0.02E \qquad \text{[see Eq. (5.29)]}.$$

The maximum elevation for which this equation applies is about $E = \sigma_\alpha = 0.1$ rad, because the intense reflecting region begins to merge into the diffuse glistening surface above this value. Thus, the low-elevation error will vary directly with elevation angle up to a maximum of about 0.002 rad error, which is close to the value obtained earlier for a target at the upper edge of the beam. The azimuth multipath error can thus be neglected at all target elevations for our case. It would become an important limitation if monopulse processing were used in an attempt to reduce scintillation error and improve the overall search-radar accuracy.

Step 8: Tropospheric Fluctuation. This error component is negligible in almost all search applications.

Step 9: Residual Refraction Bias. This step can be omitted in 2D radar, where elevation data are not gathered.

Step 10: Summary of Coverage. The detection coverage of the search radar is plotted first, to indicate the volume within which measurements are possible. In the example given here, the error in azimuth data will remain almost constant, between 0.0035 and 0.005 rad within this region, except for track-while-scan lag which increases within 20 km range.

Step 11: Discrete Interference. Since search radar measurement to $0.1\,\theta_a$ requires the same ratio S/I as does target detection, the ill effects of interference on a given scan can be either to prevent detection or to produce a false target detection near enough the true position to confuse the track-while-scan system. Loss of detection merely aggravates the problem of data rate, where a "coasting time" is already allowed for target fading. In the presence of interference, the lag error increases and the probability of losing track also increases. False targets may capture the tracking loop, or may introduce enough noise to break the track. Neither of these effects is susceptible to a generalized analysis.

Application to 3D Search Radar

Rather than give detailed analyses of each type of 3D system, we will note the major differences between such systems and the 2D system used in our earlier example. As an initial case, we should note that the scanning height-finder, used in conjunction with conventional 2D radar to provide the third coordinate on selected targets, can be treated as a special case of the search radar. It uses a narrow elevation beamwidth and scans in the elevation plane, but otherwise has most of the characteristics of the system analyzed above. Scintillation error may be reduced by repeating the scans more rapidly, so that several readings can be averaged (with independent scintillation errors). Analysis of elevation multipath and tropospheric refraction errors will usually be required, using the same techniques given earlier for the tracking radars. The normalized error sensitivity $k_{p2}/\sqrt{L_{p2}} = 1.4$ should be substituted for k_m in calculating multipath and clutter errors.

Other 3D techniques, and their important differences relative to 2D and tracking radars are as follows.

(a) Stacked-Beam Radar. In this system, the instrumental error in elevation measurement is larger than in trackers, because the beamshapes vary over the elevation coverage, and because relative channel gains are less precisely matched than in monopulse trackers. Scintillation error can be as low as in monopulse trackers, however, if the Δ/Σ ratio process is instrumented properly. Most measurements are made off the axes defined by crossover points between adjacent beams, so the appropriate off-axis error equations should be used for thermal noise, clutter and multipath.

(b) Raster-Scanning Pencil Beams. These radars can be analyzed as search radars or sequential-processing trackers in both angular coordinates, noting the different time intervals between samples and the resulting differences in scintillation errors. In the rapid-scan coordinate, the target is usually correlated over the measurement interval ($t_o/t_c \ll 1$), while in the other coordinate it is uncorrelated ($t_o/t_c \gg 1$). Frequency scan may introduce decorrelation in the target amplitude even in the rapid-scan coordinate, and the ratio $\Delta f/f_c$ should be used instead of t_o/t_c in entering Fig. 6.10. The frequency change corresponding to one beamwidth of scan is related to the electrical length of the frequency-sensitive network in the antenna. If the antenna beamwidth and the scanning coefficient C_f rad/Hz are known, the electrical length S' can be calculated:

$$S' = \left(\frac{C_f}{\theta_3}\right) \times 3 \times 10^8 \text{ m.}$$

For example, a 20-mr beam which is scanned 2 mr per MHz (or one beamwidth in 10 MHz) has an electrical length of 30 m. The ratio $\Delta f/f_c$ is equal to $2L_r/S'$, where L_r is the effective radial extent of the target. Thus, if the target has a length of 10 m, the ratio $\Delta f/f_c = (n_e - 1) = 0.67$, and the scintillation error (Fig. 6.10) is $0.11\,\theta_3 = 2.2$ mr. Most frequency-scan networks have electrical lengths comparable to target dimensions, and hence errors near $0.1\,\theta_3$ are to be expected.

(c) V-Beam Radar. In the V-beam radar, elevation angle is measured by the displacement in target azimuth between a vertical fan beam and a second beam slanted at an angle ϕ from the vertical (Fig. 8.13). In the coverage region, generally

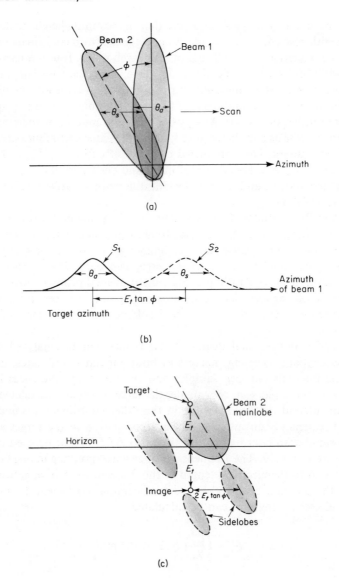

Fig. 8.13 V-beam measurement principles: (a) V-beam patterns; (b) signal wave-forms; (c) multipath effect.

below 30 deg elevation, the contributions of thermal noise, scintillation, clutter, and multipath errors in the azimuth measurements to derived elevation data are given by

$$\sigma_E = \sigma_A \sqrt{2} \cot \phi.$$

Here, the factor $\sqrt{2}$ represents the increase in error associated with taking the difference between two azimuth readings whose errors are uncorrelated, while the cot ϕ term gives the sensitivity of elevation data to errors in azimuth difference. Because ϕ is normally about 30 deg, the elevation errors are two to three times as great as those in azimuth. Tropospheric bias in elevation is the same as for any measuring system, but is often masked by large instrumental errors, as are tropospheric fluctuations in both angular coordinates.

The elevation multipath error from specular reflection in V-beam systems is determined by the sidelobes of the slanted beam at an angle $2E_t \tan \phi$ from the axis, in azimuth, and at elevation $-E_t$ from the horizon [see Fig. 8.13(c)]. In attempting to locate the azimuth angle at which S_2 reaches its peak value, surface reflections from this sidelobe will interfere with the main response, causing a shift in the apparent azimuth:

$$\sigma_A = \frac{\theta_s \rho \sqrt{L_{p2}}}{k_{p2}\sqrt{2G_{sr}n_e}} = \frac{0.49\theta_s \rho}{\sqrt{G_{sr}}} \quad \text{[see Eq. (5.31)].}$$

This error is converted into an elevation error

$$\sigma_E = \frac{\theta_s \rho \cot \phi \sqrt{L_{p2}}}{k_{p2}\sqrt{2G_{sr}n_e}} = \frac{0.49\theta_s \rho \cot \phi}{\sqrt{G_{sr}}}.$$

The same multipath effect would be observed if a beam of width $\theta_s \cot \phi$ were scanned in elevation as a height-finder. Use of an azimuth sidelobe level measured $2E_t \tan \phi$ from the axis is equivalent to scaling the entire pattern by cot ϕ and using sidelobes $2E_t$ from the axis, as would be done with the elevation-scanning beam. Not included in the above expressions is the additional error which can result from azimuth multipath components in both beams, when the surface is rough.

8.4 Range Error Analysis

The components of range error which are discussed elsewhere in this handbook are:

    (a) Thermal noise (Chap. 3);
    (b) Range-Doppler cross-talk (Chap. 4);
    (c) Clutter and interference (Chap. 5);
    (d) Multipath reflections (Chap. 5);
    (e) Glint and scintillation (Chap. 6);
    (f) Quantization (Chap. 7);
    (g) Dynamic lag (App. C); and
    (h) Atmospheric propagation (App. D).

The remaining errors in a practical system are associated with details of component and circuit design. These include:

    (i) Receiver time delay instability;

(j) Time discriminator alignment and stability;

(k) Servo loop noise;

(l) Error in converting measured delay to output data (apart from quantizing noise); and

(m) Reference oscillator frequency stability.

In a specific system, these errors add to an rms value which is almost independent of target dynamics and tracking conditions. A particular type of target will contribute a further error in the form of glint, which may be added to the circuit errors to find the instrumental error, σ_e. The instrumental error sets a limit on the accuracy of the system for this class of target, and we can use it to define a threshold of concern for variable error components, according to the general procedure of Sec. 8.2.

We will illustrate the application of this procedure to the range coordinate with an example: a digital range tracker operating out to 900-km range with a radar system having the following parameters:

Transmitted pulse width, $\tau_a = 40$ μsec,

Transmitted bandwidth, $B_a = 1.1$ MHz,

Receiver weighting: 30-db Taylor,

Pulse broadening factor, $K_h = 1.25$,

Compression ratio, $K_c = 40$,

Compressed pulse width, $\tau_{3x} = 1.0$ μsec,

Pulse compression loss, $1/\eta_f = 1.17$, or 0.7 db,

Target: small aircraft, $\bar{\sigma} = 1.0$ m^2, $L_{nr} = 6$ m, $a_t = 5$ m/sec^2, $v_t = 300$ m/sec,

Other radar parameters identical to those of the AN/FPS-16, given in Sec. 8.3. We will calculate range errors in meters, converting from time delay in μsec when necessary.

Step 1: Optimum Accuracy. Assume that the range is measured to the zero-crossing point on the derivative of each received pulse, by counting the nearest whole number of clock pulses at 20 MHz and averaging the readings in a servo bandwidth $\beta_n = 4$ Hz. The quantizing error at the output will be a fraction of the range corresponding to the interval between clock pulses, $\Delta_r = 7.5$ m:

$$\sigma_q = \frac{\Delta_r}{\sqrt{12n}} = \frac{7.5 \text{ m}}{\sqrt{12 \times 20}} = 0.48 \text{ m}.$$

The glint error can be estimated from the effective radial length of the target, $L_{nr} = 6$ m:

$$\sigma_r = 0.35 \, L_{nr} \qquad \text{[see Eq. (6.7)]}.$$

Other contributions to instrumental error can be assumed about equal to the glint error (e.g., receiver and compression network instability totaling about 15 nsec or 2.3 m rms). The total instrumental error, independent of target location and dynamics, is then

$$\sigma_e = 3 \text{ m}.$$

The range-Doppler cross-talk term, a maximum of 59 m for the target velocity of 300 m/sec, will not be included as an error because it is readily correctable once a track has been established.

Step 2: Required S/N. The thermal noise error for a pulse compression system using a uniform transmitted spectrum is given in Table 3.6, on the basis of single-pulse error, σ_{t1}:

$$\sigma_{t1} = \frac{\tau_a(\beta_h/\beta_{h1})}{1.607 K_c \sqrt{\eta_f \mathscr{R}_1}}.$$

The ratio β_h/β_{h1} is found by reference to Fig. 2.4, using the analogy to the ratio $\mathscr{L}_s/\mathscr{L}_\theta = 1.32/1.63 = 0.81$ for the 30-db Taylor distribution. The error for a tracking loop is less by the factor $\sqrt{n} = \sqrt{f_r/2\beta_n}$, and the required S/N for noise equal to instrumental error is found from

$$\sigma_t = \frac{40(0.81)}{1.607 \times 40\sqrt{(S/N) \times 40}} = \frac{2\sigma_e}{c} = 0.02 \ \mu\text{sec},$$

$$\frac{S}{N} = 8.1, \text{ or } 9.1 \text{ db.}$$

Step 3: Long-Range Transition. The range for unity S/N ratio is greater than calculated in Sec. 8.3 because of the greater average power:

$$R_o = \left[\frac{P_t G^2 \lambda^2 \bar{\sigma} \tau_a \eta_f}{(4\pi)^3 k T_o NF_o L} \right]^{1/4} = 810 \text{ km, or } 435 \text{ naut mi,}$$

$$R_2 = R_o \left(\frac{S}{N} \right)^{-1/4} = 475 \text{ km, or } 255 \text{ naut mi.}$$

Step 4: Linear Error at Target. All the range errors are expressed in linear measure, so we need calculate here only the range-dependent dynamic lag error, caused by geometrical acceleration.

$$(\ddot{R})_{max} = -\frac{v_t^2}{R}. \qquad \text{(Barton, 1964, p. 370)}$$

For a servo bandwidth $\beta_n = 4$ Hz, we have $K_a = 40$ sec^{-2}, and the peak lag error caused by geometrical acceleration is

$$\Delta_a = -\frac{v_t^2}{RK_a} = -\frac{2250}{R} \text{ m.}$$

In addition to this, of course, there will be an rms lag of $\sigma_r = a_t/K_a = 0.12$ m caused by the actual target acceleration of 5 m/sec^2 rms. This will not add significantly to instrumental error.

Step 5: Short-Range Transition. The range at which the lag equals the instrumental error is

$$R_1 = \frac{v_t^2}{3K_a} = 750 \text{ m.}$$

Step 6: Extended Clutter Error. Figure 8.6 can be used to find the elevation $E_1 = 1.4$ deg for $S/I = 8.1$, and the low-elevation tracking limit $E_{min} = 0.9$ deg for $S/I = 1$, assuming in both cases $R_c = 20$ km. In rain of 1 mm/hr, $S/I = 8.1$ at a range of about 20 km, and rain volumes beyond that range must be excluded from the region of optimum accuracy.

Step 7: Multipath Error. From Eq. (5.36), the maximum multipath error in range can be found, for a very high-sited radar, as follows:

$$\sigma_t = \frac{\rho \tau_{3x}}{\sqrt{8G_{sr}}} = \frac{0.1 \ \mu\text{sec}}{\sqrt{G_{sr}}},$$

$$\sigma_r = \frac{15 \text{ m}}{\sqrt{G_{sr}}}.$$

If the concentrated scattering is assumed to be approximately at the horizon [(Fig. 5.10(d)], the low-elevation region will begin at $E_2 = 1.3$ deg, where $G_{sr} = 25$ (see Fig. 8.6). For most radar sites, the range multipath error will remain much smaller at all elevations. For example, if the antenna is at a height $h = 100$ m above the reflecting surface, Eq. (5.34) gives, for $E_t = 0.8$ deg,

$$\sigma_r = \frac{\rho h \sin E_t}{\sqrt{2G_{sr}}} = \frac{0.3 \times 100 \times 0.014}{\sqrt{2 \times 4.3}} = 0.14 \text{ m.}$$

The quantity $\sin E_t / \sqrt{G_{sr}}$ falls off at higher angles because of the rapid increase in G_{sr} in the sidelobe region, while below this angle there is a decrease because of the $\sin E_t$ term. Even over water, where ρ may approach unity, the error cannot exceed 0.5 m for $h = 100$ m.

Step 8: Tropospheric Fluctuation. This component is negligible when compared to $\sigma_e = 3$ m.

Step 9: Residual Refraction Bias. At zero elevation and maximum range for precision tracking ($R = R_2 = 475$ km), the tropospheric bias (Fig. D.2) is about 90 m. The required accuracy of refractive correction is thus about three percent, if the multipath error is small and if tracking to the horizon is desired. If the correction can only be relied on to five percent, a contour extending from $R = 200$ km at $E = 0$ to $R = 475$ km at $E_t = 1$ deg will bound the region where residual bias is held within 3 m.

Step 10: Coverage Summary. Figure 8.14 shows the contour bounding the region of optimum accuracy, within which the total error $\sigma_r < \sqrt{2} \ \sigma_e = 4.3$ m. Also shown are the elevation limits for high-sited radars and for operation at short

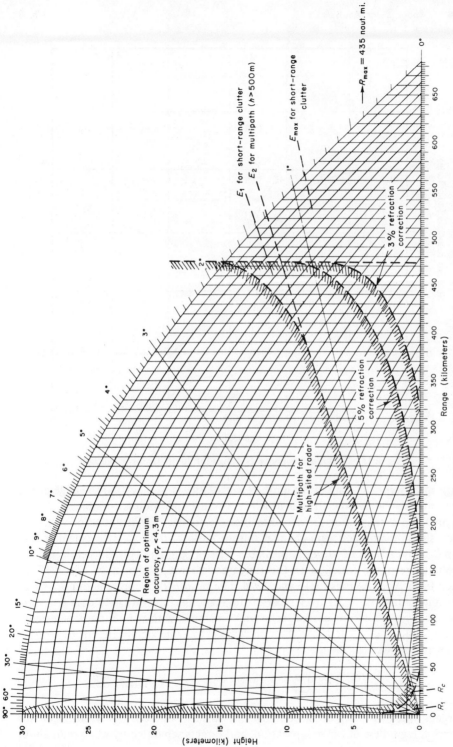

Fig. 8.14   Typical range error contours (drawn on Naval Res. Lab. chart; see Blake, 1968).

233

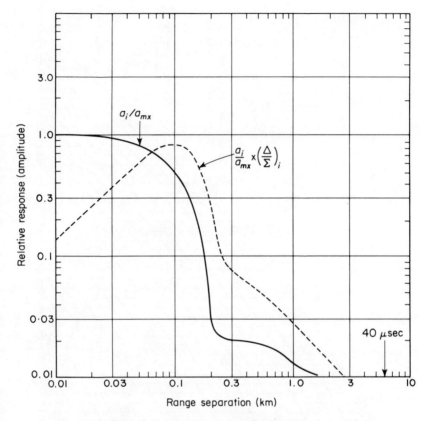

Fig. 8.15   Smoothed response vs. range for 30-db Taylor weighting.

range in clutter. The long-range limit, $R = 810$ km, is well beyond the edge of the chart, and also above the altitude at which aircraft operation is possible.

Step 11: Discrete Interference. For discrete echoes, significant error will arise when the differentiated sum-channel output ratio $(S/I_\Delta)n_e \leq 325$, calculated in Step 2 above. Loss of target is probable when this quantity falls below 6. The smoothed response function in the range coordinate is shown in Fig. 8.15, as taken from App. A, Figs. A.15 and A. 26. The ratio S/I_Δ averages about 0.2 times the sum-channel ratio S/I over most of the response interval (± 40 μsec from the tracked target). A strong surface echo ($\sigma_i = 10^4$ m²), with $n_e = 1$, must be attenuated by a factor of 1.7×10^7 in the sum channel to avoid significant error as a 1 m² target passes over it:

$$G_{sr}^2\left(\frac{a_{mx}}{a_i}\right)^2 \geq \left(\frac{\sigma_i}{\bar{\sigma}}\right) \times 325 \times 5 = 1.7 \times 10^7, \text{ or 72 db.}$$

Targets which lie 7 deg or more above the surface, where $G_{sr} > 4100$, meet this criterion regardless of range separation, and targets separated by more than about

200 m from strong clutter (where $a_{mx}/a_i > 32$) must lie above 2 deg (where $G_{sr} >$ 130). Targets within one beamwidth of the surface ($G_{sr} < 12$) cannot meet the criterion unless they are more than 6 km (40 μsec) from the strong clutter.

To ensure continuance of track,

$$G_{sr}^2\left(\frac{a_{mx}}{a_i}\right)^2 \geq \left(\frac{\sigma_i}{\bar{\sigma}}\right) \times 6 \times 5 = 3 \times 10^5, \text{ or 55 db.}$$

Targets above 3 deg meet this criterion at any range separation, as do targets above 1 deg elevation which are separated by more than 200 m in range from the clutter. In most cases, the motion of the target over the ground can be expected to increase n_e above unity, leading to further relaxation of these requirements. Target velocity will also cause a range displacement at the rate of 0.2 m per m/sec, but this effect merely shifts the range region in which clutter can interfere with the target.

Random-pulse interference is dispersed by the compression filter, so that a 1-μsec pulse is stretched over 40 μsec at the receiver output, and attenuated by 16 db. To cause a significant error, a single such pulse must be passed to the compression network without limiting, at a power level such that $(S/I)n^2K_c < 325$, or $S/I < 0.02$.

8.5 Doppler Error Analysis

The components of Doppler error which are discussed elsewhere in this handbook are:
   (a) Thermal noise (Chap. 4);
   (b) Range-Doppler cross-talk (Chap. 4);
   (c) Clutter and interference (Chap. 5);
   (d) Multipath reflection (Chap. 5);
   (e) Glint and scintillation (Chap. 6);
   (f) Quantization error (Chap. 7);
   (g) Dynamic lag (App. C); and
   (h) Atmospheric propagation (App. D).
Other errors in a practical system are dependent upon details of components and circuit design:
   (i) Receiver time delay or phase variation;
   (j) Doppler discriminator stability;
   (k) Servo loop noise;
   (l) Error in converting frequency to output data (other than quantization); and
   (m) Reference oscillator stability.
As with range and angular error analysis, we can find an rms error level σ_e which is essentially independent of target position and dynamics, and use this instru-

mental error to define a threshold of concern, beyond which other components of error are significant.

As an example, we will use the AN/FPS-16 type of radar which was discussed earlier, and will assume that coherent operation has been achieved at the repetition rate of 160 pps. Presence or absence of pulse compression is not of concern, as long as the frequency stability of the radar is maintained from pulse to pulse. Hence we will assume that the small jet aircraft ($L_{nx} = 10$ m, $\omega_a = 0.01$ rad/sec) is being tracked with the compressed pulse (Sec. 8.4 above). Errors will be calculated in hertz, converted later to m/sec.

Step 1: Optimum Accuracy. The Doppler loop might typically use a frequency discriminator matched to the expected width of the received spectral lines (a width controlled by radar and target stability). A voltage-controlled oscillator, a frequency divider, and a counter will then provide output data in the form of cycle counts at some integral multiple of the actual Doppler shift (e.g., $4f_d$, counted over intervals of 0.1 sec). The smallest data increment for this case is $\Delta_f = 2.5$ Hz, and the quantizing error is

$$\sigma_q = \frac{\Delta_f}{\sqrt{12}} = \frac{2.5}{\sqrt{12}} = 0.7 \text{ Hz.}$$

The glint error component, from Eq. (6.7) and Fig. 6.7, is

$$\sigma_f = 0.35\left[\frac{2\omega_a L_{nx}}{\lambda}\right] = \frac{0.35 \times 2 \times 0.01 \times 10}{0.055} = 1.3 \text{ Hz.}$$

If we allow an additional error of 1.5 Hz for system instabilities, the instrumental error level will be

$$\sigma_e = 2 \text{ Hz, or } 0.05 \text{ m/sec.}$$

Step 2: Required S/N. The thermal noise error is

$$\sigma_f = \frac{\sqrt{\eta_f}}{K_f \sqrt{2(S/N)_f}} \qquad \text{[see Eq. (4.41)].}$$

To find K_f, we must estimate the width of the coherent signal lines in the spectrum. If this width is based on target rotation alone, it would be

$$B_n = \frac{2\omega_a L_{nx}}{\lambda} = 3.6 \text{ Hz} \qquad \text{(see Fig. 6.7).}$$

This, however, is a minimum value which could be expected only with a rigid airframe. Vibration and jet engine modulation will broaden the lines and introduce spectral components up to several hundred hertz from the airframe return (Hynes and Gardner, 1967). As an example of moderate broadening, we will assume a

3-db bandwidth $B_{3a} = 10$ Hz. From Fig. 4.4, we find $K_f \cong 1/B_{3a}$ for a near-matched system:

$$\sigma_f = \frac{B_{3a}}{\sqrt{2(S/N)_f}} = \frac{10 \text{ Hz}}{\sqrt{2(S/N)_f}} = 2 \text{ Hz},$$

$$\left(\frac{S}{N}\right)_f = 12.5, \text{ or } 11 \text{ db.}$$

The actual value of $(S/N)_f \cong (S/N)(f_r/B_{na}) \cong 16(S/N)$ will always exceed this minimum value under conditions which permit angle and range tracking $(S/N > 1)$. Thus there will be no long-range region, as far as Doppler errors are concerned.

Step 3: Long-Range Transition. This step is omitted, and the maximum tracking range $R_o = 810$ km can be used as the limit of accurate Doppler tracking.

Step 4: Linear Error at Target. All Doppler errors are expressed in linear measure, so we need only calculate the range-dependent dynamic error caused by geometrical acceleration in the Doppler data (third derivative of range).

$$(\dddot{R})_{max} = 0.85 \frac{v_t^3}{R^2}, \qquad \text{(Barton, 1964, p. 371)}$$

$$(\Delta_a)_{max} = \frac{(\dddot{R})_{max}}{K_a} = 0.85 \frac{v_t^3}{R^2 K_a}.$$

For a Doppler loop bandwidth $\beta_n = 2$ Hz, we have $K_a = 10 \text{ sec}^{-2}$, and the lag error for a 300 m/sec straight-line course is

$$(\Delta_a)_{max} = \frac{0.85 \times 2.7 \times 10^7}{10 R_1^2} = \frac{2.3 \times 10^6}{R_1^2}.$$

Step 5: Short-Range Transition. Setting the lag error equal to instrumental error, we find

$$R_1 = \left[\frac{2.3 \times 10^6}{2}\right]^{1/2} = 1070 \text{ m.}$$

The Doppler loop in this case will function at its optimum accuracy without increase in its bandwidth at ranges shorter than those which can be tracked in angle (Sec. 8.3, Step 10). There will also be a lag error caused by actual target third derivative (unspecified in the example), but this will be independent of range.

Step 6: Clutter Error. Under conditions which permit the range and angle loops to maintain track, the Doppler loop error will not exceed its instrumental value, even if the target and the clutter are in the same filter bandwidth. This is because the instrumental error is such a large fraction of the filter bandwidth.

Step 7: Multipath Error. For a tracker with a narrow beam and reasonable side-lobe rejection, multipath error in Doppler measurement is negligible (see Chap. 5, Sec. 4).

Step 8: Tropospheric Fluctuation. Data in App. D, Sec. D.2, indicate that the range-rate fluctuation should not exceed 0.01 m/sec, even when the tracking beam moves rapidly through the troposphere.

Step 9: Residual Refraction Bias. The radial velocity errors caused by uncorrected elevation and azimuth errors are given by Eqs. (D.8) and (D.9). In each case, the error is the product of residual angle error, target velocity, and the ratio of tropospheric path length to total range. If the angular errors are held to 0.1 mr and the entire path is in the troposphere, the maximum error for our target is

$$\sigma_v = 10^{-4} v_t = 0.03 \text{ m/sec.}$$

This is about half the instrumental error, and it could be significantly larger for faster targets or for angular data which are not so well corrected. This particular error is only of importance if an attempt is made to relate the measured Doppler component to some external frame of reference, where its exact direction must be known.

Step 10: Coverage Summary. Because none of the error components discussed in the preceding steps is large enough to add significantly to instrumental error, the region of optimum Doppler accuracy extends over the entire coverage volume of the radar. This assumes, of course, that the proper spectral line can be identified to eliminate ambiguity at the repetition rate. The difficulty in establishing range rate within the tolerances required for resolution of the Doppler ambiguity is discussed in Sec. 8.6 below.

Step 11: Discrete Interference. As with thermal noise and clutter, active interference and strong fixed targets will cause loss of angle or range tracking before they affect Doppler accuracy.

The example used above does not make extensive use of the theoretical description of Doppler errors, but it does illustrate a case where the practical limitations and target effects are dominant. An analysis based on theoretical considerations for a point target could lead to gross overestimates of system accuracy.

8.6 Errors in Differentiated Data

The subjects of differentiation and errors in the resulting smoothed velocity data are discussed in App. C. It is shown there that knowledge of the frequency spectrum of position error, as well as magnitude, is needed to calculate accurately the velocity error. However, even in the absence of spectral information, Eq. (C. 8) may be used to obtain a first approximation of error in the derivative of any coordinate, x.

$$\sigma_{\dot{x}2} = \frac{\sigma_1}{\sqrt{2} \, t_d},$$

where t_d is the delay or lag in differentiated data and σ_1 is the single-sample rms error in x, evaluated within an interval less than t_d. For example, considering the range tracker discussed in Sec. 8.4 above, a first approximation of range-rate accuracy for one-second data delay would be (in the region of optimum accuracy)

$$\sigma_{v2} = \frac{3}{\sqrt{2}} = 2.1 \text{ m/sec.}$$

More refined error estimates can be made by applying the curves of Figs. C.5 and C.6 to individual components of position error, and then forming the rss sum of the resulting velocity error components.

The exact form of the differentiator weighting function is seen to be of small importance compared to the spectral parameters of the position error. Table 8.1 lists the several components of range error which have been evaluated, along with

Table 8.1

TYPICAL VELOCITY ERROR COMPUTATION

Range Error Component	*Spectrum*	σ_r (m)	σ_{v2} (m/sec)	C_v	σ_v (m/sec)
a. Quantization	White, $\beta_n = 4$ Hz	0.48	0.34	0.6	0.3
b. Glint	Markoffian, $f_3 \cong 1.6$ Hz	2.	1.4	0.7	1.0
c. Circuit drift	Markoffian, $f_3 \ll 0.1$ Hz	2.25	1.6	0	0
d. Thermal noise (at R_2)	White, $\beta_n = 4$ Hz	3.	2.1	0.6	1.3
e. Target acceleration	Markoffian, $f_3 \cong 0.1$ Hz	0.5	0.35	1.0	0.35
f. Geometrical acceleration	Sinusoid at $f_c \cong 0.015$ Hz	3	2.1	0.14	0.3
g. Ground clutter (at E_1)	White, $\beta_n = 4$ Hz	3	2.1	0.6	1.3
h. Multipath (at $E = 0.4$ deg)	Sinusoid at $f_c \cong 3$ Hz	0.07	0.05	0.16	0
Instrumental velocity error (a, b, c)			2.1		1.0
Long-range velocity error (a, b, c, d, e)			3.0		1.7
Short-range velocity error (a, b, c, e, f)			3.0		1.1
Low-elevation velocity error (a, b, c, e, g)			3.0		1.7
Differentiator lag error, $t_d a_t$					5.0

(Linear-odd differentiator function, Fig. C.4(a), $t_d = 1$ sec)

a rough description of their spectral parameters. Because the instrumental error included some long-term drift components, which contribute very little to rate error, and because the glint error is reduced appreciably by smoothing, the instrumental contribution to rate error is smaller than its first approximation would show. Total error at the boundaries of the region of optimum accuracy is also less than the approximation.

This example illustrates the difficulty in using differentiated range data to resolve the Doppler ambiguity of a Doppler tracker in a low-prf radar. The rate error of 2 m/sec corresponds to a frequency error of

$$\sigma_f = \frac{2\sigma_v}{\lambda} = 73 \text{ Hz} \qquad (\text{at } \lambda = 0.055 \text{ m}).$$

Such accuracy is inadequate to resolve ambiguity at $f_r = 160$ pps, and the delay of one second is also excessive for a target which has 5 m/sec^2 variable acceleration. Referring to Table C.4, we find that an endpoint estimator (which would avoid the 5 m/sec lag error under constant acceleration conditions) would have approximately sixteen times the variance of the midpoint estimator assumed in the velocity error analysis (or four times the rms error), given a noise input. This means that noise and clutter errors would increase from 2 m/sec to almost 8 m/sec if the output data were updated to remove lag. A simple reduction in smoothing time by a factor of 2, giving 2.5 m/sec lag, would increase the noise error by only a factor of 1.4. The total error would then be about 3.8 m/sec, which would be adequate to resolve ambiguities at $f_r > 6\sigma_f = 830$ pps.

Differentiated velocity error estimates can be made for all radar coordinates by using the same procedure illustrated here for range. The method which starts with a first approximation based on Eq. (C.8) is useful in that it establishes an upper bound for error components whose spectral parameters are uncertain. If any particular component appears to be of critical importance, its spectrum can be calculated more carefully by referring to the detailed material given in the preceding chapters of this handbook or in the references.

8.7 Coordinate Transformations

Radar data, gathered in spherical coordinates, must be transformed into other coordinate systems (e.g., Cartesian) for transmission, further processing, and display. The ellipsoid of radar error, whose semi-axes are the errors $R\sigma_E$, $R\sigma_A \cos E_t$, and σ_r for an azimuth-elevation radar system, is not changed by coordinate transformation. Errors in x-, y-, and z-coordinates, therefore, can assume any values between the smallest radar error (usually σ_r) and the largest (usually $R\sigma_E$), depending on target location. A two-dimensional example is shown in Fig. 8.16. The radar errors σ_r and σ_E are uncorrelated, and the ellipse is elongated in the elevation direction. Obviously, the error ellipse in this case cannot be described by σ_x and σ_y without considering the correlation between errors in these Cartesian coordinates. This correlation reduces the area of the ellipse below what would be found if σ_x and σ_y were used as semi-axes.

Another effect of coordinate transformation may be to eliminate geometrical accelerations from the Cartesian data, when the target is following a straight path in space. If the radar tracking loops are operated with sufficiently wide bandwidths, lag errors in the radar data can be minimized, and smoothing may then

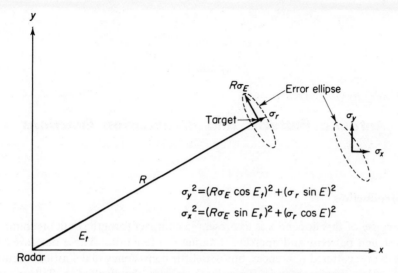

Fig. 8.16 Error relationships in coordinate transformation.

be applied without significant lag in those portions of the system following coordinate transformation. An alternative approach uses cross-coupling of the rate data in the three radar tracking loops to compensate for geometrical accelerations, so that the error signals need only take out the true target accelerations.

The ultimate solution to this type of problem is to transform the data into a system of coordinates in which the unknown target parameters are expressed as constants over the entire period of measurement. Smoothing may then be carried out on the constants with no lag errors. The best examples of this technique are in the area of satellite orbit calculations, where radar data may be fitted to the orbital parameters. The so-called "Kalman-Bucy" filter extends this technique to any problem in which the state of the radar-target system may be described by constants in a set of simultaneous equations. Although these methods offer considerable refinement in smoothing techniques, and may also make it possible to measure each unknown with the most accurate coordinate of the radar data, they must be applied with great caution. In some cases, the unknown parameter being estimated may be very sensitive to errors introduced by the coordinate transformation, and a thorough error analysis of the transformation itself is needed. Also, the usual assumption of white noise (independent from one radar sample to the next) can lead to serious errors when the smoothing interval is long. Small drift errors, which may be insignificant in the radar position data, begin to dominate the output error as soon as the smoothing interval is extended to about ten times the radar observation interval used in range and angle error analysis.

Appendix A

Antenna Patterns and Illumination Functions

A.1 Introduction

The purpose of this appendix is to present in compact form the data on commonly used antenna patterns and aperture illumination functions. These data are available in many scattered references, but with little consistency of definition, accuracy, normalization, or presentation. To unify these data, the results for all illumination functions have been recomputed to engineering accuracy, and plots have been prepared on several different scales.

Three classes of functions are considered: even, odd, and circularly-symmetrical functions. The first two types apply to rectangular apertures and line-source antennas, as well as to analogous signal waveforms. The circular functions are peculiar to circular apertures with radial symmetry. For each illumination function considered, we have computed the antenna pattern and a number of parameters which are useful to describe the function or its pattern. Table A.1 lists the illumination functions considered.

Table A.1

ILLUMINATION FUNCTIONS

Even Functions

$g(x') = \cos^n(\pi x')$; $n = 1$ to 6

$g(x') = [1 - 4(x')^2]^n$; $n = 1$ to 6

$g(x')$ is $a = 0, 0.1, 0.2, 0.3, 0.4, 0.5$

$g(x') = \exp[-1.382(nx')^2]$; $n = 1.0, 1.7, 2.4, 2.8, 3.2$

$g(x') = k + (1 - k)\cos(\pi x')$; $k = 0.5, 0.3, 0.2, 0.1, 0.05, 0.03$

$g(x') = $ Taylor linear weighting function, for sidelobe levels and $\bar{n}$ as follows:

Sidelobe level (db)	20	25	30	35	40	45
$\bar{n}$		2	3	4	5	6	8

$g(x') = $ Hamming function, $0.54 + 0.48\cos(2\pi x')$

Table A.1—*Cont.*

$g(x') = |\sin(2\pi x')|$

Odd Functions

$g(x') = x'\cos^n(\pi x');\ n = 1$ to 6

$g(x') = 2.74(nx')\exp[-1.382(nx')^2];\ n = 1.2, 2.18, 2.52, 3.34, 3.87$

$g(x')$ is $a = 0, 0.1, 0.2, 0.3, 0.4$

$g(x')$ is $a = 0, 0.1, 0.2, 0.3, 0.4, 0.5$

$g(x') = x'$ [Taylor linear function]; same cases as used in even functions, above.

$g(x') = \sin(2\pi x')$

$g(x') = 4x'\sqrt{1 - 4(x')^2}$

Circular Functions

$g(r') = [1 - 4(r')^2]^n;\ n = 0$ to 4

$g(r') = \exp[-1.382(nr')^2];\ n = 1.0, 1.7, 2.4, 2.8, 3.2$

$g(r') =$ Taylor circular weighting function, for same sidelobe levels and $\bar{n}$ as used in set of even functions, above.

Section A.2 contains the definitions and normalization procedures used for the computed data. This is followed in Sec. A.3 by a set of tables containing the descriptive parameters for all computed functions. The last section, A.4, consists of plots of selected illumination functions and their corresponding antenna patterns.

A.2 Definitions and Normalization Procedure

This section contains the definitions, normalization procedures, and interrelationships of the parameters used in the remainder of this appendix. Definitions are given for the antenna application only, while analogies to waveforms and spectra are shown in App. B. Certain general rules were followed in notation, normalization, and use of subscripts, and primary attention has been given to consistency within the handbook itself.

Normalization is used for convenience in both aperture and pattern coordinates to eliminate aperture size from the data. A primed symbol indicates a normalized, dimensionless variable (e.g., $x' = x/w$, where x is the coordinate in the aperture plane and w is the total width in that coordinate). Subscripts indicate particular values of a function or a parameter:

Subscript o indicates the largest possible value;

Subscript m indicates the largest value for a particular case;

Subscript r indicates value relative to the largest possible value.

For example:

G_o = peak directive gain for the uniformly illuminated aperture;

G_m = peak directive gain for the illumination considered;

K_r = voltage slope at beam center (for an odd function), relative to the slope for linear-odd illumination.

Antenna Parameter Definitions

$$\text{\textit{Aperture Illumination Function}} \equiv g(x, y)$$

where x and y are rectangular coordinates defining the aperture plane.

$$\text{\textit{Beam Pattern}} \equiv F(u, v) = \int_{-h/2}^{h/2} \int_{-w/2}^{w/2} g(x, y)\, e^{-j2\pi(xu+yv)}\, dx\, dy.$$

Angular Coordinate Relationships:

Transform	Spherical	Polar Transform	Direction Cosine
u	$= (1/\lambda)\sin\theta\cos\phi$	$= U\cos\phi$	$= (1/\lambda)\cos\alpha$
v	$= (1/\lambda)\sin\theta\sin\phi$	$= U\sin\phi$	$= (1/\lambda)\cos\beta$
$du\,dv$	$= (1/2\lambda^2)\lvert\sin(2\theta)\rvert\, d\theta\, d\phi =$	$U\,dU\,d\phi =$	$(1/\lambda^2)\lvert\sin\alpha\sin\beta\rvert\, d\alpha\, d\beta$

The spherical coordinates θ and ϕ are defined in Fig. A.1 with respect to aperture-plane coordinates.

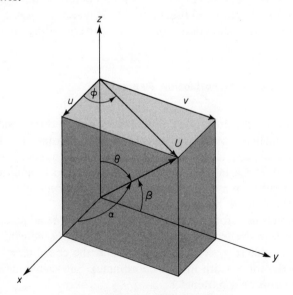

Fig. A.1 Antenna pattern coordinates.

Inverse transform relationship of illumination to pattern:

$$g(x, y) = \int\!\!\int\limits_{-\infty}^{\infty} F(u, v)e^{j2\pi(ux+vy)} \, du \, dv.$$

$$Directivity \equiv G(u, v) = \frac{\text{Far-field power density for antenna}}{\text{Far-field power density for isotropic radiator}}.$$

$$G(u, v) = \frac{4\pi \, |F(u, v)|^2}{\lambda^2 \displaystyle\int\!\!\int\limits_{-\infty}^{\infty} |F(u, v)|^2 \, du \, dv} = \frac{4\pi \, |F(u, v)|^2}{\lambda^2 \displaystyle\int\!\!\int\limits_{A} |g(x, y)|^2 \, dx \, dy}.$$

(The wavelength λ is expressed in the same units as x and y.)

$$Maximum \ value \ of \ G \equiv G_m = \frac{4\pi \left| \displaystyle\int\!\!\int\limits_{A} g(x, y) \, dx \, dy \right|^2}{\displaystyle\int\!\!\int\limits_{A} |g(x, y)|^2 \, dx \, dy}.$$

Largest possible value of $G_m \equiv G_o = 4\pi A/\lambda^2$ for $g(x, y)$ constant over aperture.

$$Aperture \ area \equiv A.$$

$$Effective \ aperture \ area \equiv A_r = \frac{G_m \lambda^2}{4\pi}.$$

$$Aperture \ efficiency \equiv \eta_a \equiv \frac{A_r}{A} = \frac{G_m}{G_o}.$$

Half-power (or 3-db) beamwidths, θ_{u3} and θ_{v3} are as defined in Fig. A.2.

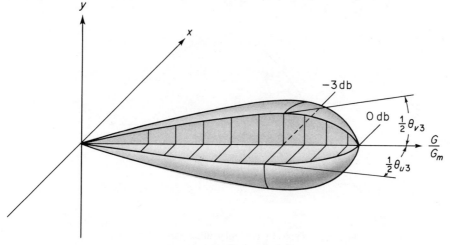

Fig. A.2 Typical main lobe.

For a rectangular aperture, the coordinates are easily normalized to the aperture dimensions. Normalized functions and variables, denoted by primed symbols, are (see Fig. A.3):

$$x' = \frac{x}{w}, \qquad u' = uw = \frac{w}{\lambda} \sin \theta \cos \phi,$$

$$y' = \frac{y}{h}, \qquad v' = vh = \frac{h}{\lambda} \sin \theta \sin \phi,$$

$$F(u', v') = wh \int_{-1/2}^{1/2} \int_{-1/2}^{1/2} g(x', y') \, e^{-j2\pi(x'u'+y'v')} \, dx' \, dy',$$

$$G(u', v') = \frac{4\pi wh}{\lambda^2} \times \frac{|F(u', v')|^2}{\displaystyle\iint_{-\infty}^{\infty} |F(u', v')|^2 \, du' \, dv'}$$

$$= \frac{4\pi}{\lambda^2 wh} \times \frac{|F(u', v')|^2}{\displaystyle\int_{-1/2}^{1/2} \int_{-1/2}^{1/2} |g(x', y')|^2 \, dx' \, dy'},$$

$$G_o = \frac{4\pi wh}{\lambda^2},$$

$$\eta_a = \frac{|F(0, 0)|^2}{\displaystyle\iint_{-\infty}^{\infty} |F(u', v')|^2 \, du' \, dv'} = \frac{\left| \displaystyle\int_{-1/2}^{1/2} \int_{-1/2}^{1/2} g(x', y') \, dx' \, dy' \right|^2}{\displaystyle\int_{-1/2}^{1/2} \int_{-1/2}^{1/2} |g(x', y')|^2 \, dx' \, dy'}.$$

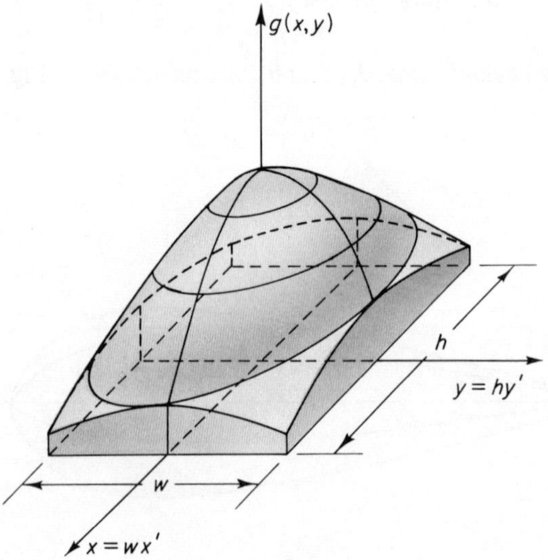

Fig. A.3 Typical rectangular illumination function.

For a rectangular aperture with separable illumination functions in x and y, we have:

$$g(x', y') = g(x')g(y'),$$

$$F(u', v') = F(u')F(v'),$$

$$F(u') = w \int_{-1/2}^{1/2} g(x')e^{-j2\pi x' u'} \, dx',$$

$$F(v') = h \int_{-1/2}^{1/2} g(y')e^{-j2\pi y' v'} \, dy',$$

$$G(u')G(v') = \frac{4\pi hw}{\lambda^2} \times \frac{|F(u')|^2}{\int_{-\infty}^{\infty} |F(u')|^2 \, du'} \times \frac{|F(v')|^2}{\int_{-\infty}^{\infty} |F(v')|^2 \, dv'}.$$

There are several parameters defined in terms of $F(u)$ or $g(x)$, for one-dimensional illumination functions, which are useful in describing properties of antenna patterns. These are:

Pattern maximum,

$$F_m \equiv F(0) = w \int_{-1/2}^{1/2} g(x') \, dx',$$

Total power radiated,

$$C \equiv \int_{-\infty}^{\infty} |F(u)|^2 \, du = \int_{-w/2}^{w/2} |g(x)|^2 \, dx$$

$$= w \int_{-1/2}^{1/2} |g(x')|^2 \, dx',$$

Aperture efficiency,

$$\eta_x = \frac{|F_m|^2}{\int_{-\infty}^{\infty} |F(u')|^2 \, du'} = \frac{\left| \int_{-1/2}^{1/2} g(x') \, dx' \right|^2}{\int_{-1/2}^{1/2} |g(x')|^2 \, dx'},$$

Noise beamwidth,

$$\theta_n \equiv \frac{\lambda \int_{-\infty}^{\infty} |F(u')|^2 \, du'}{w |F_m|^2} = \frac{\lambda}{\eta_x w} \quad \text{(radians)},$$

Noise aperture width,

$$w_n \equiv \frac{w \int_{-1/2}^{1/2} |g(x')|^2 \, dx'}{|g_m|^2} \quad \text{(units of } x\text{)},$$

Rms aperture width (for illumination power),

$$\mathscr{L}_s \equiv 2\pi w \left[\frac{\int_{-1/2}^{1/2} [x']^2 |g(x')|^2 \, dx'}{\int_{-1/2}^{1/2} |g(x')|^2 \, dx'} \right]^{1/2} \qquad \text{(units of } x),$$

Rms beamwidth,

$$\Theta \equiv \frac{\lambda}{w} \left[\frac{\int_{-1/2}^{1/2} \left| \frac{dg(x')}{dx'} \right|^2 \, dx'}{\int_{-1/2}^{1/2} |g(x')|^2 \, dx'} \right]^{1/2} \qquad \text{(radians)},$$

$$\frac{\text{Voltage slope at beam center}}{\text{Slope for linear-odd illum.}} \equiv K_r = \frac{\sqrt{12} \int_{-1/2}^{1/2} x' g(x') \, dx'}{\left[\int_{-1/2}^{1/2} |g(x')|^2 \, dx' \right]^{1/2}}.$$

For a circular aperture, with illumination which is a function of radius from the center of the aperture, it is more convenient to use polar coordinates for both the pattern and the illumination function. In this case:

$$F(U) = 2\pi \int_0^{D/2} g(r) J_o(2\pi r U) r \, dr,$$

where J_o is the zero-order Bessel function.

Here, $U = \sqrt{u^2 + v^2} = \dfrac{\sin \theta}{\lambda}$,

$r = \sqrt{x^2 + y^2}$,

D = aperture diameter,

r, D, and λ are in the same units.

$$G(U) = \left(\frac{2}{\lambda^2} \right) \frac{|F(U)|^2}{\int_0^\infty |F(U)|^2 U \, dU} = \left(\frac{2}{\lambda^2} \right) \frac{|F(U)|^2}{\int_0^{D/2} |g(r)|^2 r \, dr},$$

$$G_o = \frac{\pi^2 D^2}{\lambda^2} = \frac{4\pi A}{\lambda^2}.$$

Normalizing to D, the aperture diameter, we have

$$r' = \frac{r}{D}, \qquad U' = DU = \frac{D \sin \theta}{\lambda},$$

$$F(U') = \left(\frac{\pi D^2}{4} \right) 8 \int_0^{1/2} g(r') J_o(2\pi r' U') r' \, dr',$$

$$G(U') = \left(\frac{2D^2}{\lambda^2} \right) \times \frac{|F(U')|^2}{\int_0^\infty |F(U')|^2 U' \, dU'} = \left(\frac{2}{\lambda^2 D^2} \right) \times \frac{|F(U')|^2}{\int_0^{1/2} |g(r')|^2 r' \, dr'}.$$

Further properties of the circular aperture are:

$$F_m = 2\pi \left| \int_0^{D/2} g(r)\, r\, dr \right| = \left(\frac{\pi D^2}{4} \right) 8 \left| \int_0^{1/2} g(r')\, r'\, dr' \right|,$$

$$C = 2\pi \int_0^{D/2} |g(r)|^2 \, r\, dr = \left(\frac{\pi D^2}{4} \right) 8 \int_0^{1/2} |g(r')|^2 \, r'\, dr',$$

$$G_m = \frac{4\pi (F_m)^2}{\lambda^2 C} = \frac{\pi^2 D^2}{\lambda^2} \eta_a = G_o \eta_a.$$

A.3 Tabular Data

The following tables contain the descriptive parameters for all the illumination functions listed in Table A.1, as well as certain pattern parameters for each case. For convenience in use, to avoid repeated reference to the definitions of Sec. A.2, a brief definition of each parameter is included along with the entry.

LIST OF TABLES

Taylor Illumination Function

$$g_T(x') = \frac{1 + 2 \sum_{m=1}^{\bar{n}-1} F_m \cos (2\pi m x')}{1 + 2 \sum_{m=1}^{\bar{n}-1} F_m}$$

$$F_m = \frac{(-1)^{m+1} \prod_{n=1}^{\bar{n}-1}[1 - m^2/\sigma^2[A^2 + (n - \frac{1}{2})^2]]}{2 \prod_{n=1}^{\bar{n}-1}[1 - (m^2/n^2)]_{m \neq n}}$$

$$\sigma^2 = \frac{\bar{n}^2}{A^2 + (\bar{n} - \frac{1}{2})^2}$$

$$SLL = -20 \log[\cosh(\pi A)]$$

Providing that: $\bar{n} \geq 2A^2 + \frac{1}{2}$.

SLL	A	$\bar{n}$ (min)
20	0.9527	2
25	1.1367	3
30	1.3197	4
35	1.5034	5
40	1.6864	6
45	1.8701	8

Circular Taylor Illumination Function, $g_T(r')$

$$g_T(r') = \frac{\sum_{m=1}^{\bar{n}-1} \frac{J_o(2\pi\mu_m r')F(\mu_m, A, \bar{n})}{[J_o(\pi\mu_m)]^2} + 1}{\sum_{m=1}^{\bar{n}-1} \frac{F(\mu_m, A, \bar{n})}{[J_o(\pi\mu_m)]^2} + 1}$$

$$F(\mu_m, A, \bar{n}) = 0, \qquad m = 0$$

$$F(\mu_m, A, \bar{n}) = \frac{-J_o(\pi\mu_m) \prod_{n=1}^{\bar{n}-1}\left[1 - \frac{\mu_m^2}{\sigma^2[A^2 + (n - \frac{1}{2})^2]}\right]}{\prod_{\substack{n=1 \\ n \neq m}}^{\bar{n}-1}\left[1 - \frac{\mu_m^2}{\mu_n^2}\right]}$$

$$\sigma^2 = \frac{\mu_{\bar{n}}^2}{A^2 + (\bar{n} - \frac{1}{2})^2} \qquad SLL = -20 \log[\cosh(\pi A)]$$

$$\bar{n} \geq 2A^2 + \frac{1}{2}$$

SLL	A	$\bar{n}$ (min)	n	μ_n
20	0.9527	2	0	0
25	1.1376	3	1	1.21967
30	1.3197	4	2	2.23313
35	1.5034	5	3	3.23832
40	1.6864	6	4	4.24106
45	1.8701	8	5	5.24276
			6	6.24392
			7	7.24475
			8	8.24539

Table A.2

EVEN ILLUMINATION FUNCTIONS; $g(x') = \cos^n (\pi x')$

$g(x')$	$\cos (\pi x')$	$\cos^2 (\pi x')$	$\cos^3 (\pi x')$	$\cos^4 (\pi x')$	$\cos^5 (\pi x')$	$\cos^6 (\pi x')$
Figure number	A.5	A.6	A.7			
$A = \int_{-1/2}^{1/2} g(x')\,dx'$	0.630	0.495	0.420	0.371	0.336	0.309
$20 \log (A)$	−4.01	−6.11	−7.53	−8.61	−9.47	−10.19
$C = \int_{-1/2}^{1/2} g^2(x')\,dx'$	0.495	0.371	0.309	0.271	0.244	0.223
$10 \log (C)$	−3.05	−4.30	−5.09	−5.67	−6.13	−6.51
$D = \int_{-1/2}^{1/2} \left[\frac{dg(x')}{dx'}\right]^2 dx'$	4.93	4.93	5.55	6.16	6.74	7.28
$G = \int_{-1/2}^{1/2} [x'g(x')]^2\,dx'$	0.0161	0.0074	0.0044	0.0030	0.0022	0.0017
$H = \int_{-1/2}^{1/2} (x')^2 g(x')\,dx'$	0.030	0.0161	0.0104	0.0074	0.0056	0.0044
$\eta_x = A^2/C$	0.802	0.660	0.571	0.509	0.463	0.429
$10 \log (\eta_x)$	−0.956	−1.804	−2.44	−2.93	−3.34	−3.68
$\theta_n w/\lambda = C/A^2$	1.246	1.515	1.752	1.964	2.16	2.33
$\theta_3 w/\lambda$	1.189	1.441	1.659	1.849	2.03	2.09
$\Theta w/\lambda = \sqrt{D/C}$	3.16	3.65	4.23	4.77	5.26	5.71
$\mathscr{L}_s/w = 2\pi\sqrt{G}/C$	1.136	0.888	0.753	0.665	0.602	0.554
$10 \log [G_{sr}]$	+23	+31.5	+39	+47	+54	+61.3

$w =$ Aperture width
$\lambda =$ Wavelength
$\eta_x =$ Aperture efficiency
$\theta_n =$ Noise beamwidth (rad)

$\theta_3 =$ 3-db beamwidth (rad)
$\Theta =$ Rms beamwidth (rad)
$\mathscr{L}_s =$ Rms aperture width
$G_{sr} = \dfrac{\text{Gain of main beam}}{\text{Gain of highest sidelobe}}$

Table A.3

EVEN ILLUMINATION FUNCTIONS, $g(x') = [1 - 4(x')^2]^n$

$g(x')$	$[1-4(x')^2]$	$[1-4(x')^2]^2$	$[1-4(x')^2]^3$	$[1-4(x')^2]^4$	$[1-4(x')^2]^5$	$[1-4(x')^2]^6$
Figure number	A.8					
$A = \int_{-1/2}^{1/2} g(x')\,dx'$	0.660	0.528	0.453	0.403	0.366	0.338
$20\log(A)$	-3.61	-5.55	-6.88	-7.91	-8.74	-9.43
$C = \int_{-1/2}^{1/2} g^2(x')\,dx'$	0.528	0.402	0.338	0.297	0.268	0.246
$10\log(C)$	-2.77	-3.95	-4.72	-5.28	-5.73	-6.09
$D = \int_{-1/2}^{1/2}\left[\dfrac{dg(x')}{dx'}\right]^2 dx'$	5.33	4.87	5.32	5.82	6.30	6.76
$G = \int_{-1/2}^{1/2} [x'g(x')]^2\,dx'$	0.0188	0.0091	0.0056	0.0039	0.0029	0.0022
$H = \int_{-1/2}^{1/2} (x')^2 g(x')\,dx'$	0.0329	0.0188	0.0125	0.0091	0.0070	0.0056
$\eta_x = A^2/C$	0.825	0.693	0.607	0.546	0.500	0.464
$10\log(\eta_x)$	-0.836	-1.592	-2.17	-2.63	-3.01	-3.34
$\theta_n w/\lambda = C/A^2$	1.212	1.443	1.648	1.832	2.00	2.16
$\theta_3 w/\lambda$	1.179	1.365	1.568	1.731	1.885	2.028
$\Theta w/\lambda = \sqrt{D/C}$	3.18	3.48	3.97	4.43	4.85	5.25
$\mathscr{L}_s/w = 2\pi\sqrt{G/C}$	1.187	0.947	0.811	0.721	0.655	0.605
$10\log[G_{sr}]$	+21.3	+27.5	+34.7	+38.5	+43.4	+48

w = Aperture width
λ = Wavelength
η_x = Aperture efficiency
θ_n = Noise beamwidth (rad)

θ_3 = 3-db beamwidth (rad)
Θ = Rms beamwidth (rad)
$\mathscr{L}_s$ = Rms aperture width
$G_{sr} = \dfrac{\text{Gain of main beam}}{\text{Gain of highest sidelobe}}$

Table A.4

EVEN ILLUMINATION FUNCTIONS, $g(x') = $ *Trapezoidal*

$g(x')$					
Figure number	A.9				A.4
$A = \int_{-1/2}^{1/2} g(x')\,dx'$	0.495	0.594	0.693	0.792	0.891	1.000
$20 \log (A)$	−6.02	−4.52	−3.18	−2.02	−1.002	0
$C = \int_{-1/2}^{1/2} g^2(x')\,dx'$	0.330	0.462	0.594	0.726	0.858	1.000
$10 \log (C)$	−4.81	−3.35	−2.26	−1.389	−0.663	0
$D = \int_{-1/2}^{1/2} \left[\frac{dg(x')}{dx'}\right]^2 dx'$	3.92	5.00	6.66	10.00	20.00	—
$G = \int_{-1/2}^{1/2} [x'g(x')]^2\,dx'$	0.0082	0.0128	0.0209	0.0342	0.0542	0.0850
$H = \int_{-1/2}^{1/2} (x')^2 g(x')\,dx'$	0.0206	0.0257	0.0334	0.0448	0.0608	0.0850
$\eta_x = A^2/C$	0.742	0.764	0.808	0.864	0.925	1.000
$10 \log (\eta_x)$	−1.293	−1.171	−0.924	−0.635	−0.338	0
$\theta_n w/\lambda = C/A^2$	1.347	1.309	1.237	1.157	1.081	1.000
$\theta_3 w/\lambda$	1.273	1.242	1.172	1.076	0.976	0.886
$\Theta w/\lambda = \sqrt{D/C}$	3.45	3.29	3.35	3.71	4.83	—
$\mathscr{L}_s/w = 2\pi\sqrt{G/C}$	0.993	1.046	1.179	1.364	1.579	1.832
$10 \log [G_{sr}]$	+26.5	+28.7	+19	+15.1	+13.6	+13.3

w = Aperture width
λ = Wavelength
η_x = Aperture efficiency
θ_n = Noise beamwidth (rad)

θ_3 = 3-db beamwidth (rad)
Θ = Rms beamwidth (rad)
$\mathscr{L}_s$ = Rms aperture width
$G_{sr} = \dfrac{\text{Gain of main beam}}{\text{Gain of highest sidelobe}}$

Table A.5

EVEN ILLUMINATION FUNCTIONS, TRUNCATED GAUSSIAN

n	1.0	1.7	2.4	2.8	3.2
Edge illumination (db)	−3.0	−8.67	−17.29	−23.5	−30.7
Figure number		A.10	A.11	A.12	
$A = \int_{-1/2}^{1/2} g(x')\,dx'$	0.894	0.7433	0.595	0.523	0.463
$20 \log (A)$	−0.974	−2.58	−4.51	−5.63	−6.69
$C = \int_{-1/2}^{1/2} g^2(x')\,dx'$	0.807	0.594	0.438	0.377	0.330
$10 \log (C)$	−0.929	−2.26	−3.59	−4.24	−4.82
$D = \int_{-1/2}^{1/2} \left[\frac{dg(x')}{dx}\right]^2 dx'$	—	—	—	—	—
$G = \int_{-1/2}^{1/2} [x'\,g(x')]^2\,dx'$	0.0566	0.0290	0.0132	0.0085	0.0058
$H = \int_{-1/2}^{1/2} (x')^2 g(x')\,dx'$	0.0690	0.0478	0.0291	0.0212	0.0154
$\eta_x = A^2/C$	0.990	0.930	0.808	0.727	0.650
$10 \log (\eta_x)$	−0.0445	−0.313	−0.928	−1.387	−1.869
$\theta_n w/\lambda = C/A^2$	1.010	1.075	1.238	1.376	1.538
$\theta_3 w/\lambda$	0.920	1.025	1.167	1.296	1.444
$\Theta w/\lambda = \sqrt{D/C}$	—	—	—	—	—
$\mathscr{L}_s/w = 2\pi\sqrt{G/C}$	1.664	1.389	1.091	0.949	0.834
$10 \log [G_{sr}]$	+15.5	+20.8	+32.1	+37	+47.5

w = Aperture width
λ = Wavelength
η_x = Aperture efficiency
$g(x') = \exp[-1.382\,(nx')^2]$
θ_n = Noise beamwidth (rad)

θ_3 = 3-db beamwidth (rad)
Θ = Rms beamwidth (rad)
$\mathscr{L}_s$ = Rms aperture width
$G_{sr} = \dfrac{\text{Gain of main beam}}{\text{Gain of highest sidelobe}}$

Table A.6

EVEN ILLUMINATION FUNCTIONS, COSINE-ON-PEDESTAL

k	0.5	0.3	0.2	0.1	0.05	0.03
Figure number			A.13			
$A = \int_{-1/2}^{1/2} g(x')\,dx'$	0.815	0.741	0.704	0.667	0.649	0.641
$20\log(A)$	−1.775	−2.60	−3.05	−3.51	−3.76	−3.86
$C = \int_{-1/2}^{1/2} g^2(x')\,dx'$	0.689	0.597	0.559	0.524	0.509	0.503
$10\log(C)$	−1.619	−2.24	−2.53	−2.80	−2.93	−2.98
$D = \int_{-1/2}^{1/2}\left[\dfrac{dg(x')}{dx}\right]^2 dx'$	−	−	−	−	−	−
$G = \int_{-1/2}^{1/2}[x'\,g(x')]^2\,dx'$	0.0402	0.0281	0.0232	0.0193	0.0176	0.0170
$H = \int_{-1/2}^{1/2}(x')^2 g(x')\,dx'$	0.0574	0.0463	0.0408	0.0353	0.0325	0.0314
$\eta_x = A^2/C$	0.965	0.920	0.888	0.849	0.827	0.817
$10\log(\eta_x)$	−0.1569	−0.363	−0.516	−0.711	−0.827	−0.877
$\theta_n w/\lambda = C/A^2$	1.037	1.087	1.126	1.178	1.210	1.224
$\theta_3 w/\lambda$	0.996	1.028	1.069	1.121	1.151	1.165
$\Theta w/\lambda = \sqrt{D/C}$	−	−	−	−	−	−
$\mathscr{L}_s/w = 2\pi\sqrt{G}/C$	1.518	1.363	1.283	1.206	1.170	1.156
$10\log[G_{sr}]$	+17.8	+20.5	+21.8	+22.9	+23.1	+23.1

$g(x') = k + (1-k)\cos(\pi x')$

w = Aperture width
λ = Wavelength
η_x = Aperture efficiency
θ_n = Noise beamwidth (rad)

θ_3 = 3-db beamwidth (rad)
Θ = Rms beamwidth (rad)
$\mathscr{L}_s$ = Rms aperture width
$G_{sr} = \dfrac{\text{Gain of main beam}}{\text{Gain of highest sidelobe}}$

Table A.7

EVEN ILLUMINATION FUNCTIONS, TAYLOR*

$\bar{n}$	2	3	4	5	6	8
SLL	20	25	30	35	40	45
Figure number	A.14		A.15		A.16	
$A = \int_{-1/2}^{1/2} g(x')\,dx'$	0.756	0.689	0.638	0.596	0.562	0.532
$20\log (A)$	-2.43	-3.23	-3.90	-4.49	-5.01	-5.48
$C = \int_{-1/2}^{1/2} g^2(x')\,dx'$	0.601	0.528	0.478	0.442	0.414	0.390
$10\log (C)$	-2.21	-2.78	-3.20	-3.55	-3.83	-4.09
$D = \int_{-1/2}^{1/2} \left[\dfrac{dg(x')}{dx'}\right]^2 dx'$	—	—	—	—	—	—
$G = \int_{-1/2}^{1/2} [x'g(x')]^2\,dx'$	0.0325	0.0224	0.0167	0.0131	0.0107	0.0090
$H = \int_{-1/2}^{1/2} (x')^2 g(x')\,dx'$	0.0517	0.0417	0.0344	0.0289	0.0247	0.0213
$\eta_x = A^2/C$	0.951	0.900	0.850	0.804	0.763	0.726
$10\log (\eta_x)$	-0.218	-0.455	-0.704	-0.948	-1.178	-1.390
$\theta_n w/\lambda = C/A^2$	1.051	1.111	1.176	1.244	1.311	1.377
$\theta_3 w/\lambda$	0.983	1.049	1.115	1.179	1.250	1.301
$\Theta w/\lambda = \sqrt{D/C}$	—	—	—	—	—	—
$\mathscr{L}_s/w = 2\pi\sqrt{G/C}$	1.462	1.297	1.176	1.085	1.015	0.959
$10\log [G_{sr}]$	$+20.9$	$+25.9$	$+30.9$	$+35.9$	$+40.9$	$+45.9$

$g(x')$

$w =$ Aperture width
$\lambda =$ Wavelength
$\eta_x =$ Aperture efficiency
$\theta_n =$ Noise beamwidth (rad)

$\theta_3 =$ 3-db beamwidth (rad)
$\Theta =$ Rms beamwidth (rad)
$\mathscr{L}_s =$ Rms aperture width
$G_{sr} = \dfrac{\text{Gain of main beam}}{\text{Gain of highest sidelobe}}$

*Defined on page 249f.

Table A.8

EVEN ILLUMINATION FUNCTIONS, SPECIAL

$g(x')$	Hamming $0.54 + 0.46 \cos(2\pi x')$	$|\sin(2\pi x')|$
Figure number	A.17	
$A = \int_{-1/2}^{1/2} g(x')\, dx'$	0.536	0.640
$20 \log (A)$	-5.43	-3.88
$C = \int_{-1/2}^{1/2} g^2(x')\, dx'$	0.394	0.505
$10 \log (C)$	-4.05	-2.97
$D = \int_{-1/2}^{1/2} \left[\dfrac{dg(x')}{dx'}\right]^2 dx'$	—	—
$G = \int_{-1/2}^{1/2} [x'\, g(x')]^2\, dx'$	0.0092	0.0349
$H = \int_{-1/2}^{1/2} (x')^2 g(x')\, dx'$	0.0216	0.0468
$\eta_x = A^2/C$	0.729	0.811
$10 \log (\eta_x)$	-1.375	-0.909
$\theta_n w/\lambda = C/A^2$	1.373	1.233
$\theta_3 w/\lambda$	1.296	0.936
$\Theta w/\lambda = \sqrt{D/C}$	—	—
$\mathscr{L}_s/w = 2\pi\sqrt{G/C}$	0.961	1.654
$10 \log [G_{sr}]$	$+42.8$	$+7.9$

$g(x')$ curve, with axis markings $-\tfrac{1}{2}$, 0, $\tfrac{1}{2}$, $x' = x/w$.

w = Aperture width
λ = Wavelength
η_x = Aperture efficiency
θ_n = Noise beamwidth (rad)

θ_3 = 3-db beamwidth (rad)
Θ = Rms beamwidth (rad)
$\mathscr{L}_s$ = Rms aperture width
G_{sr} = $\dfrac{\text{Gain of main beam}}{\text{Gain of highest sidelobe}}$

Table A.9

ODD ILLUMINATION FUNCTIONS, $g(x') = x' \cos^n (\pi x')$

$g(x')$	$x' \cos (\pi x')$	$x' \cos^2 (\pi x')$	$x' \cos^3 (\pi x')$	$x' \cos^4 (\pi x')$	$x' \cos^5 (\pi x')$	$x' \cos^6 (\pi x')$
Figure number	A.19	A.20				
$C = \int_{-1/2}^{1/2} g^2(x')\,dx'$	0.507	0.432	0.379	0.340	0.310	0.288
$10 \log (C)$	-2.95	-3.64	-4.22	-4.69	-5.08	-5.41
$D = \int_{-1/2}^{1/2} \left[\frac{dg(x')}{dx'} \right]^2 dx'$	0.641	0.329	0.247	0.205	0.1785	0.1596
$F = 2\pi \int_{-1/2}^{1/2} x' g(x')\,dx'$	1.050	0.775	0.605	0.494	0.416	0.360
$G = \int_{-1/2}^{1/2} [x' g(x')]^2\,dx'$	0.0398	0.0226	0.0148	0.0105	0.0080	0.0063
$H = \int_{-1/2}^{1/2} (x')^2 \|g(x')\|\,dx'$	0.0513	0.0326	0.0227	0.0168	0.0131	0.0105
$K_r = \dfrac{\sqrt{12}F}{2\pi\sqrt{C}}$	0.808	0.647	0.539	0.465	0.410	0.368
$20 \log (K_r)$	-1.847	-3.79	-5.36	-6.66	-7.75	-8.69
$10 \log (G_r)$	-2.56	-3.24	-3.82	-4.30	-4.69	-5.02
$10 \log \left[\dfrac{G_{se}}{\eta_x} \right]$	$+18.6$	$+26.1$	$+33.4$	$+40.1$	$+47.3$	$+54$

w = Aperture width
$K_r = \dfrac{\text{Voltage slope at beam center}}{\text{Slope for linear odd illumination}}$
$G_r = \dfrac{\text{Peak gain}}{\text{Peak gain for uniform illumination}}$
$\dfrac{G_{se}}{\eta_x} = \dfrac{\text{Peak gain for uniform illumination}}{\text{Gain of highest sidelobe}}$

Table A.10

ODD ILLUMINATION FUNCTIONS, TRUNCATED RAYLEIGH

n	1.2	2.175	2.52	2.88	3.34	3.87
Edge illumination (db)	0	−4.84	−8.87	−13.02	−20.3	−30.5
Figure number		A.21	A.22		A.23	
$C = \int_{-1/2}^{1/2} g^2(x')\,dx'$	0.520	0.608	0.551	0.493	0.428	0.371
$10 \log(C)$	−2.84	−2.20	−2.59	−3.07	−3.68	−4.31
$D = \int_{-1/2}^{1/2} \left[\dfrac{dg(x')}{dx'}\right]^2 dx'$	—	—	—	—	—	—
$F = 2\pi \int_{-1/2}^{1/2} x'g(x')\,dx'$	1.309	1.287	1.148	0.984	0.791	0.611
$G = \int_{-1/2}^{1/2} [x'g(x')]^2\,dx'$	0.0699	0.0564	0.0430	0.0311	0.0206	0.0133
$H = \int_{-1/2}^{1/2} (x')^2\,|g(x')|\,dx'$	0.0761	0.0681	0.0579	0.0467	0.0343	0.0236
$K_r = \dfrac{\sqrt{12}F}{2\pi\sqrt{C}}$	0.995	0.909	0.848	0.769	0.663	0.551
$20 \log(K_r)$	−0.0395	−0.829	−1.433	−2.28	−3.57	−5.18
$10 \log(G_r)$	−2.24	−2.25	−2.46	−2.79	−3.32	−3.95
$10 \log\left[\dfrac{G_{se}}{\eta_x}\right]$	+12.6	+20.	+26.	+30.8	+40.6	+52.7

w = Aperture width

$g(x') = 2.74nx' \exp\left[-1.382(nx')^2\right]$

$K_r = \dfrac{\text{Voltage slope at beam center}}{\text{Slope for linear odd illumination}}$

$G_r = \dfrac{\text{Peak gain}}{\text{Peak gain for uniform illumination}}$

$\dfrac{G_{se}}{\eta_x} = \dfrac{\text{Peak gain for uniform illumination}}{\text{Gain of highest sidelobe}}$

Table A.11

ODD ILLUMINATION FUNCTIONS, RECTANGULAR FAMILY

$g(x')$				
Figure number	A.24			
$C = \int_{-1/2}^{1/2} g^2(x') \, dx'$	1	0.800	0.600	0.400	0.200
$10 \log (C)$	0	-0.997	-2.22	-3.98	-6.99
$D = \int_{-1/2}^{1/2} \left[\dfrac{dg(x')}{dx'}\right]^2 dx'$	—	—	—	—	—
$F = 2\pi \int_{-1/2}^{1/2} x' g(x') \, dx'$	1.586	1.518	1.325	1.008	0.566
$G = \int_{-1/2}^{1/2} [x'g(x')]^2 \, dx'$	0.0850	0.0842	0.0793	0.0662	0.0411
$H = \int_{-1/2}^{1/2} (x')^2 |g(x')| \, dx'$	0.0850	0.0842	0.0793	0.0662	0.0411
$K_r = \dfrac{\sqrt{12}\,F}{2\pi\sqrt{C}}$	0.874	0.935	0.943	0.878	0.698
$20 \log (K_r)$	-1.166	-0.580	-0.511	-1.127	-3.13
$10 \log (G_r)$	-2.76	-2.39	-2.87	-4.23	-7.07
$10 \log \left[\dfrac{G_{se}}{\eta_x}\right]$	$+13.3$	$+15.5$	$+8.44$	$+6.03$	$+7.42$

w = Aperture width

K_r = Voltage slope at beam center

 = Slope for linear odd illumination

$G_r = \dfrac{\text{Peak gain}}{\text{Peak gain for uniform illumination}}$

$\dfrac{G_{se}}{\eta_x} = \dfrac{\text{Peak gain for uniform illumination}}{\text{Gain of highest sidelobe}}$

Table A.12

ODD ILLUMINATION FUNCTIONS, TRIANGULAR FAMILY

$g(x')$	(0.5)	(0.5, 0.1)	(0.5, 0.2)	(0.5, 0.3)	(0.5, 0.4)	(0.5)
Figure number						A.18
$C = \int_{-1/2}^{1/2} g^2(x')\,dx'$	0.333	0.333	0.333	0.333	0.333	0.333
$10\log(C)$	−4.78	−4.78	−4.78	−4.78	−4.78	−4.78
$D = \int_{-1/2}^{1/2}\left[\dfrac{dg(x')}{dx'}\right]^2 dx'$	4.00	25.00	16.67	16.67	25.00	4.00
$F = 2\pi\int_{-1/2}^{1/2} x'g(x')\,dx'$	0.518	0.622	0.726	0.829	0.933	1.068
$G = \int_{-1/2}^{1/2} [x'g(x')]^2\,dx'$	0.0082	0.0125	0.0188	0.0270	0.0373	0.0520
$H = \int_{-1/2}^{1/2} (x')^2\,|g(x')|\,dx'$	0.0206	0.0255	0.0321	0.0404	0.0503	0.0643
$K_r = \dfrac{\sqrt{12}F}{2\pi\sqrt{C}}$	0.502	0.594	0.693	0.792	0.890	1
$20\log(K_r)$	−5.98	−4.53	−3.19	−2.03	−1.01	0
$10\log(G_r)$	−5.09	−3.85	−3.08	−2.69	−2.53	−2.43
$10\log\left[\dfrac{G_{se}}{\eta_x}\right]$	—	+31.5	+25.	+14.1	+11.5	+10.7

$g(x')$ shown against axis $x' = x/w$, with values marked at $-\tfrac{1}{2}$, $\tfrac{1}{2}$, and 1.

$w = $ Aperture width

$K_r = $ Voltage slope at beam center / Slope for linear odd illumination

$G_r = \dfrac{\text{Peak gain}}{\text{Peak gain for uniform illumination}}$

$\dfrac{G_{se}}{\eta_x} = \dfrac{\text{Peak gain for uniform illumination}}{\text{Gain of highest sidelobe}}$

Table A.13

ODD ILLUMINATION FUNCTIONS, TAYLOR DERIVATIVE

$\bar{n}$	2	3	4	5	6	8
SLL	20	25	30	35	40	45
Figure number	A.25		A.26		A.27	
$C = \int_{-1/2}^{1/2} g^2(x')\,dx'$	0.486	0.682	0.608	0.553	0.517	0.489
$10 \log (C)$	−3.13	−1.660	−2.16	−2.57	−2.86	−3.11
$D = \int_{-1/2}^{1/2} \left[\dfrac{dg(x')}{dx'}\right]^2 dx'$	–	–	–	–	–	–
$F = 2\pi \int_{-1/2}^{1/2} x'\,g(x')\,dx'$	1.258	1.444	1.303	1.177	1.074	0.985
$G = \int_{-1/2}^{1/2} [x'\,g(x')]^2\,dx'$	0.0628	0.0759	0.0579	0.0455	0.0374	0.0316
$H = \int_{-1/2}^{1/2} (x')^2\,|g(x')|\,dx'$	0.0723	0.0801	0.0695	0.0604	0.0530	0.0468
$K_r = \dfrac{\sqrt{12F}}{2\pi\sqrt{C}}$	0.989	0.959	0.918	0.868	0.819	0.773
$20 \log (K_r)$	−0.0968	−0.366	−0.755	−1.227	−1.731	−2.24
$10 \log (G_r)$	−2.21	−2.18	−2.25	−2.40	−2.57	−2.76
$10 \log \left[\dfrac{G_{se}}{\eta_x}\right]$	+14.2	+17.3	+21.3	+25.0	+29.9	+34.

w = Aperture width

$g(x') = x'\,g_T(x')$ [See Table A.7 for definition of $g_T(x')$]

$K_r = \dfrac{\text{Voltage slope at beam center}}{\text{Slope for linear odd illumination}}$

$G_r = \dfrac{\text{Peak gain}}{\text{Peak gain for uniform illumination}}$

$\dfrac{G_{se}}{\eta_x} = \dfrac{\text{Peak gain for uniform illumination}}{\text{Gain of highest sidelobe}}$

Table A.14

ODD ILLUMINATION FUNCTIONS, SPECIAL

$g(x')$	$\sin(2\pi x)$	$4x'\sqrt{1-4(x')^2}$
Figure number	A.28	
$C = \int_{-1/2}^{1/2} g^2(x')\,dx'$	0.495	0.528
$10 \log(C)$	-3.05	-2.77
$D = \int_{-1/2}^{1/2} \left[\dfrac{dg(x')}{dx'}\right]^2 dx'$	0	0
$F = 2\pi \int_{-1/2}^{1/2} x'\,g(x')\,dx'$	0.990	1.216
$G = \int_{-1/2}^{1/2} [x'\,g(x')]^2\,dx'$	0.0349	0.0565
$H = \int_{-1/2}^{1/2} (x')^2\,|g(x')|\,dx'$	0.0468	0.0655
$K_r = \dfrac{\sqrt{12F}}{2\pi\sqrt{C}}$	0.772	0.918
$20 \log(K_r)$	-2.25	-0.741
$10 \log(G_r)$	-2.70	-2.29
$10 \log\left[\dfrac{G_{se}}{\eta_x}\right]$	$+21.$	$+14.1$

w = Aperture width

$K_r = \dfrac{\text{Voltage slope at beam center}}{\text{Slope for linear odd illumination}}$

$G_r = \dfrac{\text{Peak gain}}{\text{Peak gain for uniform illumination}}$

$\dfrac{G_{se}}{\eta_x} = \dfrac{\text{Peak gain for uniform illumination}}{\text{Gain of highest sidelobe}}$

Table A.15

CIRCULAR ILLUMINATION FUNCTIONS, $g(r') = [1 - (2r')^2]^n$

n	(*Uniform*) 0	1	2	3	4
Figure number	A.29	A.30			
$A = 8\int_0^{1/2} r' g(r')\, dr'$	1	0.490	0.327	0.245	0.1960
$20 \log (A)$	0	−6.20	−9.72	−12.22	−14.15
$C = 8\int_0^{1/2} r' g^2(r')\, dr'$	1	0.327	0.1960	0.1399	0.1088
$10 \log (C)$	0	−4.86	−7.08	−8.54	−9.63
$\eta_a = A^2/C$	1	0.735	0.545	0.429	0.353
$10 \log (\eta_a)$	0	−1.338	−2.64	−3.68	−4.52
$\theta_n D/\lambda = C/A^2$	1	1.361	1.836	2.33	2.83
$\theta_3 D/\lambda$	1.016	1.267	1.467	1.681	1.889
$10 \log [G_{sr}]$	+17.6	+24.5	+30.5	+35.8	+40.6

D = Aperture diameter θ_n = Noise beamwidth (rad)

λ = Wavelength θ_3 = 3-db beamwidth (rad)

η_a = Aperture efficiency $G_{sr} = \dfrac{\text{Gain of main beam}}{\text{Gain of highest sidelobe}}$

Table A.16

CIRCULAR ILLUMINATION FUNCTIONS, TRUNCATED GAUSSIAN

n	1.	1.7	2.4	2.8	3.2
Edge illumination (db)	-3.	-8.67	-17.29	-23.5	-30.7
Figure number		A.31	A.32	A.33	
$A = 8 \int_0^{1/2} r' g(r')\, dr'$	0.843	0.627	0.428	0.339	0.269
$20 \log (A)$	-1.486	-4.05	-7.37	-9.40	-11.39
$C = 8 \int_0^{1/2} r' g^2(r')\, dr'$	0.718	0.427	0.242	0.1801	0.1383
$10 \log (C)$	-1.441	-3.70	-6.16	-7.44	-8.59
$\eta_a = A^2/C$	0.990	0.922	0.757	0.638	0.525
$10 \log (\eta_a)$	-0.0445	-0.354	-1.212	-1.952	-2.80
$\theta_n D/\lambda = C/A^2$	1.010	1.085	1.322	1.567	1.905
$\theta_3 D/\lambda$	1.049	1.120	1.224	1.338	1.469
$10 \log [G_{sr}]$	+19.2	+23.3	+34.5	+43.3	+49.1

D = Aperture diameter
λ = Wavelength
$g(r') = \exp\left[-1.382(nr')^2\right]$
η_a = Aperture efficiency

θ_n = Noise beamwidth (rad)
θ_3 = 3-db beamwidth (rad)
$G_{sr} = \dfrac{\text{Gain of main beam}}{\text{Gain of highest sidelobe}}$

Table A.17

CIRCULAR ILLUMINATION FUNCTIONS, CIRCULAR TAYLOR*

$\bar{n}$	2	3	4	5	6	8
SLL	20	25	30	35	40	45
Figure number	A.34		A.35		A.36	
$A = 8\int_0^{1/2} r' g(r')\,dr'$	0.681	0.558	0.503	0.438	0.394	0.359
$20\log(A)$	-3.33	-5.06	-5.97	-7.17	-8.09	-8.93
$C = 8\int_0^{1/2} r' g^2(r')\,dr'$	0.480	0.341	0.299	0.248	0.220	0.1968
$10\log(C)$	-3.18	-4.67	-5.24	-6.06	-6.58	-7.06
$\eta_a = A^2/C$	0.966	0.914	0.846	0.774	0.706	0.651
$10\log(\eta_a)$	-0.1490	-0.389	-0.725	-1.113	-1.510	-1.867
$\theta_n D/\lambda = C/A^2$	1.035	1.094	1.182	1.292	1.416	1.537
$\theta_3 D/\lambda$	1.076	1.120	1.164	1.234	1.287	1.334
$10\log[G_{sr}]$	$+22.0$	$+26.2$	$+31.2$	$+36.6$	$+41.0$	$+45.0$

D = Aperture diameter
λ = Wavelength
η_a = Aperture efficiency
θ_n = Noise beamwidth (rad)

θ_3 = 3-db beamwidth (rad)
$G_{sr} = \dfrac{\text{Gain of main beam}}{\text{Gain of highest sidelobe}}$

*Defined on page 250.

A.4 Plotted Data

This section contains the complete data describing selected illumination functions, listed in Table A.18, and their corresponding patterns. These are divided into three groups of functions: even, odd, and circular. Each function is described by a set of four figures, as follows:

(a) A plot of the illumination function, $g(x')$. Curves for g, g^2, and $20 \log (g)$ are plotted versus normalized aperture distance, x', from 0 to 0.5. These plots cover only positive values of x' because the functions have either even or odd symmetry about the axis. The curves are more easily read on the larger scale.

(b) Plots of normalized antenna pattern function. Curves for the voltage pattern, F/F_m, and the power pattern, $(F/F_m)^2$, are plotted versus normalized angle, u'. The region near the main lobe is plotted on an expanded scale in angle.

(c) Plots of the normalized antenna pattern, on a db-linear scale. The region near the main lobe is again expanded for greater reading accuracy.

(d) Plots of the normalized pattern, on a db-log scale. Two logarithmic cycles for angle are combined on one, by using separate scales at top and bottom. The portion of the main lobe $0.1 < u' < 1.0$ is also plotted on an expanded vertical scale to give greater accuracy near its peak.

Table A.18

LIST OF PLOTTED FUNCTIONS

Figure Number	Illumination Function	Figure Number	Illumination Function
A.4	Uniform rectangular	A.21	Truncated Rayleigh, $n = 2.17$
A.5	Cosine	A.22	" " $n = 2.88$
A.6	Cosine^2	A.23	" " $n = 3.34$
A.7	Cosine^3	A.24	Rectangular odd
A.8	Parabolic	A.25	Taylor derivative, 20-db sidelobes
A.9	Triangular	A.26	" " 30-db sidelobes
A.10	Truncated Gaussian, $n = 1.7$	A.27	" " 40-db sidelobes
A.11	" " $n = 2.4$	A.28	Sine $(\pi x')$
A.12	" " $n = 2.8$	A.29	Uniform circular
A.13	Cosine-on-pedestal, $k = 0.2$	A.30	Parabolic circular
A.14	Taylor, $\bar{n} = 2$, 20-db sidelobes	A.31	Truncated Gaussian circular, $n = 1.7$
A.15	Taylor, $\bar{n} = 4$, 30-db sidelobes	A.32	" " " $n = 2.4$
A.16	Taylor, $\bar{n} = 6$, 40-db sidelobes	A.33	" " " $n = 2.8$
A.17	Hamming	A.34	Taylor circular, $\bar{n} = 2$, 20-db sidelobes
A.18	Linear odd	A.35	" " $\bar{n} = 4$, 30-db "
A.19	$x' \cos (\pi x')$	A.36	" " $\bar{n} = 6$, 40-db "
A.20	$x' \cos^2 (\pi x')$		

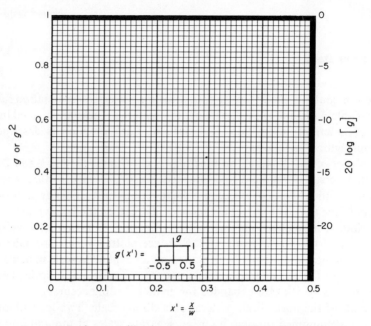

Fig. A.4(a)   Aperture illumination function. [Uniform illumination.]

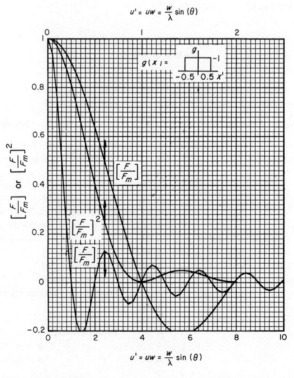

Fig. A.4(b)   Normalized antenna pattern. [Uniform illumination.]

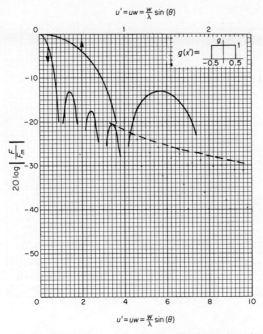

Fig. A.4(c) Normalized antenna pattern. [Uniform illumination.]

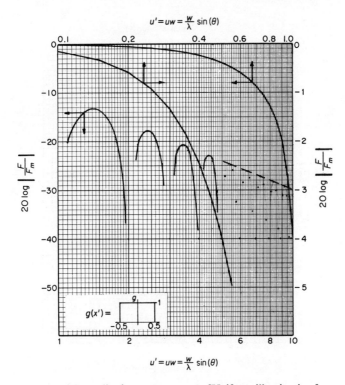

Fig. A.4(d) Normalized antenna pattern. [Uniform illumination.]

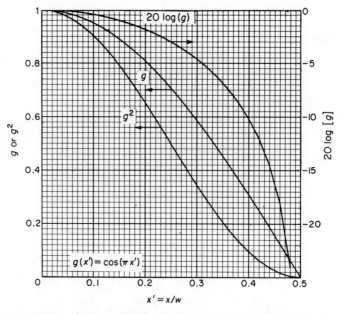

Fig. A.5(a) Aperture illumination function. [Cosine illumination.]

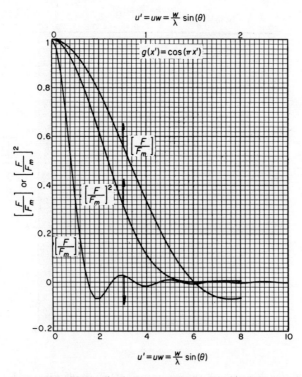

Fig. A.5(b) Normalized antenna pattern. [Cosine illumination.]

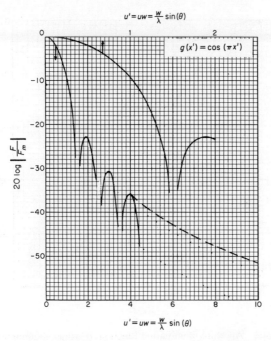

Fig. A.5(c)   Normalized antenna pattern. [Cosine illumination.]

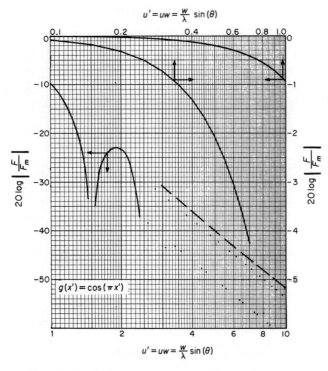

Fig. A.5(d)   Normalized antenna pattern. [Cosine illumination.]

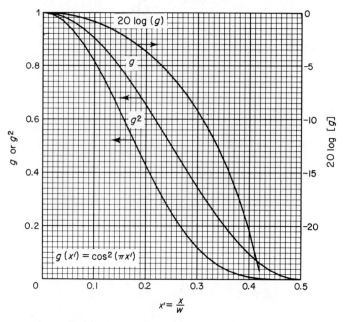

Fig. A.6(a)   Aperture illumination function. [Cosine² illumination.]

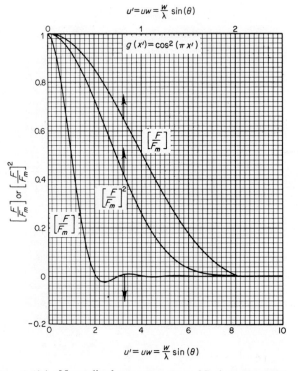

Fig. A.6(b)   Normalized antenna pattern. [Cosine² illumination.]

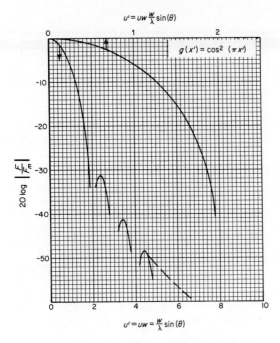

$u' = uw \dfrac{w}{\lambda} \sin(\theta)$

$g(x') = \cos^2(\pi x')$

$20 \log \left| \dfrac{F}{F_m} \right|$

$u' = uw = \dfrac{w}{\lambda} \sin(\theta)$

Fig. A.6(c)   Normalized antenna pattern. [Cosine² illumination.]

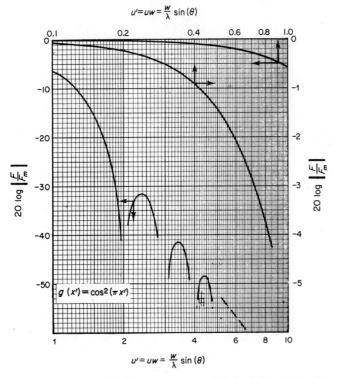

$u' = uw = \dfrac{w}{\lambda} \sin(\theta)$

$20 \log \left| \dfrac{F}{F_m} \right|$

$20 \log \left| \dfrac{F}{F_m} \right|$

$g(x') = \cos^2(\pi x')$

$u' = uw = \dfrac{w}{\lambda} \sin(\theta)$

Fig. A.6(d)   Normalized antenna pattern. [Cosine² illumination.]

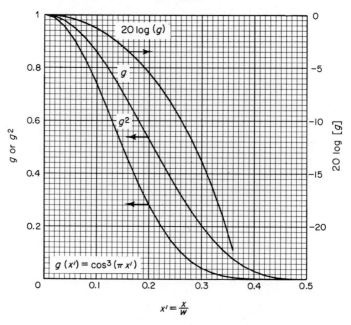

Fig. A.7(a)   Aperture illumination function. [Cosine³ illumination.]

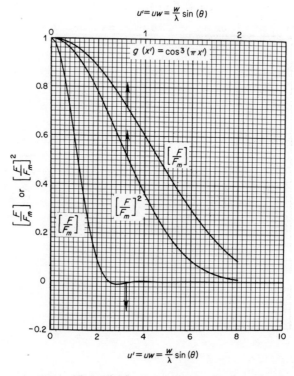

Fig. A.7(b)   Normalized antenna pattern. [Cosine³ illumination.]

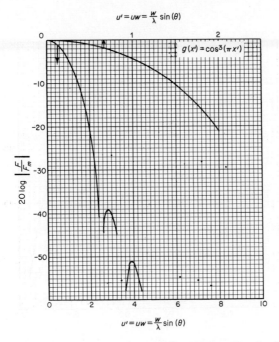

Fig. A.7(c)   Normalized antenna pattern. [Cosine³ illumination.]

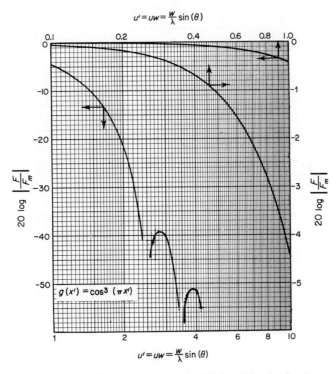

Fig. A.7(d)   Normalized antenna pattern. [Cosine³ illumination.]

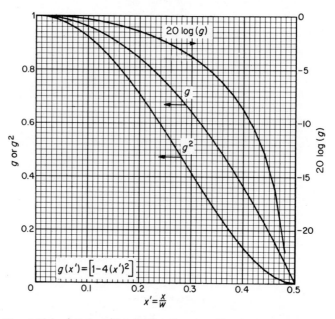

Fig. A.8(a) Aperture illumination function. [Parabolic illumination.]

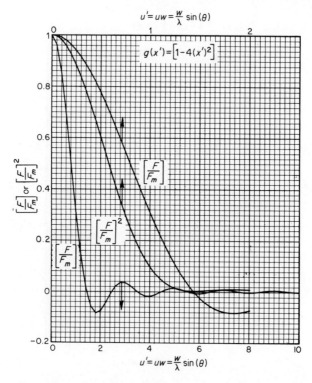

Fig. A.8(b) Normalized antenna pattern. [Parabolic illumination.]

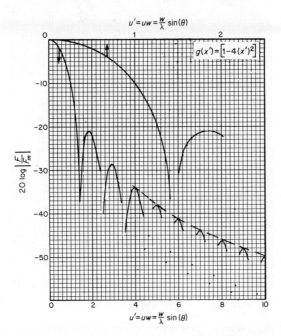

Fig. A.8(c) Normalized antenna pattern. [Parabolic illumination.]

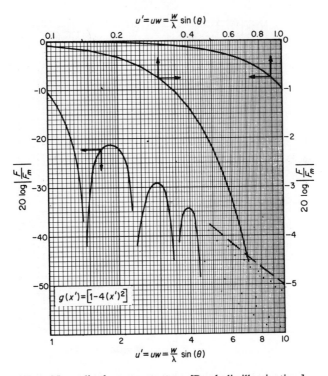

Fig. A.8(d) Normalized antenna pattern. [Parabolic illumination.]

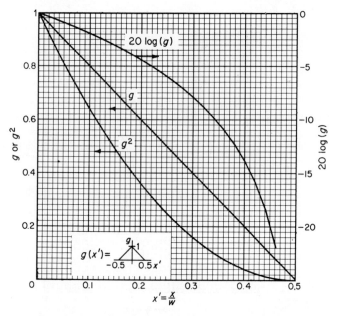

Fig. A.9(a) Aperture illumination function. [Triangular illumination.]

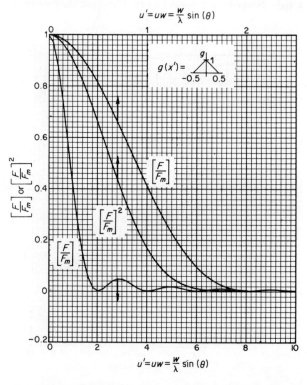

Fig. A.9(b) Normalized antenna pattern. [Triangular illumination.]

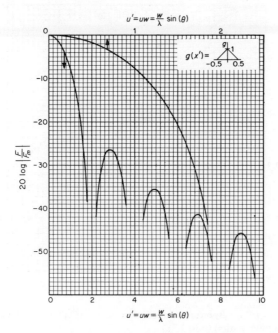

Fig. A.9(c)   Normalized antenna pattern. [Triangular illumination.]

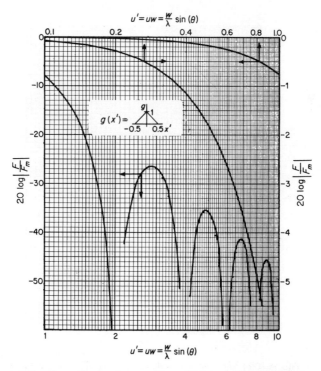

Fig. A.9(d)   Normalized antenna pattern. [Triangular illumination.]

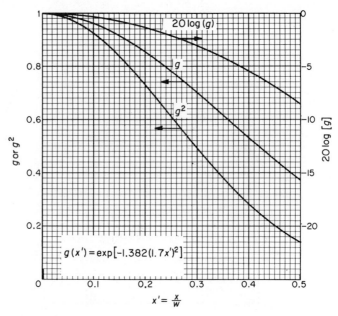

Fig. A.10(a) Aperture illumination function. [Truncated Gaussian illumination ($n = 1.7$).]

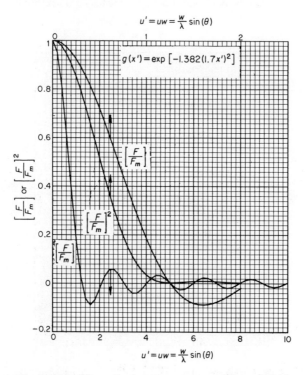

Fig. A.10(b) Normalized antenna pattern. [Truncated Gaussian illumination ($n = 1.7$).]

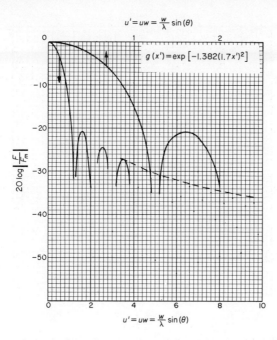

Fig. A.10(c) Normalized antenna pattern. [Truncated Gaussian illumination ($n = 1.7$).]

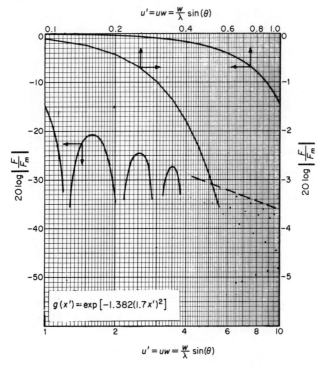

Fig. A.10(d) Normalized antenna pattern. [Truncated Gaussian illumination ($n = 1.7$).]

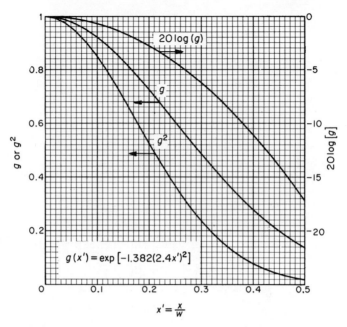

Fig. A.11(a) Aperture illumination function. [Truncated Gaussian illumination ($n = 2.4$).]

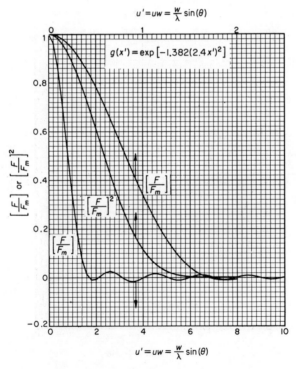

Fig. A.11(b) Normalized antenna pattern. [Truncated Gaussian illumination ($n - 2.4$).]

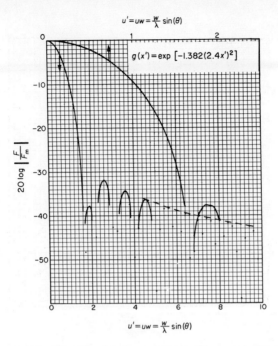

Fig. A.11(c) Normalized antenna pattern. [Truncated Gaussian illumination ($n = 2.4$).]

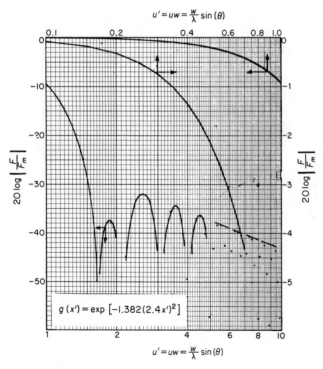

Fig. A.11(d) Normalized antenna pattern. [Truncated Gaussian illumination ($n = 2.4$).]

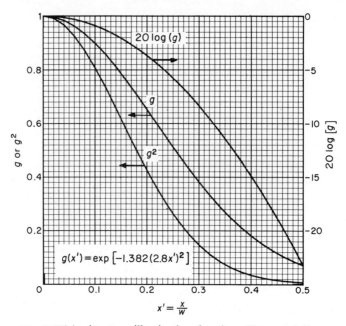

Fig. A.12(a) Aperture illumination function. [Truncated Gaussian illumination ($n = 2.8$).]

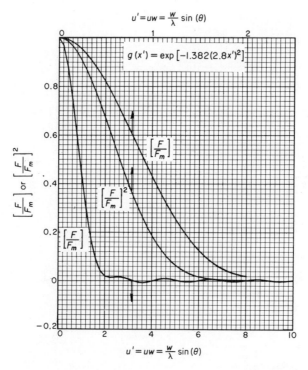

Fig. A.12(b) Normalized antenna pattern. [Truncated Gaussian illumination ($n = 2.8$).]

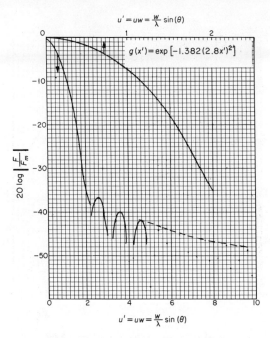

Fig. A.12(c) Normalized antenna pattern. [Truncated Gaussian illumination ($n = 2.8$).]

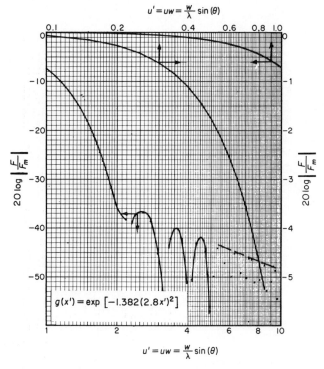

Fig. A.12(d) Normalized antenna pattern. [Truncated Gaussian illumination ($n = 2.8$).]

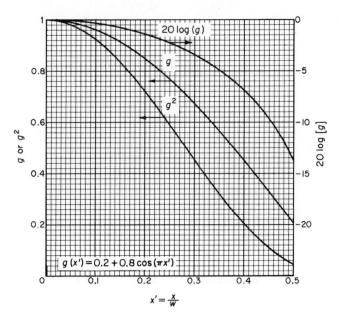

Fig. A.13(a) Aperture illumination function. [Cosine-on-pedestal illumination.]

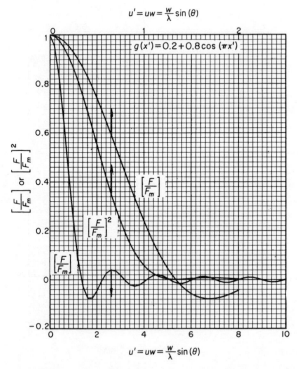

Fig. A.13(b) Normalized antenna pattern. [Cosine-on-pedestal illumination.]

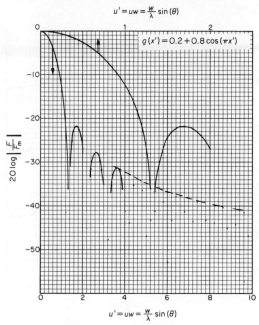

Fig. A.13(c) Normalized antenna pattern. [Cosine-on-pedestal illumination.]

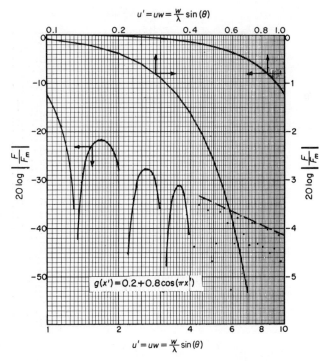

Fig. A.13(d) Normalized antenna pattern. [Cosine-on-pedestal illumination.]

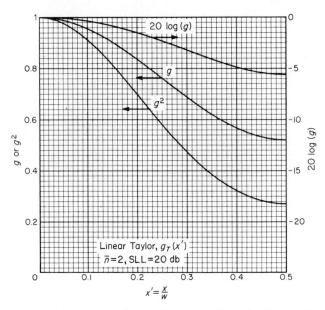

Fig. A.14(a) Aperture illumination function. [Taylor illumination (SLL = 20 db).]

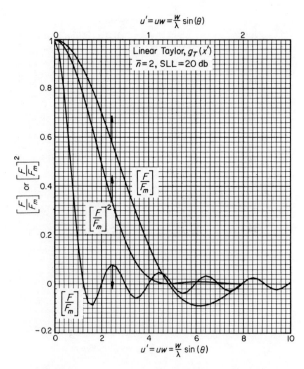

Fig. A.14(b) Normalized antenna pattern. [Taylor illumination (SLL = 20 db).]

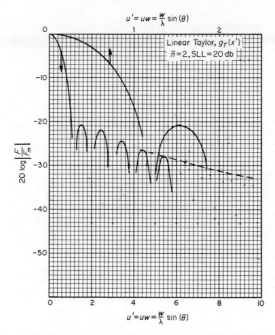

Fig. A.14(c) Normalized antenna pattern. [Taylor illumination (SLL = 20 db).]

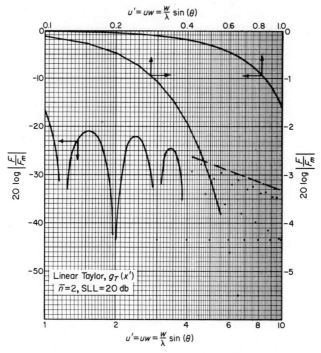

Fig. A.14(d) Normalized antenna pattern. [Taylor illumination (SLL = 20 db).]

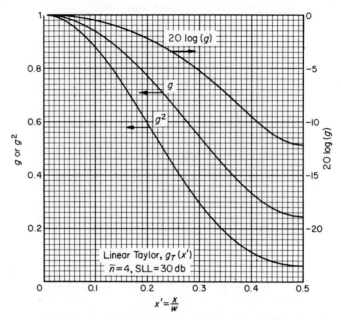

Fig. A.15(a) Aperture illumination function. [Taylor illumination (SLL = 30 db).]

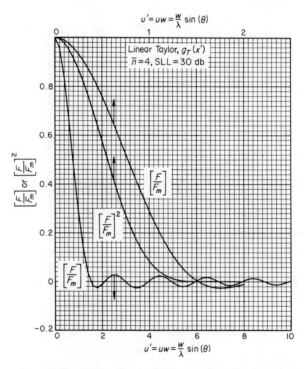

Fig. A.15(b) Normalized antenna pattern. [Taylor illumination (SLL = 30 db).]

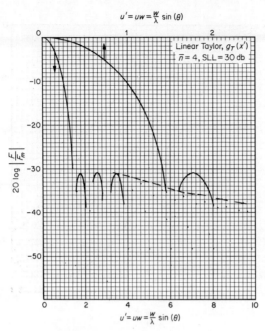

Fig. A.15(c) Normalized antenna pattern. [Taylor illumination (SLL = 30 db).]

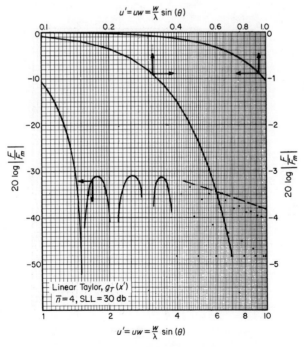

Fig. A.15(d) Normalized antenna pattern. [Taylor illumination (SLL = 30 db).]

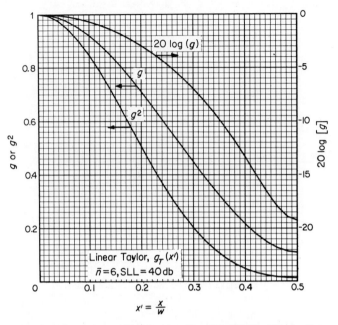

Fig. A.16(a) Aperture illumination function. [Taylor illumination (SLL = 40 db).]

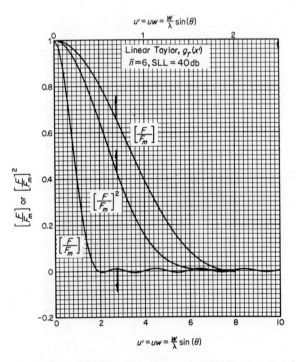

Fig. A.16(b) Normalized antenna pattern. [Taylor illumination (SLL = 40 db).]

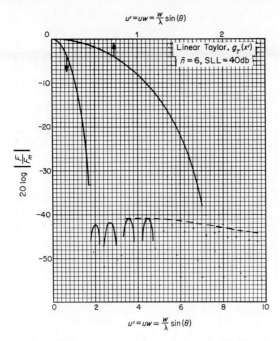

Fig. A.16(c) Normalized antenna pattern. [Taylor illumination (SLL = 40 db).]

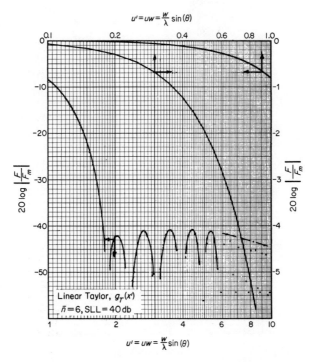

Fig. A.16(d) Normalized antenna pattern. [Taylor illumination (SLL = 40 db).]

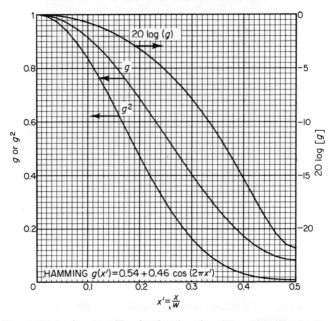

Fig. A.17(a) Aperture illumination function. [Hamming illumination.]

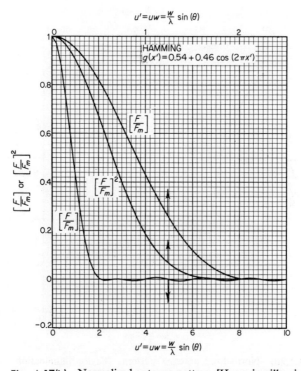

Fig. A.17(b) Normalized antenna pattern. [Hamming illumination.]

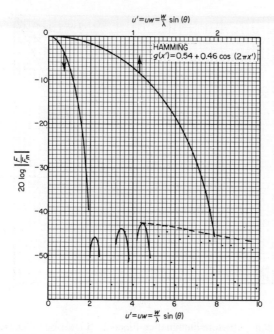

Fig. A.17(c) Normalized antenna pattern. [Hamming illumination.]

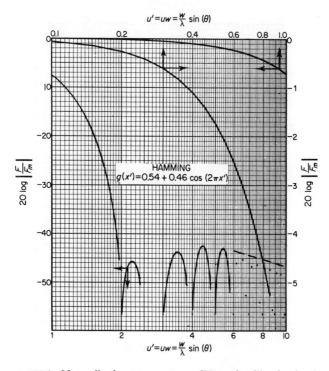

Fig. A.17(d) Normalized antenna pattern. [Hamming illumination.]

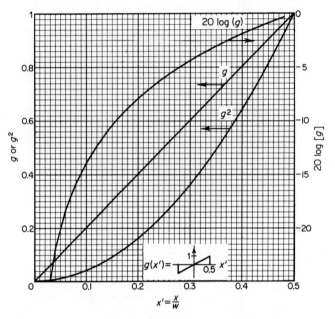

Fig. A.18(a) Aperture illumination function. [Linear-odd illumination.]

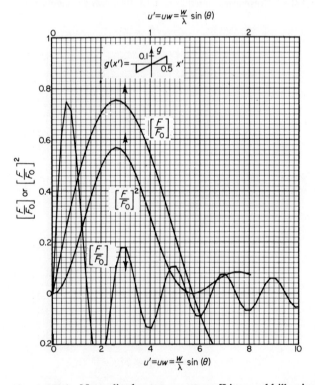

Fig. A.18(b) Normalized antenna pattern. [Linear-odd illumination.]

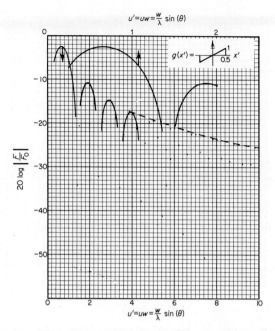

Fig. A.18(c) Normalized antenna pattern. [Linear-odd illumination.]

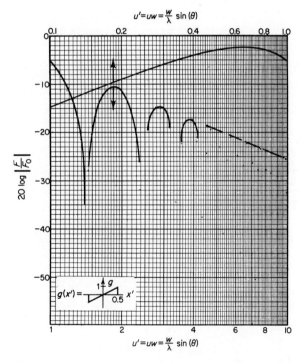

Fig. A.18(d) Normalized antenna pattern. [Linear-odd illumination.]

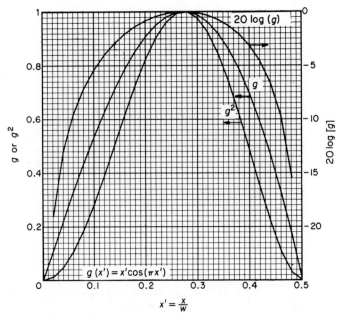

Fig. A.19(a) Aperture illumination function. [$x' \cos (\pi x')$ illumination.]

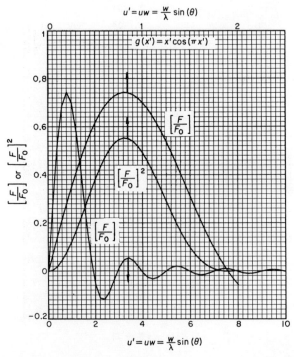

Fig. A.19(b) Normalized antenna pattern. [$x' \cos (\pi x')$ illumination.]

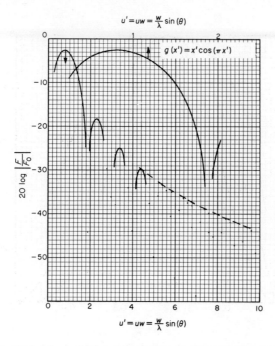

Fig. A.19(c) Normalized antenna pattern. [$x'\cos(\pi x')$ illumination.]

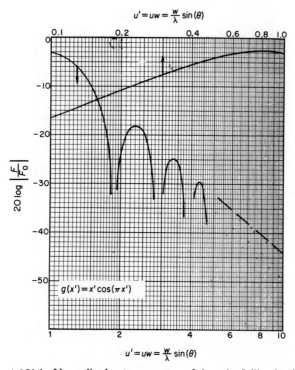

Fig. A.19(d) Normalized antenna pattern. [$x'\cos(\pi x')$ illumination.]

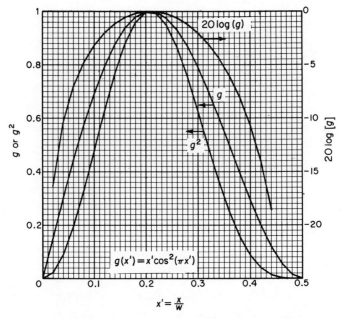

Fig. A.20(a) Aperture illumination function. [$x' \cos^2 (\pi x')$ illumination.]

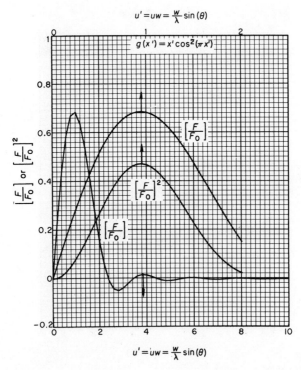

Fig. A.20(b) Normalized antenna pattern. [$x' \cos^2 (\pi x')$ illumination.]

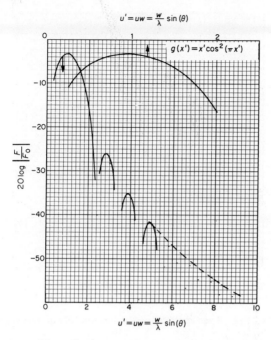

Fig. A.20(c) Normalized antenna pattern. [$x' \cos^2 (\pi x')$ illumination.]

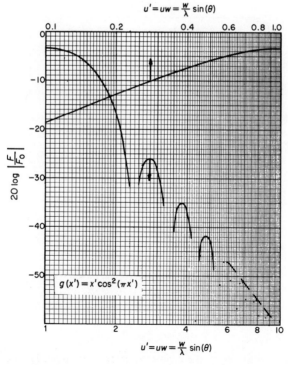

Fig. A.20(d) Normalized antenna pattern. [$x' \cos^2 (\pi x')$ illumination.]

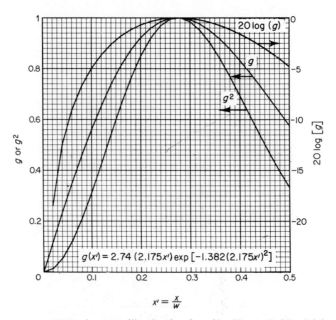

$$x' = \frac{x}{w}$$

Fig. A.21(a) Aperture illumination function. [Truncated Rayleigh illumination ($n = 2.17$).]

$$u' = uw = \frac{w}{\lambda} \sin(\theta)$$

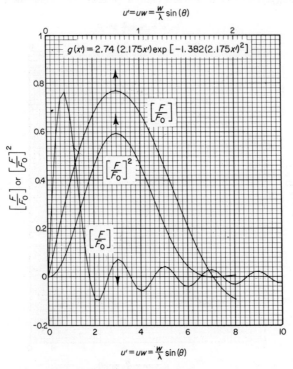

$$u' = uw = \frac{w}{\lambda} \sin(\theta)$$

Fig. A.21(b) Normalized antenna pattern. [Truncated Rayleigh illumination ($n = 2.17$).]

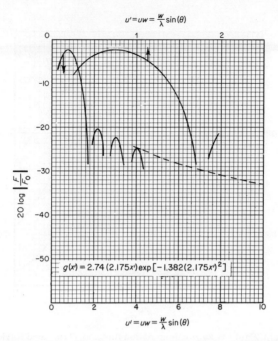

Fig. A.21(c) Normalized antenna pattern. [Truncated Rayleigh illumination $(n = 2.17)$.]

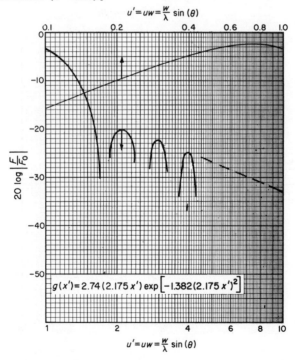

Fig. A.21(d) Normalized antenna pattern. [Truncated Rayleigh illumination $(n = 2.17)$.]

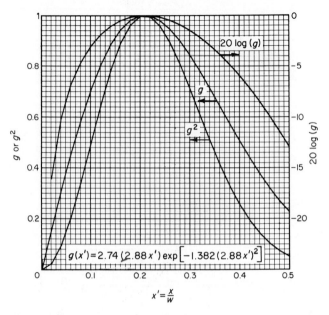

Fig. A.22(a) Aperture illumination function. [Truncated Rayleigh illumination ($n = 2.88$).]

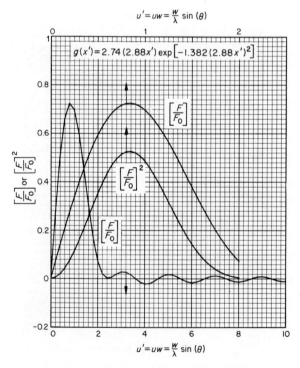

Fig. A.22(b) Normalized antenna pattern. [Truncated Rayleigh illumination ($n = 2.88$).]

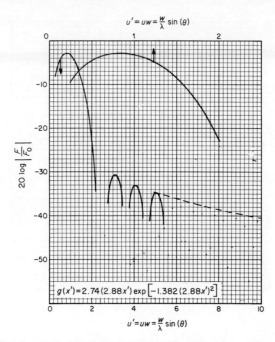

Fig. A.22(c) Normalized antenna pattern. [Truncated Rayleigh illumination ($n = 2.88$).]

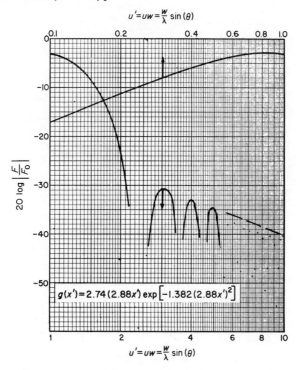

Fig. A.22(d) Normalized antenna pattern. [Truncated Rayleigh illumination ($n = 2.88$).]

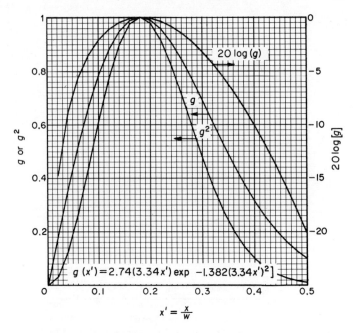

Fig. A.23(a) Aperture illumination function. [Truncated Rayleigh illumination ($n = 3.34$).]

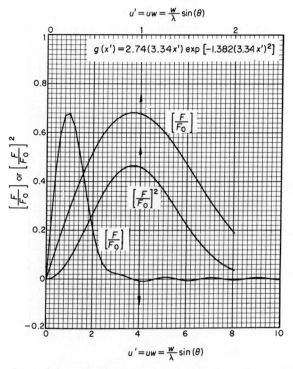

Fig. A.23(b) Normalized antenna pattern. [Truncated Rayleigh illumination ($n = 3.34$).]

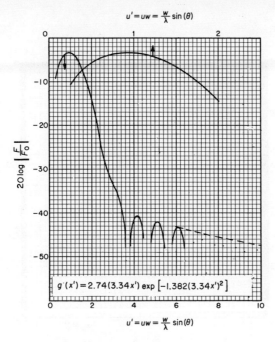

Fig. A.23(c) Normalized antenna pattern. [Truncated Rayleigh illumination ($n = 3.34$).]

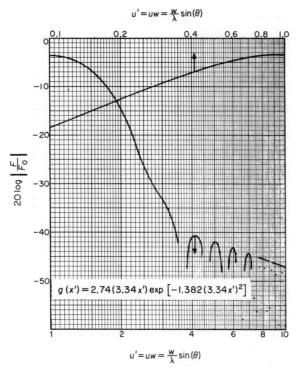

Fig. A.23(d) Normalized antenna pattern. [Truncated Rayleigh illumination ($n = 3.34$).]

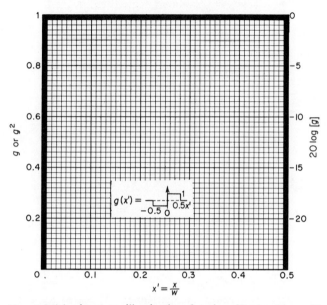

Fig. A.24(a) Aperture illumination function [Rectangular odd illumination.]

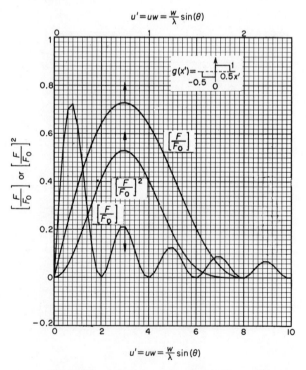

Fig. A.24(b) Normalized antenna pattern. [Rectangular odd illumination.]

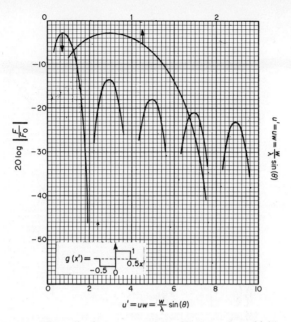

Fig. A.24(c) Normalized antenna pattern. [Rectangular odd illumination.]

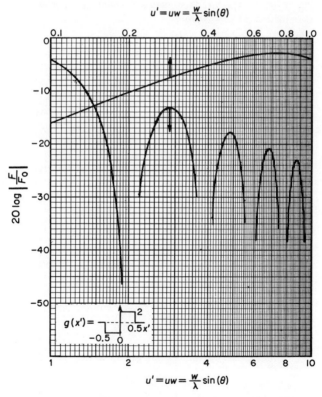

Fig. A.24(d) Normalized antenna pattern. [Rectangular odd illumination.]

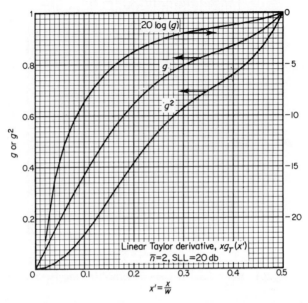

Fig. A.25(a) Aperture illumination function. [Taylor derivative illumination (SLL = 20 db).]

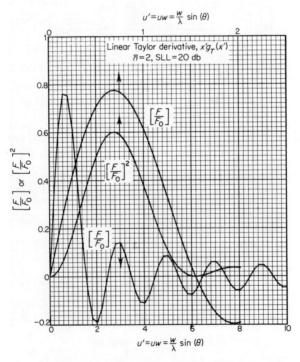

Fig. A.25(b) Normalized antenna pattern. [Taylor derivative illumination (SLL = 20 db).]

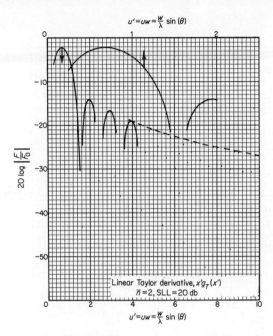

Fig. A.25(c) Normalized antenna pattern. [Taylor derivative illumination (SLL = 20 db).]

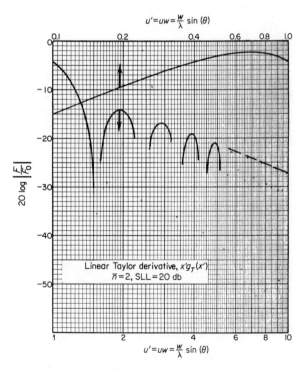

Fig. A.25(d) Normalized antenna pattern. [Taylor derivative illumination (SLL = 20 db).]

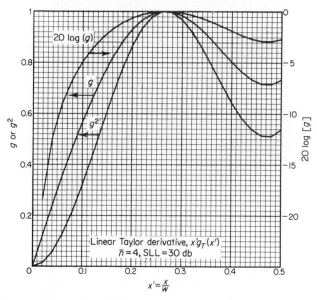

Fig. A.26(a) Aperture illumination function. [Taylor derivative illumination (SLL = 30 db).]

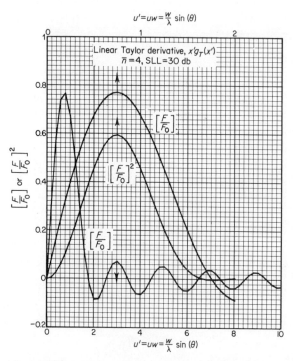

Fig. A.26(b) Normalized antenna pattern. [Taylor derivative illumination (SLL = 30 db).]

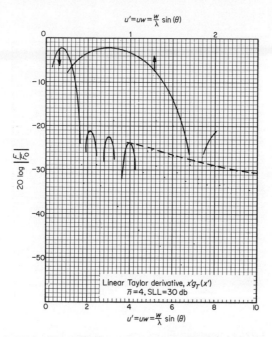

Fig. A.26(c) Normalized antenna pattern. [Taylor derivative illumination (SLL = 30 db).]

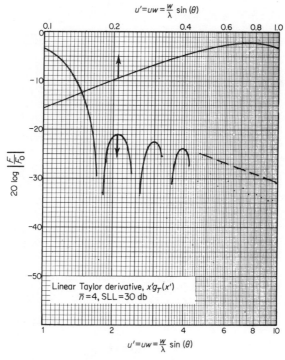

Fig. A.26(d) Normalized antenna pattern. [Taylor derivative illumination (SLL = 30 db).]

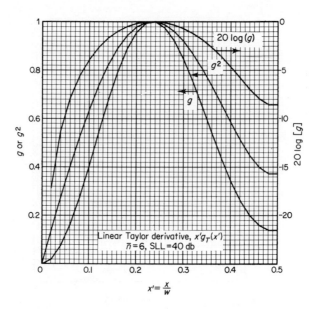

Fig. A.27(a) Aperture illumination function. [Taylor derivative illumination (SLL = 40 db).]

$$u' = uw = \frac{w}{\lambda}\sin(\theta)$$

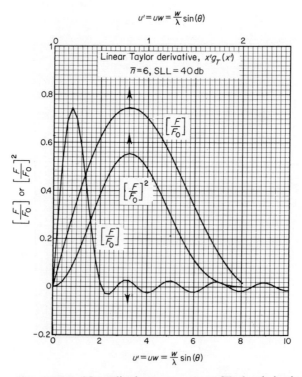

Fig. A.27(b) Normalized antenna pattern. [Taylor derivative illumination (SLL = 40 db).]

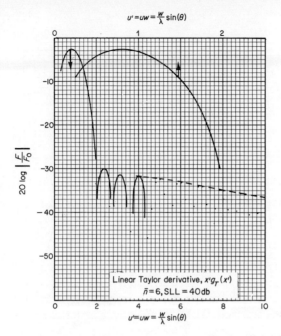

Fig. A.27(c) Normalized antenna pattern. [Taylor derivative illumination (SLL = 40 db).]

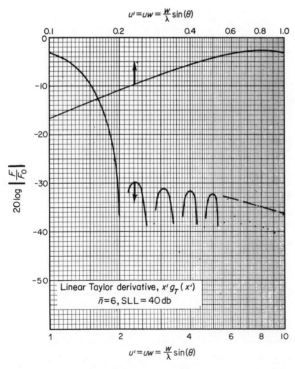

Fig. A.27(d) Normalized antenna pattern. [Taylor derivative illumination (SLL = 40 db).]

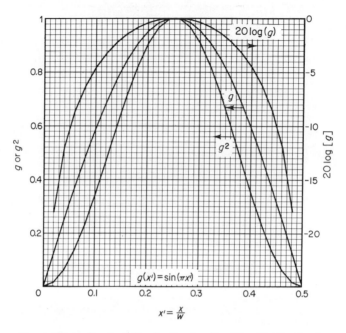

Fig. A.28(a)   Aperture illumination function [Sine illumination.]

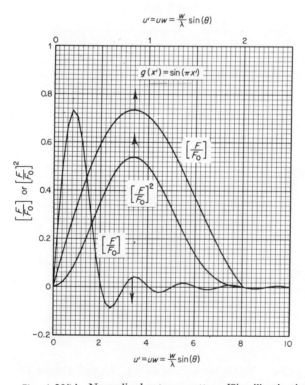

Fig. A.28(b)   Normalized antenna pattern. [Sine illumination.]

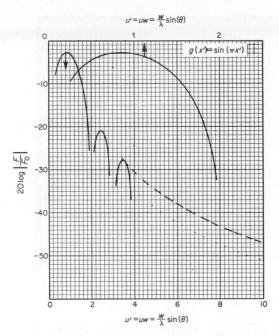

Fig. A.28(c) Normalized antenna pattern. [Sine illumination.]

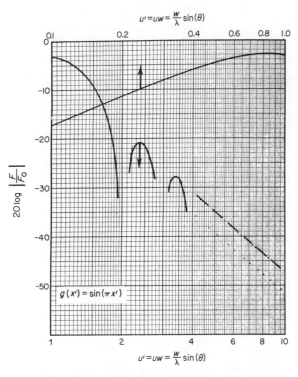

Fig. A.28(d) Normalized antenna pattern. [Sine illumination.]

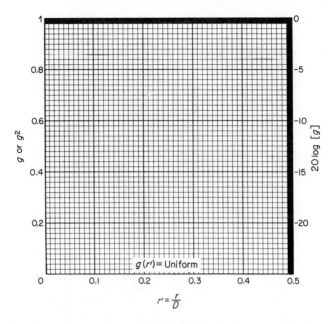

Fig. A.29(a) Aperture illumination function. [Uniform circular illumination.]

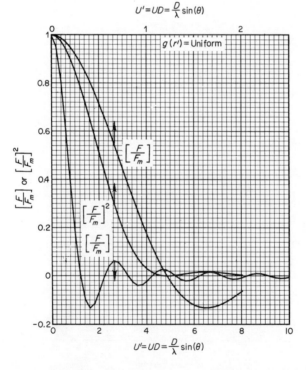

Fig. A.29(b) Normalized antenna pattern. [Uniform circular illumination.]

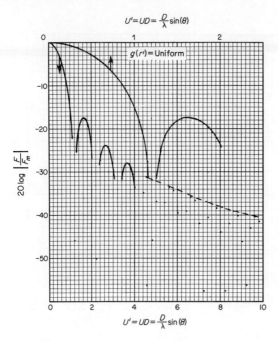

$$U' = UD = \frac{D}{\lambda}\sin(\theta)$$

Fig. A.29(c) Normalized antenna pattern. [Uniform circular illumination.]

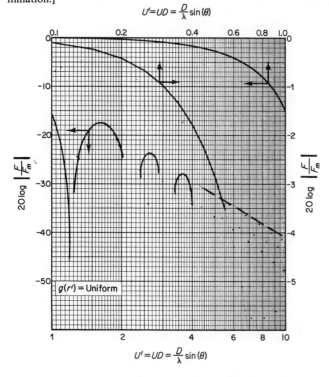

$$U' = UD = \frac{D}{\lambda}\sin(\theta)$$

Fig. A.29(d) Normalized antenna pattern. [Uniform circular illumination.]

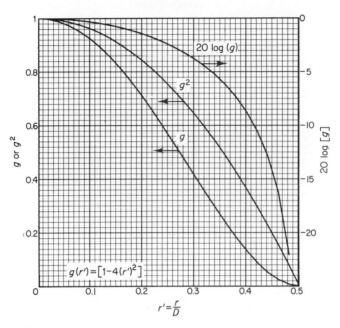

Fig. A.30(a) Aperture illumination function. [Parabolic circular illumination.]

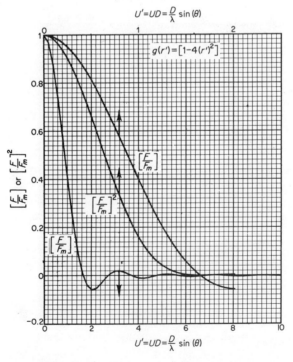

Fig. A.30(b) Normalized antenna pattern. [Parabolic circular illumination.]

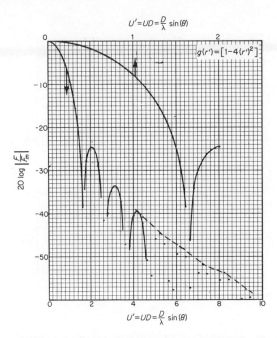

Fig. A.30(c) Normalized antenna pattern. [Parabolic circular illumination.]

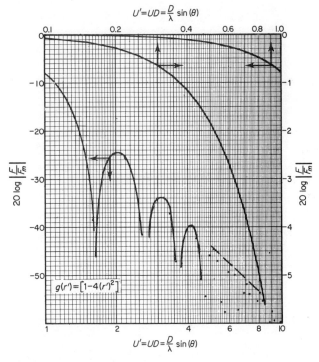

Fig. A.30(d) Normalized antenna pattern. [Parabolic circular illumination.]

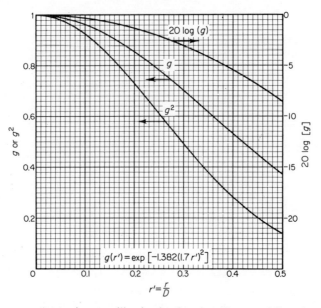

Fig. A.31(a) Aperture illumination function. [Truncated Gaussian circular illumination ($n = 1.7$).]

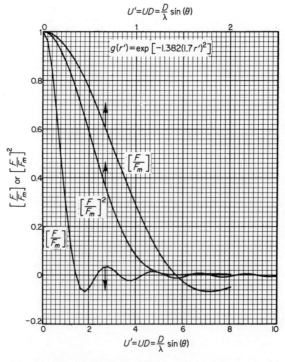

Fig. A.31(b) Normalized antenna pattern. [Truncated Gaussian circular illumination ($n = 1.7$).]

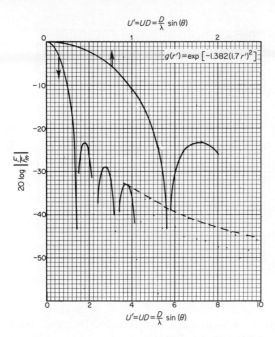

Fig. A.31(c) Normalized antenna pattern. [Truncated Gaussian circular illumination ($n = 1.7$).]

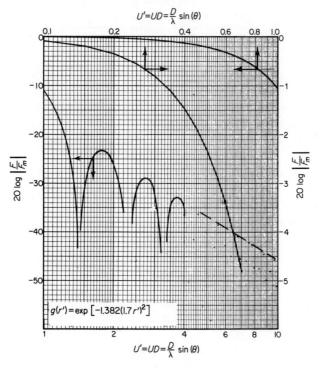

Fig. A.31(d) Normalized antenna pattern. [Truncated Gaussian circular illumination ($n = 1.7$).]

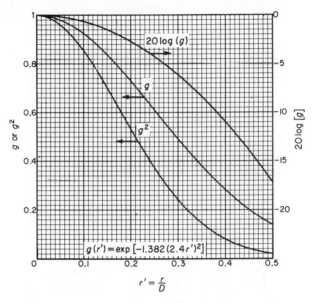

Fig. A.32(a) Aperture illumination function. [Truncated Gaussian circular illumination ($n = 2.4$).]

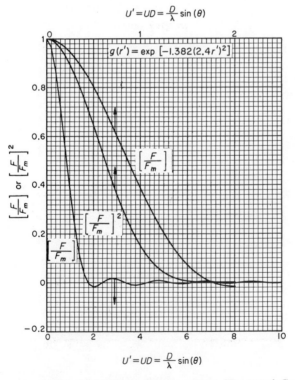

Fig. A.32(b) Normalized antenna pattern. [Truncated Gaussian circular illumination ($n = 2.4$).]

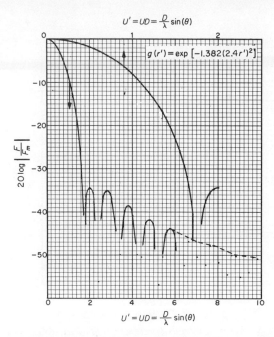

Fig. A.32(c) Normalized antenna pattern. [Truncated Gaussian circular illumination ($n = 2.4$).]

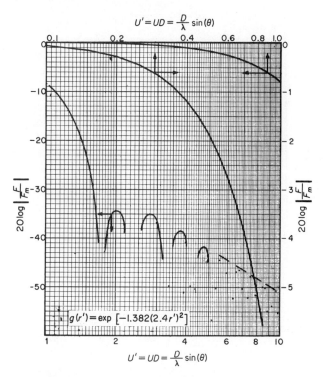

Fig. A.32(d) Normalized antenna pattern. [Truncated Gaussian circular illumination ($n = 2.4$).]

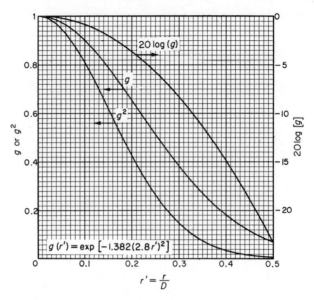

Fig. A.33(a)   Aperture illumination function. [Truncated Gaussian circular illumination ($n = 2.8$).]

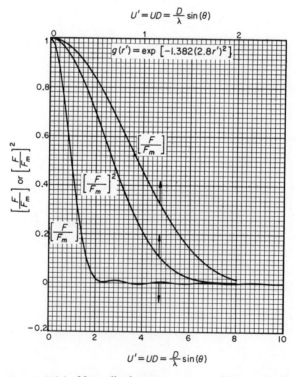

Fig. A.33(b)   Normalized antenna pattern. [Truncated Gaussian circular illumination ($n = 2.8$).]

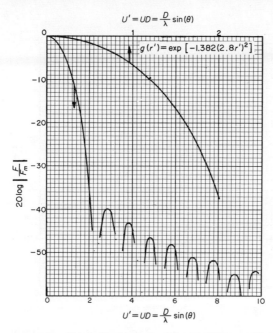

Fig. A.33(c) Normalized antenna pattern. [Truncated Gaussian circular illumination ($n = 2.8$).]

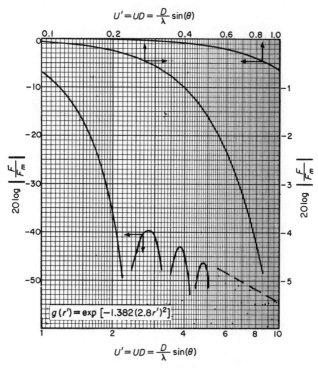

Fig. A.33(d) Normalized antenna pattern. [Truncated Gaussian circular illumination ($n = 2.8$).]

327

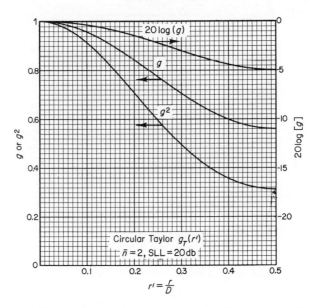

Fig. A.34(a) Aperture illumination function. [Taylor circular illumination (SLL = 20 db).]

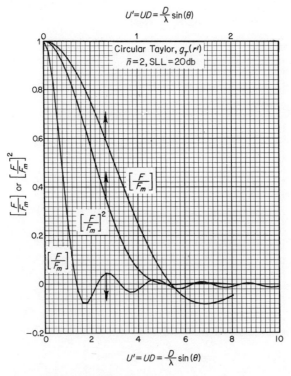

Fig. A.34(b) Normalized antenna pattern. [Taylor circular illumination (SLL = 20 db).]

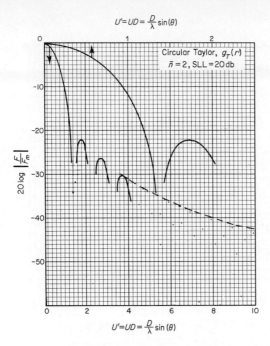

$$U' = UD = \frac{D}{\lambda}\sin(\theta)$$

Circular Taylor, $g_T(r')$
$\bar{n} = 2$, SLL = 20 db

$20 \log \left| \frac{F}{F_m} \right|$

$$U' = UD = \frac{D}{\lambda}\sin(\theta)$$

Fig. A.34(c) Normalized antenna pattern. [Taylor circular illumination (SLL = 20 db).]

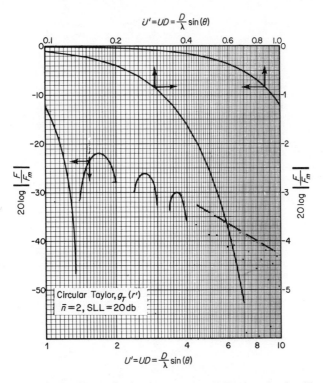

$$U' = UD = \frac{D}{\lambda}\sin(\theta)$$

$20 \log \left| \frac{F}{F_m} \right|$

Circular Taylor, $g_T(r')$
$\bar{n} = 2$, SLL = 20 db

$20 \log \left| \frac{F}{F_m} \right|$

$$U' = UD = \frac{D}{\lambda}\sin(\theta)$$

Fig. A.34(d) Normalized antenna pattern. [Taylor circular illumination (SLL = 20 db).]

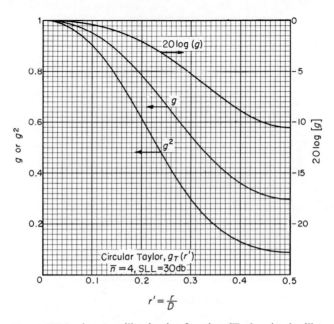

Fig. A.35(a) Aperture illumination function. [Taylor circular illumination (SLL = 30 db).]

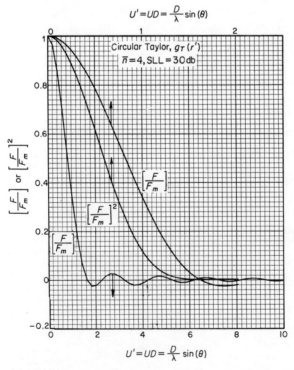

Fig. A.35(b) Normalized antenna pattern. [Taylor circular illumination (SLL = 30 db).]

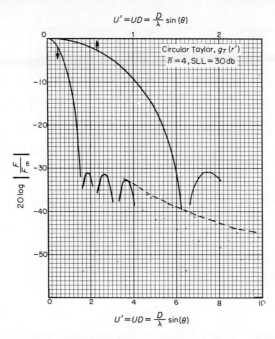

$$U'=UD=\frac{D}{\lambda}\sin(\theta)$$

Fig. A.35(c) Normalized antenna pattern. [Taylor circular illumination (SLL = 30 db).]

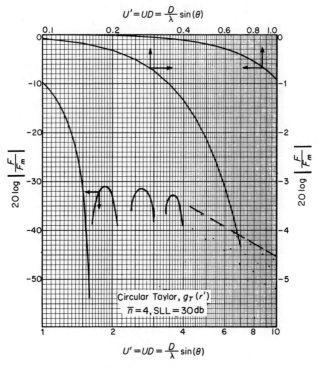

$$U'=UD=\frac{D}{\lambda}\sin(\theta)$$

Fig. A.35(d) Normalized antenna pattern. [Taylor circular illumination (SLL = 30 db).]

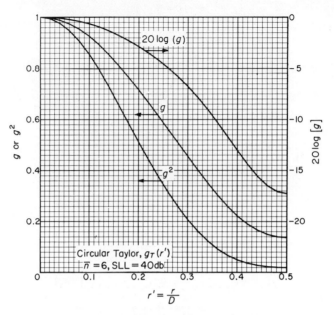

Fig. A.36(a) Aperture illumination function. [Taylor circular illumination (SLL = 40 db).]

$$U' = UD = \frac{D}{\lambda}\sin(\theta)$$

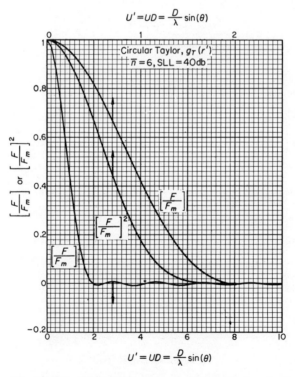

Fig. A.36(b) Normalized antenna pattern. [Taylor circular illumination (SLL = 40 db).]

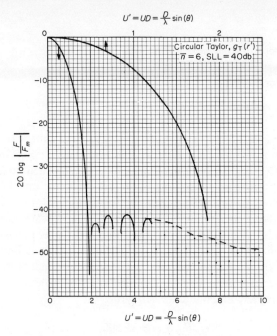

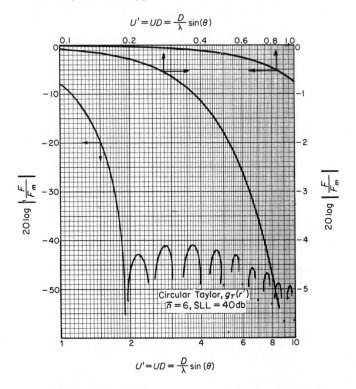

Fig. A.36(c) Normalized antenna pattern. [Taylor circular illumination (SLL = 40 db).]

Fig. A.36(d) Normalized antenna pattern. [Taylor circular illumination (SLL = 40 db).]

A.5 Antenna Relationships

Gain-Beamwidth Product

The beam formed by a directive antenna generally has a gain near the maximum which can be obtained from the available aperture area, subject to reasonable sidelobe levels. Gain falls off in all directions relative to the beam axis, with a pattern determined by aperture shape and illumination function. For almost all practical illuminations, however, the beams are similarly shaped near their centers. This similarity leads to combinations of parameters which are essentially independent of illumination function, so that the antenna designer or systems engineer may apply "rules of thumb" to estimate one parameter when others are known.

The first rule of thumb relates the on-axis gain to the product of the beamwidths in the two angular coordinates. For a fully-filled aperture, whether it be a reflector or an array of elementary antennas, we can write

$$G_m \underset{\text{(power ratio)}}{\theta_{u3}} \underset{\text{(degrees)}}{\theta_{v3}} \cong 35{,}500,$$

or

$$G_m \underset{\text{(power ratio)}}{\theta_{u3}} \underset{\text{(radians)}}{\theta_{v3}} \cong 10.75,$$

or

$$G_m \underset{\text{(power ratio)}}{\theta_{u3}} \underset{\text{(milliradians)}}{\theta_{v3}} \cong 10.75 \times 10^6.$$

It is often convenient to express these equations in decibels:

$$(G_m)_{\text{db}} + 10 \log \theta_{u3} + 10 \log \theta_{v3} \cong 45.5 \qquad (\theta \text{ in deg})$$
$$(G_m)_{\text{db}} + 10 \log \theta_{u3} + 10 \log \theta_{v3} \cong 10.5 \qquad (\theta \text{ in rad})$$
$$(G_m)_{\text{db}} + 10 \log \theta_{u3} + 10 \log \theta_{v3} \cong 70.5 \qquad (\theta \text{ in mr})$$

For a horn-fed reflector, we should subtract the spillover loss, typically 0.8 db. For a reflector which is shaped to produce a cosecant^2 beam in one coordinate, we should subtract a total of 2.3 db. For an array, there will generally be a feed distribution loss, analogous to spillover and with a magnitude which depends upon the design of the feed network.

Figure A.37 gives a plot of θ_{u3} vs θ_{v3} for various values of $(G_m)_{\text{db}}$. Table A.19 lists gain-beamwidth products for some common aperture illumination functions. The product varies by about 0.6 db between the uniform and the heavily weighted apertures. It increases as sidelobes drop, indicating that more of the energy is being concentrated in the main beam.

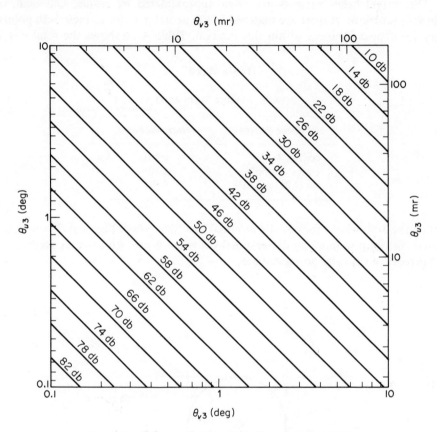

Fig. A.37   Gain as a function of half-power beamwidth.

Table A.19

GAIN-BEAMWIDTH PRODUCTS

Aperture Illumination	$G_m\theta_3^2$ (deg)	$10\log(G_m\theta_3^2)$
Uniform circular	33,700	45.28
Uniform rectangular	32,300	45.10
Rectangular, weighted:		
$\cos(\pi x)\cos(\pi y)$	37,400	45.73
$\cos^2(\pi x)\cos^2(\pi y)$	37,300	45.72
$\cos^3(\pi x)\cos^3(\pi y)$	36,800	45.66
Taylor, 20-db sidelobes	36,000	45.57
" 30-db "	36,900	45.67
" 40-db "	37,000	45.68

Power Distribution in Beam

Directional beam patterns are often approximated by cosine, Gaussian, or $(\sin x)/x$ functions. If these are matched to the actual patterns at their 3-db points, they are almost identical within this interval. Table A.20 shows the total power

Table A.20

POWER WITHIN 3-db WIDTH

Voltage Pattern	Integrated Power
$\cos(\pi u/2)$	0.805
$\exp(-1.385u^2)$	0.804
$\sin(2.783u)/2.783u$	0.817

within the 3-db widths for one-dimensional functions listed. Figure A.38 shows the power distribution in a two-dimensional Gaussian beam, with figures accurate to 0.2 percent of the total power radiated.

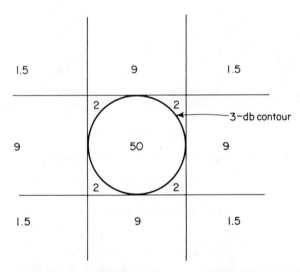

Fig. A.38 Power distribution in a two-dimensional Gaussian beam (figures in percent).

Half-Power Beamwidth

Spencer (1946) gives the following formula for the half-power beamwidth, θ_3, in terms of the moments of the aperture illumination function:

$$\theta_3 \text{ (rad)} \cong 0.26\left(\frac{\lambda}{w}\right)\sqrt{\frac{\mu_0}{\mu_2}} = 0.26\lambda\left[\frac{\int_A g(x)\,dx}{\int_A x^2 g(x)\,dx}\right]^{1/2}.$$

The constant 0.26 has been adjusted to fit a typical weighted aperture illumination. Table A.21 gives the fractional error in θ_3 for some common illumination functions. These data also appear in Chap. 2, Fig. 2.4.

Table A.21

BEAMWIDTH CONSTANTS

Aperture Illumination (rectangular)	$\dfrac{\theta_3}{0.26(\lambda/w)\sqrt{\mu_0/\mu_2}}$
Uniform	0.986
$\cos(\pi x')$	0.996
$\cos^2(\pi x')$	1.000
$\cos^3(\pi x')$	1.005
Taylor, 20-db	0.974
" 30-db	0.998
" 40-db	1.002

Beamwidth Measures

In this handbook, we have adopted the half-power beamwidth as the basic measure of angular resolution. Various other parameters are used in the literature, including the "noise beamwidth" θ_n, the rms beamwidth Θ, the angular resolution constant θ_w (analogous to Woodward's time and frequency resolution constants), and Spencer's integral approximation to half-power beamwidth. The ratios of these various measures to θ_3 are listed in Table A.22 for several illumination functions. Definitions of each measure are given at the end of the table.

Table A.22

COMPARISON OF BEAMWIDTH MEASURES

Illumination Function	First Sidelobe (db)	Spencer's θ_s/θ_3	Noise θ_n/θ_3	Rms Θ/θ_3	Woodward θ_w/θ_3
Uniform	13.3	1.1014	1.128	∞	1.129
Cosine	23	1.004	1.048	2.656	2.548
Cosine^2	31.5	1.000	1.076	2.530	2.726
Cosine^3	39	0.995	1.058	2.548	2.813
Taylor, 20-db	20.9	1.027	1.068	∞	2.399
" , 30-db	30.9	1.002	1.054	∞	2.682
" , 40-db	40.9	0.998	1.050	∞	2.783

Table A.22—*Cont.*

Definitions:

$$\theta_3 = \text{half-power beamwidth}$$

$$\theta_s = .26\frac{\lambda}{w}\left[\frac{\displaystyle\int_{-1/2}^{1/2} g(x')\,dx'}{\displaystyle\int_{-1/2}^{1/2} (x')^2\,|g(x')|\,dx'}\right]^{1/2}$$

$$\theta_n = \frac{\lambda}{w}\frac{\displaystyle\int_{-1/2}^{1/2} |g(x')|^2\,dx'}{\left[\displaystyle\int_{-1/2}^{1/2} g(x')\,dx'\right]^2}$$

$$\Theta = \frac{\lambda}{w}\left[\frac{\displaystyle\int_{-1/2}^{1/2}\left|\frac{dg(x')}{dx'}\right|^2\,dx'}{\displaystyle\int_{-1/2}^{1/2} |g(x')|^2\,dx'}\right]^{1/2}$$

$$\theta_w = \frac{\lambda}{w}\frac{\displaystyle\int_{-1/2}^{1/2} |g(x')|^4\,dx'}{\left[\displaystyle\int_{-1/2}^{1/2} |g(x')|^2\,dx'\right]^2}$$

Appendix B

Waveform Analogies

The data in App. A, although antenna oriented, are also useful for waveform analysis. In particular, the linear even and linear odd illumination functions are directly analogous to either time-limited waveforms or frequency-limited spectra. Since the waveform and spectrum are related through the Fourier transform, just as the antenna illumination function and pattern, the plots describe functions commonly used in waveform analysis.

The remainder of this section defines the terms used in waveform analysis, lists analogies, and gives examples of how the analogies may be applied.

B.1 Definitions

This section contains the definitions of waveform parameters and functions. In particular, it defines the
  (a) Fourier transform,
  (b) Voltage waveform and amplitude spectrum,
  (c) Correlation function and energy spectrum, and
  (d) Response function (ambiguity function).
Time duration and bandwidth are the normalizing parameters. Either can be analogous to the antenna aperture width if the function is zero outside the interval. This basic analogy, combined with the fact that voltage waveforms and spectra are related in the same way as the antenna pattern and illumination function, makes the antenna data also useful for waveform analysis.

Similar notation is used here:

Prime (') indicates a normalized variable,

Subscript o indicates the largest possible value,

Subscript m indicates the largest value for a particular case, and

Subscript r indicates the value relative to the largest possible value.

The following definitions apply only to finite energy waveforms because these are analogous to the antenna functions. Radar pulses or pulse groups are examples of time-limited waveforms which have finite energy and zero average power. A sine wave, a repeating pulse train, and noise are examples of infinite waveforms.

These have infinite energy and finite average power, and cannot be used with the following definitions.

Fourier Transform

Following Woodward (1953), we define the Fourier transform as

$$A(f) = \int_{-\infty}^{\infty} a(t) \exp(-j2\pi ft)\, dt \equiv \mathscr{F}[a(t)],$$

and its inverse as

$$a(t) = \int_{-\infty}^{\infty} A(f) \exp(j2\pi ft)\, df = \mathscr{F}^{-1}[A(f)].$$

Convolution (indicated by $\otimes$) is a basic Fourier operation and is defined by:

$$a(t) \otimes b(t) = \int_{-\infty}^{\infty} a(t_d) b(t - t_d)\, dt_d$$
$$= \mathscr{F}^{-1}[A(f) B(f)].$$

Table B.1

A SHORT TABLE OF TRANSFORM PAIRS

Operation	Function	Transform
	$a(t)$	$A(f)$
time reversal	$a(-t)$	$A(-f)$
conjugate	$a^*(t)$	$A^*(-f)$
derivative	$d[a(t)]/dt$	$2\pi jf\, A(f)$
derivative	$-2\pi jt\, a(t)$	$d[A(f)]/df$
translation	$a(t - \tau)$	$A(f) \exp(-2\pi jf\tau)$
translation	$a(t) \exp(2\pi j\phi t)$	$A(f - \phi)$
normalization	$a(t/T)$	$|T|\, A(fT)$
normalization	$|B|\, a(tB)$	$A(f/B)$
convolution	$a(t) \otimes b(t)$	$A(f)\, B(f)$
convolution	$a(t)\, b(t)$	$A(f) \otimes B(f)$
	$\delta(t)$	1

rect $(t) \equiv$ $\qquad$ sinc $(f) \equiv \dfrac{\sin \pi f}{\pi f}$

tri $(t) \equiv$ $\qquad$ sinc^2 (f)

$\exp(-\pi t^2)$ $\qquad\qquad\qquad$ $\exp(-\pi f^2)$
$t \exp(-\pi t^2)$ $\qquad\qquad\qquad$ $-jf \exp(-\pi f^2)$

Also,

$$A(f) \otimes B(f) = \int_{-\infty}^{\infty} A(f_d)B(f - f_d) \, df_d$$
$$= \mathcal{F}[a(t)b(t)].$$

Parsaval's Theorem equates the energies of a function and its transform:

$$\text{Energy, } E = \int_{-\infty}^{\infty} |a(t)|^2 \, dt = \int_{-\infty}^{\infty} |A(f)|^2 \, df.$$

When $a(t)$ is a modulation waveform on a carrier, the integrals equal $2E$. Additional operations and some common transform pairs are given in Table B.1 taken from Woodward.

Voltage Waveform and Amplitude Spectrum

The Fourier transform and its inverse relate the signal voltage waveform, $a(t)$, to its amplitude spectrum, $A(f)$.

The functions in App. A are analogous to low-pass waveforms. This is not a limitation, since most band-pass waveforms can be represented by a modulated carrier. The modulation is the interesting feature of the waveform, and this is generally a complex low-pass waveform.

Bandwidth describes the frequency extent of the voltage spectrum and can be defined in a variety of ways. Some of the more common are listed here:

3-db bandwidth,

$$B_{3a} \equiv \text{Width between 3-db points (in Hz) of } A(f),$$

Noise bandwidth,

$$B_{na} \equiv \frac{\int_{-\infty}^{\infty} |A(f)|^2 \, df}{|A_m|^2} \qquad \text{(Hz)}$$

$$= \frac{\int_{-\infty}^{\infty} |a(t)|^2 \, dt}{\left| \int_{-\infty}^{\infty} a(t) \, dt \right|^2} \qquad \text{(Hz)},$$

Rms bandwidth,

$$\beta_a \equiv \left[\frac{\int_{-\infty}^{\infty} (2\pi f)^2 |A(f)|^2 \, df}{\int_{-\infty}^{\infty} |A(f)|^2 \, df} \right]^{1/2} \qquad \text{(Hz)},$$

Band limit,

$$B_a \equiv \text{width of a limited band (no energy outside the interval } B_a) \qquad \text{(Hz)}.$$

Corresponding measures of waveform duration are:

3-db duration,

$$\tau_{3a} \equiv \text{width between 3-db points of } |a(t)| \quad \text{(sec)},$$

Effective (noise) duration,

$$\tau_n \equiv \frac{\displaystyle\int_{-\infty}^{\infty} |a(t)|^2 \, dt}{|a_m|^2} \quad \text{(sec)}$$

$$= \frac{\displaystyle\int_{-\infty}^{\infty} |A(f)|^2 \, df}{\left|\displaystyle\int_{-\infty}^{\infty} A(f) \, df\right|^2} \quad \text{(sec)},$$

Rms duration,

$$\alpha_a \equiv \left[\frac{\displaystyle\int_{-\infty}^{\infty} (2\pi t)^2 \, |a(t)|^2 \, dt}{\displaystyle\int_{-\infty}^{\infty} |a(t)|^2 \, dt} \right]^{1/2} \quad \text{(sec)},$$

Time limit,

$$\tau_a \equiv \text{Time interval beyond which the waveform is zero} \quad \text{(sec)}.$$

The above parameters, identified with the subscript h instead of a, also apply to filters where $a(t)$ is replaced by the filter's impulse response, $h(t)$, and $A(f)$ is replaced by the filter's frequency response, $H(f)$. If the filter is matched to the signal waveform (for maximum signal-to-noise ratio), $h(t) = a^*(-t)$ and $H(f) = A^*(f)$.

Autocorrelation Function and Energy Spectrum

The autocorrelation function, R_c, of a waveform is the integrated product of the waveform with itself delayed in time by an amount, t_d. For waveforms with finite energy,

$$R_c(t_d) = \int_{-\infty}^{\infty} a^*(t) \, a(t + t_d) \, dt \quad \text{(W-sec)}.$$

The energy spectrum, W, of a waveform is the squared magnitude of the voltage spectrum,

$$W(f_d) = |A(f_d)|^2 = A(f_d)A^*(f_d) \quad \left(\frac{\text{W-sec}}{\text{Hz}}\right),$$

where energy is equally divided between the positive and negative frequencies. These are related to each other through the Fourier transform

$$R_c(t_d) = \int_{-\infty}^{\infty} W(f_d)e^{+j2\pi f_d t_d}\, df_d = \mathcal{F}^{-1}[W(f_d)],$$

and

$$W(f_d) = \int_{-\infty}^{\infty} R_c(t_d)e^{-j2\pi f_d t_d}\, dt_d = \mathcal{F}[R_c(t_d)].$$

Notice that $R_c(t_d)$ and $W(f_d)$ are both real and even functions so that the Fourier cosine transform could be used (see, for example, Burdic, 1968).

Response Function

The response function, $\psi_o(t_d, f_d)$, describes the matched-filter output where the input waveform is displaced in time and frequency from the point to which the filter is matched. Woodward's ambiguity function is simply the squared magnitude of the response function.

The matched filter impulse response is $h(t) = a^*(-t)/C$, and the transfer function is $H(f) = A^*(f)/C$, where C is an arbitrary gain constant with the dimensions of $A(f)$. We can define $\psi_o(t_d, f_d)$ in terms of the waveform, $a(t)$, or its amplitude spectrum, $A(f)$. Some definitions are listed below:

$$\psi_o(t_d, f_d) = \frac{1}{C}\int_{-\infty}^{\infty} a^*(t)a(t - t_d)e^{j2\pi f_d t}\, dt \qquad \text{[see Eq. (1.11)]}$$

$$= \frac{1}{C}\int_{-\infty}^{\infty} A^*(f)A(f - f_d)e^{j2\pi f t_d}\, df \qquad \text{[see Eq. (1.10)]}$$

$$= \frac{1}{C}\iint_{-\infty}^{\infty} [A^*(-f)a(t)e^{j2\pi f t}]\, e^{j2\pi(-f t_d + f_d t)}\, df\, dt$$

$$= \frac{1}{C}\mathcal{F}^{-1}[A^*(-f)a(t)e^{j2\pi f t}]$$

$$= \frac{1}{C}\iint_{-\infty}^{\infty} A^*(f - f_d)a^*(-t - t_d)e^{-j2\pi f t}\, df\, dt$$

$$= \frac{1}{C}A(-f_d)a^*(-t_d)\otimes e^{j2\pi f_d t_d}.$$

Notice the symmetry between $A(f)$ and $a(t)$ in the above expressions.

Some properties of the response function are listed below:

$$|\psi_o(t_d, f_d)| \leq |\psi_o(0, 0)| = E \qquad \text{(energy in waveform)},\dagger$$

†When $a(t)$ is a modulation waveform on a carrier, $|\psi(0, 0)| = 2E$.

$$|\psi_o(t_d, f_d)| = |\psi_o(-t_d, -f_d)|,$$

$$\psi_o(t_d, 0) = R_c(t_d),$$

$$\int\int_{-\infty}^{\infty} |\psi_o(t_d, f_d)|^2 \, dt_d \, df_d = E^2 \qquad \text{(see footnote on p. 343).}$$

B.2 Analogies

Much of the data can be applied for waveform analysis by simply making the proper analogy. The antenna illumination function and pattern are related through the Fourier transform just as the voltage waveform and spectrum and the autocorrelation function and energy spectrum. In each case, one member of the Fourier transform pair usually has limited extent. The antenna is aperture-limited while

Table B.2

ANALOGIES

	Linear Antenna (Aperture-Limited)	Voltage Waveform (Time-Limited)	Amplitude Spectrum (Frequency-Limited)	Autocorrelation Function (Time-Limited)	Energy Spectrum (Frequency-Limited)
Width parameter	w (length)	τ (sec)	B (Hz)	τ (sec)	B (Hz)
Variables	x (length)	t (sec)	f (Hz)	t_d (sec)	f_d (Hz)
	$x' = x/w$	$t' = t/\tau$	$f' = f/B$	$t_d' = t_d/\tau$	$f_d' = f_d/B$
	$u\left(\dfrac{1}{\text{length}}\right)$	f (Hz)	t (sec)	f_d (Hz)	t_d (sec)
	$u' = wu$	$f' = f\tau$	$t' = tB$	$f_d' = f_d\tau$	$t_d' = t_d B$
Functions	$g(x)$	$a(t), h(t)$	$A(f), H(f)$	$R_c(t_d)$	$W(f_d)$
	$F(u)$	$A(f), H(f)$	$a(t), h(t)$	$W(f_d)$	$R_c(t_d)$
Parameters	C	$E \times \tau$	$E \times B$		
	$\dfrac{\Theta w}{\lambda}$	$\beta \tau$	αB		
	$\dfrac{\theta_3 w}{\lambda}$	$B_3 \tau$	$\tau_3 B$		
	$\dfrac{\theta_n w}{\lambda}$	$B_n \tau$	$\tau_n B$		
	$\mathscr{L}/w$	α/τ	β/B		

The various parameters may take on the subscripts a or h to distinguish waveform from filter parameters.

a waveform may be time-limited or frequency-limited. In either case, the limited function is analogous to the antenna illumination function.

A summary of the possible analogies between the antenna functions and the functions used in waveform analysis is given in Table B.2.

B.3 Examples

To illustrate how the data in App. A can be used in waveform analysis, we give here two example problems.

EXAMPLE 1: Find the amplitude spectrum, 3-db bandwidth, noise bandwidth, and rms bandwidth of a $\cos^3$ voltage pulse 2 μsec long at the base.

Solution: The $\cos^3$ voltage pulse is time-limited and, from Table B.2, is analogous to a linear even illumination function with $\tau = 2 \times 10^{-6}$ sec. A plot of the voltage pulse is obtained from the $\cos^3$ illumination function (Fig. A.7) by replacing x' with $t' = t\,(\text{sec})/2 \times 10^{-6}$, as shown in Fig. B.1.

A plot of the amplitude spectrum (in db amplitude vs. linear frequency) is found from the corresponding antenna pattern by replacing u' with $f' = 2 \times 10^{-6} f(\text{Hz})$. This is shown in Fig. B.2.

The various bandwidth measures are found in Table A.2.

$B_3\tau$ is analogous to $\theta_3 w/\lambda$ which equals 1.659 so that

$$B_3 = \frac{1.659}{2 \times 10^{-6}} = 829.5 \text{ kHz.}$$

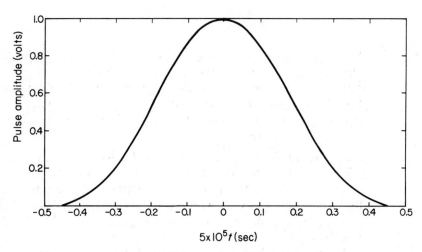

Fig. B.1   Voltage pulse of Example 1.

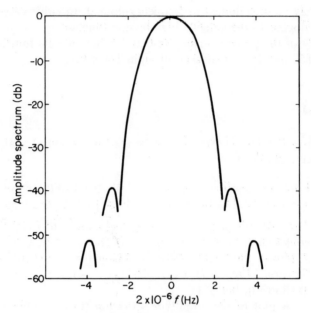

Fig. B.2   Amplitude spectrum of Example 1.

$B_n\tau$ is analogous to $\theta_n w/\lambda$ which equals 1.752 so that

$$B_n = \frac{1.752}{2 \times 10^{-6}} = 876 \text{ kHz.}$$

$\beta\tau$ is analogous to $\Theta w/\lambda$ which equals 4.23 so that

$$\beta = \frac{4.23}{2 \times 10^{-6}} = 2,115 \text{ kHz.}$$

EXAMPLE 2:   A radar transmits a coded pulse with a rectangular spectrum 10 MHz wide. In the receiver, the pulse phase coding is removed and the spectrum is weighted for low time sidelobes. The weighting filter has a Taylor 40 db, $\bar{n} = 6$ amplitude response. Find the shape of the output pulse, its rms bandwidth, and the signal-to-noise ratio loss in the weighting filter.

Solution: The pulse amplitude spectrum is band-limited and analogous to a linear even illumination function with w corresponding to B which is 10^7 Hz. A plot of the output pulse is obtained from the Taylor pattern (Fig. A.16) by replacing u' by Bt or $10^{+7}t$(sec). This is shown in Fig. B.3 on a relative voltage scale and in Fig. B.4 on a db scale.

From Table B.2, the rms bandwidth, β, is analogous to rms aperture width, $\mathscr{L}_s$. The data of Table A.7 for the linear-even Taylor function shows $\mathscr{L}_s/w$ to be 1.015 so that

$$\frac{\beta}{B} = 1.015,$$

and

$$\beta = 10.15 \text{ MHz.}$$

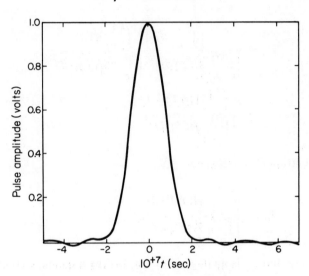

Fig. B.3 Voltage pulse of Example 2.

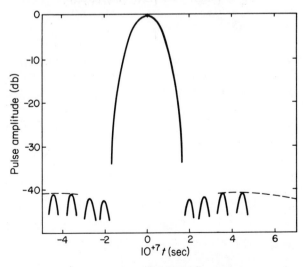

Fig. B.4 Voltage pulse of Example 2.

Signal-to-noise loss due to the weighting filter is found relative to the matched filter, which yields the maximum signal-to-noise ratio.† This is defined in terms of the weighting filter frequency response, $H(f)$, by

$$\text{Signal-to-noise loss, } L_m = \frac{(\text{signal-to-noise})_{\text{matched filter}}}{(\text{signal-to-noise})_{\text{weighted}}},$$

or

$$L_m = \frac{\left[\int_{-B/2}^{B/2} df\right]^2 \Big/ \int_{-B/2}^{B/2} df}{\left|\int_{-B/2}^{B/2} H(f)\, df\right|^2 \Big/ \int_{-B/2}^{B/2} |H(f)|^2\, df},$$

$$= \frac{B \int_{-B/2}^{B/2} |H(f)|^2\, df}{\left|\int_{-B/2}^{B/2} H(f)\, df\right|^2}.$$

Letting $f' = f/B$ yields

$$L_m = \frac{\int_{-1/2}^{1/2} |H(f')|^2\, df'}{\left|\int_{-1/2}^{1/2} H(f')\, df'\right|^2}.$$

From this, we see that L_m is analogous to $1/\eta_a$ for the antenna, so that in this case

$$L_m = (.763)^{-1} = 1.31, \text{ or } 1.178 \text{ db.}$$

---

†Signal power/noise power measured when output signal is at its peak.

Appendix C

Data Filtering and Smoothing

In this appendix, we apply the theory of linear filters to the special cases of smoothing and differentiation of data. The data, or signal, and the error, or noise, are assumed to have spectral components extending upwards from zero frequency, and bandwidths are given as "single-sided" values. Weighting functions are assumed to exist only for positive time delays (past input data).

Output errors in position and velocity data are described in terms of filter parameters and noise spectral density, and also as ratios of smoothed to unsmoothed error, and of smoothed error to a minimum error which would have been obtained with an optimized filter. This theory is not restricted to spatial coordinates of the target, but may be applied equally well to measurements of frequency and its derivatives, to signal amplitude, and to other measured quantities.

C.1 Data Smoothing

The curves and tables of App. A, with conversions to the time and frequency domain as in App. B, describe the properties of filters which are used to smooth or average radar output data. The major difference in application to smoothing is that the data and errors in any output coordinate are usually described in terms of real, low-frequency functions whose spectra occupy the positive-frequency region only. If x denotes the measured coordinate, the single-sided noise power spectrum $W(f)$ is expressed in (units of x)^2 per Hz. The output power (variance) from a filter whose transfer function is $H(f)$ is then

$$\sigma_x^2 = \int_0^\infty W(f)\,|H(f)|^2\,df = \int_0^{B_s} W(f)\,|H(f)|^2\,df, \qquad \text{(C.1)}$$

where B_s is the upper limit of filter response and σ_x is the standard deviation in units of x. In many cases, the noise density remains essentially constant over the filter bandwidth, and "white noise" relationships may be applied:

$$W(f) \cong W_n \text{ in the region } 0 < f < B_s,$$

$$\sigma_x^2 \cong W_n \int_0^{B_s} |H(f)|^2 \, df = W_n \beta_n. \tag{C.2}$$

Here, β_n is the single-sided noise bandwidth of the data filter.†

For example, in pulsed radar where a measurement sample is taken on each pulse and held until the next pulse arrives t_p seconds later, the error spectrum is of the form

$$W(f) = W_0 \left[\frac{\sin \pi f t_p}{\pi f t_p} \right]^2 \quad f \geq 0.$$

(See Fig. C.1.) The zero-frequency density W_0 is related to the single-pulse error σ_1 by

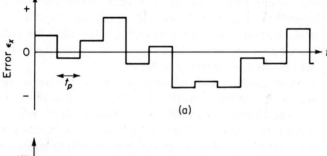

(a)

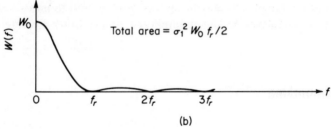

(b)

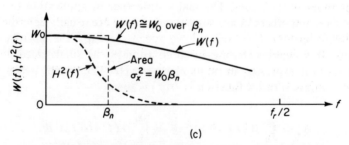

(c)

Fig. C.1   Pulsed data after sample-and-hold operation: (a) typical error waveform; (b) power spectrum of error; (c) low-frequency spectrum and response of filter.

†Referring to the filter bandwidth used in Chap. 3, $B_{nh} = 2\beta_n$.

$$\sigma_1^2 = W_0 \frac{1}{2t_p} = \frac{W_0 f_r}{2}, \tag{C.3}$$

where f_r is the pulse repetition frequency. When the filter bandwidth $B_s \ll f_r$, we may consider $W(f)$ to be uniform noise of density $W_n = W_0$ over B_s. Then we have

$$\sigma_x^2 = W_0 \beta_n = \sigma_1^2 \frac{2\beta_n}{f_r} = \frac{\sigma_1^2}{n_e}. \tag{C.4}$$

The effective number of pulses averaged by the filter is

$$n_e \equiv \frac{f_r}{2\beta_n} = f_r t_o. \tag{C.5}$$

This is the number of pulses which, averaged with equal weighting, would give the same noise output as is obtained with the actual filter. The effective averaging time of the filter is

$$t_o \equiv \frac{1}{2\beta_n} = \left[2 \int_0^\infty |H(f)|^2 \, df \right]^{-1}. \tag{C.6}$$

The minimum noise error for continuous data extending over a finite time interval t_s is obtained when all data are given equal weight in the filter:

$$h(t) = \frac{1}{t_s} \qquad 0 < t < t_s,$$

$$H(f) = \frac{\sin \pi f t_s}{\pi f t_s}.$$

For this case, $t_o = t_s$. In other cases, the ratio t_s/t_o or $2\beta_n t_s$ will be used as a measure of the extent to which the actual error variance exceeds the minimum value obtainable with any fixed smoothing interval t_s, assuming that the input noise is white. In App. B, where two-sided bandwidths are used, this same ratio is given by $B_n \tau$, which is analogous to $\theta_n w/\lambda$ in the antenna case (App. A). The performance of the filter is best, using this criterion, when $t_s/t_o = 1$, and values of this ratio will usually vary between 1 and 2 for practical, finite-memory weighting functions.

Another important characteristic of a data filter is the time delay or lag, t_d, introduced by the filter in slowly varying data. This delay is simply the median point of the weighting function, determined by

$$\int_0^{t_d} h(t) \, dt = \int_{t_d}^\infty h(t) \, dt. \tag{C.7}$$

All the filter functions described in App. A and B were symmetrical, and hence (for the even functions) $t_d = t_s/2$. In data filters, an important class of exceptions to this condition is represented by the RC (exponential) smoothing networks, often found in DC analog systems. These filters are asymmetrical and have weighting functions of infinite duration (see Fig. C.2 for typical examples).

Table C.1 lists the performance figures for several typical filter weighting functions. Values for other functions may be found from App. A, using the filter-antenna analogies of App. B. In the table, we have normalized the parameters to the delay time, t_d, of the filter. This permits comparison between the finite-memory and the infinite-memory types of filter. In two cases, the Gaussian and the $(\sin x)/x$ filters, both delay and memory are infinite, and only bandwidth ratios are compared.

The data bandwidth of a smoothing filter can be described by the one-sided, 3-db bandwidth, B_{3s}. Obviously, in the presence of white noise, it is desirable to have the ratio B_{3s}/β_n as large as possible, although it can never reach unity in a realizable filter.

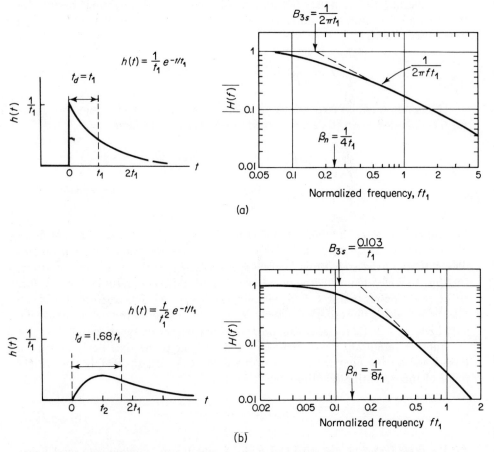

Fig. C.2 Exponential smoothing filter functions: (a) single exponential weighting; (b) double (cascaded) exponential weighting.

Table C.1

PARAMETERS OF DATA SMOOTHING FILTERS

Shape of Weighting Function	$t_s h(t)$, or $t_1 h(t)$ ($t \geq 0$, $t' = t - t_d$)	Delay t_d	$t_s/t_o = 2\beta_n t_s$	$\beta_n t_d$	$B_{3s} t_d$	B_{3s}/β_n	B_{20}/B_3 (at −20 db)
Rectangular	Unity over t_s	$t_s/2$	1.000	0.250	0.221	0.886	5.87*
Triangular	$1 - |2t'/t_s|$	$t_s/2$	1.333	0.333	0.319	0.957	2.3
Parabola on pedestal	$1 - 2t'^2/t_s^2$	$t_s/2$	1.033	0.258	0.243	0.941	3.7*
Parabolic	$1 - 4t'^2/t_s^2$	$t_s/2$	1.200	0.300	0.289	0.954	2.63
Cosine	$\cos \pi f t'/t_s$	$t_s/2$	1.246	0.312	0.297	0.955	2.72
Cosine-squared	$\cos^2 \pi f t'/t_s$	$t_s/2$	1.515	0.379	0.380	0.953	2.28
Cosine⁴	$\cos^4 \pi f t'/t_s$	$t_s/2$	1.964	0.491	0.463	0.944	2.28
Exponential	$\exp(-t/t_1)$	t_1	∞	0.250	0.159	0.637	10.
Cascaded exponential	$(t/t_1)\exp(-t/t_1)$	$1.68t_1$	∞	0.210	0.173	0.824	4.66
Gaussian	$\exp(-t^2/2\sigma_t^2)$	∞	∞	∞	∞	0.940	2.56
$(\sin x)/x$	$(\sin \pi B_s t')/\pi B_s t'$	∞	∞	∞	∞	1.000	1.00
Analogous quantity from App. A	Illumination function, $g(x)$	$w/2\lambda$	$\theta_n w/\lambda$	$\theta_n w/4\lambda$	$\theta_3 w/4\lambda$	θ_3/θ_n	θ_{20}/θ_3

*The 20-db width of these filters is determined by sidelobes which are within 20 db of the zero-frequency response.

As with antennas, there are cases where filters with large sidelobes are undesirable. Large noise components may be present at frequencies well above the desired passband, and a rapid fall-off in response may be more important than the minimization of the white-noise component. The rate of fall-off in the frequency domain depends upon the nature of the discontinuities in the weighting function. If one or more discontinuities are present in $h(t)$, the envelope of the transfer function $H(f)$ will vary inversely with f beyond some critical value (6 db per octave fall-off). If the discontinuity is in the first derivative of $h(t)$, the envelope of $H(f)$ will vary inversely with f^2 (12 db per octave), and so forth for each higher order of derivative. For example, the cosine weighting function has a 12 db per octave fall-off, the $\cos^2$ function an 18 db per octave fall-off, etc. Of the functions shown in Table C.1, only the rectangular and the parabolic-on-a-pedestal have sidelobes within 20 db of the maximum response.

Tracking Servo Performance

When the radar output filter takes the form of a closed-loop servo, its filter weighting and transfer functions can be extremely complex. Most servo systems, however, can be characterized with acceptable accuracy by a few measured or calculated parameters which are similar to those used above for passive filters:

B_{3_s}, the 3-db, closed-loop servo bandwidth, single-sided (Hz);

β_n, the equivalent noise bandwidth, single-sided (Hz);

K_v, the velocity error constant; and

K_a, the acceleration error constant.

These are related to each other and to time delay in the following ways:

$t_d = 1/K_v$, for low-frequency signals;

$t_d' \cong 1/\beta_n$, for transients and high-frequency signals;

$K_a \cong 2.5 \, \beta_n^2 \cong 6B_{3_s}^2$.

Further relationships between time constants, bandwidths, and error constants may be found in the literature of control systems theory, or in summary form in Chap. 9 of Barton (1964). The two time delays, t_d and t_d', are necessary to describe the usual, complex servo response in its low-frequency and high-frequency regions. The inclusion of an integrator or of equivalent compensation within the servo amplifier can increase K_v to any arbitrary figure (even to infinity), but at the expense of long settling times. The transient response is better described by the "rise time" t_d', which is determined by servo bandwidth.

C.2 Differentiation of Data

Filters used for obtaining derivatives of radar data in any coordinate will have characteristics similar to those of the difference channels used for measurement. The differentiating filter functions have odd symmetry, and may be described in terms of a smoothing filter with even functions $H(f)$ and $h(t)$, in cascade with an ideal differentiator. The time delay and 3-db bandwidth for velocity data are those of the smoothing filter.

A simple differentiator weighting function consists of a pair of impulses of opposite polarity, separated by t_s in time [Fig. C.3(a)]. The transfer function in the

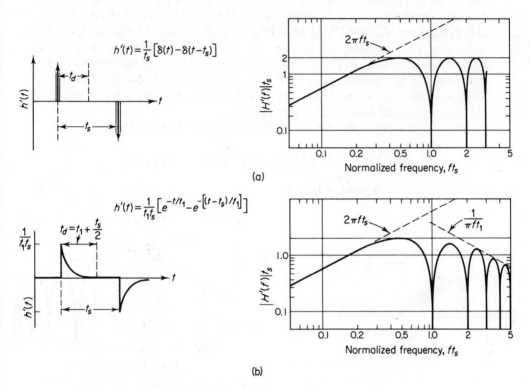

Fig. C.3 Simple differentiating filter functions: (a) two-point differentiating function; (b) smoothed two-point differentiating function.

low-frequency region ($f t_s \ll 1$) is that of the ideal differentiator, $H(f) = j2\pi f$. Above about $f t_s = 0.5$, the response follows a pattern of uniformly spaced lobes of amplitude two, extending to infinite frequency. The weighting function will be recognized as the derivative of the rectangular smoothing filter, and the transfer function is $j2\pi f$ times the $(\sin x)/x$ response of the rectangular smoothing filter. In practice, the two sampling impulses will be stretched by the limited response of portions of the measurement system [Fig. C.3(b)], and the high-frequency response will begin to fall off above some frequency $f t_s > 1$. If we represent the error in x, as smoothed over the sampling pulse, as σ_1, and assume this error to be independent over the interval t_s, the velocity error of the simple, two-point differentiator will be

$$\sigma_{x2} = \frac{\sqrt{2}\,\sigma_1}{t_s} = \frac{\sigma_1}{\sqrt{2}\,t_d}. \tag{C.8}$$

The velocity data will be delayed by $t_s/2$ from the input through the filter, for signal frequencies within the bandpass of the differentiator.

Consider, now, the linear-odd weighting function, which is optimum for estimating the first derivative within a restricted time period t_s, with white noise and no high-order derivatives in the input signal. This function [Fig. C.4(a)] is equivalent to the parabolic smoothing filter [Fig. C.4(b)] in cascade with an ideal differentiator. The time delay in following a slow change in velocity is again $t_d = t_s/2$. The 3-db bandwidth is $B_{3s} = 0.289/t_d$, indicating that velocity fluctuations at this frequency will be reproduced at 0.707 times their actual amplitudes. Note, from Table C.1, that this response extends to almost twice the frequency of either of the two exponential filters, with a given delay.

The noise output of the differentiating filter, for white noise input, is

$$\sigma_{\dot{x}}^2 = W_n \int_0^{B_s} |H'(f)|^2 \, df = W_n \int_0^{B_s} (2\pi f)^2 \, |H(f)|^2 \, df$$
$$= W_n \beta_n \beta_h^2 = (\sigma_x \beta_h)^2. \tag{C.9}$$

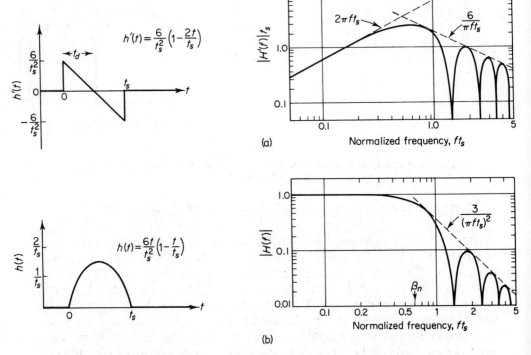

Fig. C.4 Optimum differentiator function: (a) linear-odd weighting; (b) equivalent smoothing functions.

The parameter β_h is the rms bandwidth, defined in Chap. 3, and σ_x is the rms error in position at the output of the equivalent smoothing filter. Table C.2 gives the bandwidth parameters for several types of filter, in terms of delay time.

If the input data consist of many individual pulses, each with rms error σ_1, the output error can be expressed as

$$\sigma_{\dot{x}} = \frac{\sigma_1 \beta_h}{\sqrt{n_e}} = \frac{\sigma_1 \beta_h}{\sqrt{f_r t_o}} = \sigma_1 \beta_h \sqrt{\frac{2\beta_n}{f_r}}. \tag{C.10}$$

For the optimum (linear-odd) filter in white noise, we can find from Tables C.1 and C.2 the following figures:

$$\beta_n t_d = 0.300, \qquad \beta_h t_d = 1.58.$$

Hence, using Eq. (C.9),

$$(\sigma_{\dot{x}})_{min} = \sqrt{\frac{3W_n}{4t_d^3}} = \frac{\sqrt{3}\,(\sigma_x)_{min}}{t_d} = \frac{\sqrt{12}\,(\sigma_x)_{min}}{t_s}, \tag{C.11}$$

where $(\sigma_x)_{min}$ is the error in x after smoothing with the rectangular filter over $t_s = 2t_d$ seconds. If the input data consist of $n = f_r t_s$ pulses ($n \gg 1$), with independent single-pulse position error σ_1, this expression can be combined with Eq. (C.3) to give

$$(\sigma_{\dot{x}})_{min} = \sigma_1 \sqrt{\frac{3}{2f_r t_d^3}} = \sigma_1 \sqrt{\frac{12}{f_r t_s^3}} = \frac{\sigma_1}{t_s} \sqrt{\frac{12}{n}}. \tag{C.12}$$

For filters other than the linear-odd, the factor $\sqrt{12/n}$ is replaced by $4\beta_h t_d \sqrt{\beta_n t_d}$, which is four times the quantity listed in the sixth column of Table C.2.

The velocity error of any filter may be compared to that of a simple, two-point differentiator, and the ratio of the two errors may be represented by the "velocity error ratio," C_v, defined as

$$C_v \equiv \frac{\sigma_{\dot{x}}}{\sigma_{\dot{x}2}} = \frac{\sqrt{2}\,t_d \sigma_{\dot{x}}}{\sigma_1}. \tag{C.13}$$

For the linear-odd filter on pulsed data,

$$C_v = \sqrt{\frac{3}{f_r t_d}} = \sqrt{\frac{6}{n}} \quad (n > 12). \tag{C.14}$$

For this same filter on white noise which has been prefiltered to a bandwidth B_{n1}, we have

$$\sigma_1 = \sqrt{W_n B_{n1}},$$

Table C.2

PARAMETERS OF DIFFERENTIATING FILTERS

(The differentiator weighting function is the first derivative of the smoothing filter $h(t)$ listed below.)

Shape of Smoothing Weighting Function	$t_s h(t)$, or $t_1 h(t)$ ($t \geq 0$, $t' = t - t_d$)	Delay t_d	$\beta_h t_d$	β_h/B_{3s}	$\beta_h t_d \sqrt{\beta_n t_d} = \sigma_{\dot{x}}\sqrt{t_d^3/W_n}$	$\beta_h\sqrt{\beta_n/B_{3s}} = \sigma_x/\sqrt{W_n B_{3s}^3}$
Rectangular	Unity over t_s	$t_s/2$	∞	∞	∞	∞
Triangular	$1 - \|2t'/t_s\|$	$t_s/2$	1.73	5.43	1.000	5.56
Parabola on pedestal	$1/2t'^2/t_{\frac{s}{2}}^2$	$t_s/2$	∞	∞	∞	∞
Parabolic	$1 - 4t'^2/t_s^2$	$t_s/2$	1.58	5.48	0.865	5.61
Cosine	$\cos \pi f t'/t_s$	$t_s/2$	1.57	5.30	0.876	5.44
Cosine-squared	$\cos^2 \pi f t'/t_s$	$t_s/2$	1.92	5.06	1.18	5.19
Cosine^4	$\cos^4 \pi f t'/t_s$	$t_s/2$	2.38	5.15	1.67	5.30
Exponential	$\exp(-t/t_1)$	t_1	∞	∞	∞	∞
Cascaded exponential	$(t/t_1)\exp(-t/t_1)$	$1.68t_1$	1.68	9.75	0.772	10.7
Gaussian	$\exp(-t'^2/2\sigma_t^2)$	∞	∞	2.67	∞	2.76
$(\sin x)/x$	$(\sin \pi B_s t')/\pi B_s t'$	∞	∞	3.62	∞	3.62
Analogous quantity from App. A	Illumination function, $g(x)$	$\dfrac{w}{2\lambda}$	$\dfrac{\Theta w}{2\lambda}$	$\dfrac{2\Theta}{\theta_3}$	$\dfrac{\Theta w}{4\lambda}\sqrt{\dfrac{\theta_n w}{\lambda}}$	$\dfrac{2\Theta}{\theta_3}\sqrt{\dfrac{\theta_n}{\theta_3}}$

and, for $B_{n1}t_d \gg 1$,

$$C_v = \sqrt{\frac{2}{B_{n1}t_d}} (\beta_h t_d \sqrt{\beta_n t_d}), \tag{C.15}$$

where the quantity in parentheses is the parameter listed in column six of Table C.2. The expression in Eq. (C.15) is applicable to all types of filter, and may be evaluated from the data in Table C.2 or from analogous terms in App. A.

Velocity Error for Nonwhite Noise

Integrals of the form of Eq. (C.9) for nonwhite noise can become quite complex, since the term $W(f)$ must be included within the integral. It is seldom necessary, however, to evaluate these integrals exactly, if only because the magnitude and the spectral characteristics of radar errors cannot be predicted accurately. Often, the error $\sigma_{\dot{x}2}$ for a two-point estimator, given in Eq. (C.8), yields adequate results when only the rms value σ_1 of individual position error samples and the smoothing interval t_s are known. To show how the error of the two-point estimate compares with the velocity error of the linear-odd filter, in three different noise cases, we have plotted in Fig. C.5 the velocity error ratio C_v, defined in Eq. (C.13). The following properties are shown:

(a) When the position noise error has a Markoffian spectrum,

$$W(f) = W_0 \frac{f_a^2}{f_a^2 + f^2} \qquad (f_a \text{ is half-power spectral width}),$$

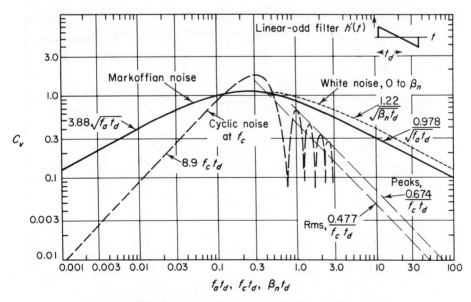

Fig. C.5 Velocity error ratio vs. normalized spectral parameter.

the velocity error is very close to $\sigma_{\dot{x}2}$ for f_a beween $0.03/t_d$ and $2/t_d$. The ratio C_v varies with the square root of $f_a t_d$ or $1/f_a t_d$ beyond this region. In calculating C_v for this noise spectrum, σ_1 is taken as the position error under the entire curve of $W(f)$.

(b) For white noise which has been passed through a filter of bandwidth $\beta_n \gg 1/t_d$, the ratio C_v varies as $(\beta_n t_d)^{-1/2}$. The curve would be the same as for Markoffian noise, but has been displaced upwards in normalized frequency by the factor $\pi/2$ which relates noise bandwidth to half-power bandwidth of the Markoffian spectrum.

(c) For sinusoidal errors of the form

$$\epsilon_x = A \sin 2\pi f_c t,$$

the rms position error is $\sigma_1 = 0.707A$. The C_v curve is simply the transfer function of the filter, multiplied by the factor $1.414t_d$. The worst velocity error occurs when the period of the error is twice the smoothing interval, or four times the delay. For error periods less than t_d (frequencies greater than $1/t_d$), the ratio C_v varies as $(f_c t_d)^{-1}$. The lobe and null structure in this region results from the sharp discontinuities in the idealized weighting function, and will not be present in practical filters. For most cases, then, the rms approximation shown in Fig. C.5 should be used to estimate error magnitudes.

Nonoptimum filters for differentiation of position data can also be characterized by a factor C_v, which will be near unity over a wide range of error spectral widths. Curves in Fig. C.6 show this ratio for a cascaded exponential differentiator with

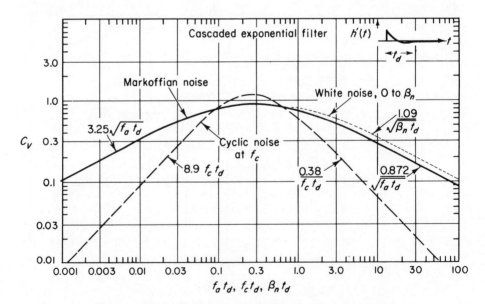

Fig. C.6 Velocity error ratio vs. normalized spectral parameter.

Markoffian, sinusoidal, and band-limited white noise. As before, the filter performance stays close to that of the two-point estimator of equal delay as long as the frequency parameters of the noise are within two or three octaves of $0.25/t_d$. The lobing structure which characterized the finite-memory filter is absent here, but the curves are otherwise almost indistinguishable from the previous case.

Time Lag and Bandwidth

Since each differentiating filter is equivalent to an ideal differentiator in cascade with a smoothing filter, the delay-bandwidth relationships can be obtained from Table C.1. Under conditions of constant acceleration, the velocity output data will lag the input by the delay t_d. From Table C.2, we find that the linear-odd differentiator (derivative of the parabolic) has the smallest noise output of the finite-memory filters for a given delay and input noise density. The derivative of the cosine weighting function is nearly as good in every respect, and the infinite-memory cascaded-exponential filter is somewhat better than either parabolic or cosine, except that its data bandwidth is only about half as great, for equal delay.

C.3 Smoothing and Estimation of Polynomial Signals

In certain cases, it is convenient to represent a small segment of the signal or data in the form of a polynomial in time t, where the coefficients represent the velocity and higher-order derivatives of target motion:

$$x(t) = x_0 + \dot{x}_0 t + \tfrac{1}{2}\ddot{x}_0 t^2 + \tfrac{1}{6}\dddot{x}_0 t^3 + \cdots . \tag{C.16}$$

The coefficients are to be estimated on the basis of data gathered over a period t_s seconds, which may be centered at the time for which the estimate is made, or may end at that time. The optimum weighting functions are listed in Table C.3 and illustrated in Fig. C.7 for the case of white noise. The output noise from an optimum estimator will vary with the degree of the polynomial to which the data are being fitted, as shown in Table C.4. The higher the degree, the greater the output noise for each estimate. The increases are especially marked when the estimate is updated to the end of the interval, because the updating process requires that the next higher derivative be estimated and added into the output. For example, the velocity error for the first-degree case (constant velocity) is the same as given earlier in Eq. (C.11) for the linear-odd differentiator. When an endpoint estimate of velocity is to be made with constant acceleration, however, the rms error is larger by a factor of four.

The advantage of high-order curve fitting and endpoint estimation, of course, is that these procedures reduce or eliminate lag errors which otherwise would arise in the smoothing process. This compromise between noise and lag errors was discussed by Nesline (1961), and is another example of the need to choose an optimum system bandwidth on the basis of expected target dynamics (Barton, 1964, pp. 306–310).

Table C.3

WEIGHTING FUNCTIONS FOR ESTIMATION OF POLYNOMIALS

(from Nesline)

Type of Estimate	*Degree of Polynomial*	*Weighting Function* $(u = t/t_s, 0 \leq u \leq 1)$
Midpoint Estimate		
Position	0, 1	$t_s h = 1$
Position	2, 3	$t_s h = -1.5(1 - 10u + 10u^2)$
Velocity	1, 2	$t_s^2 h' = 6(1 - 2u)$
Velocity	3	$t_s^2 h' = -15(1 - 16u + 42u^2 - 28u^3)$
Acceleration	2, 3	$t_s^3 h'' = 60(1 - 6u + 6u^2)$
Endpoint Estimate		
Position	0	$t_s h = 1$
Position	1	$t_s h = 4(1 - \frac{3}{2}u)$
Position	2	$t_s h = 9(1 - 4u + \frac{10}{3}u^2)$
Position	3	$t_s h = 16(1 - \frac{15}{2}u + 15u^2 - \frac{35}{4}u^3)$
Velocity	1	$t_s^2 h' = 6(1 - 2u)$
Velocity	2	$t_s^2 h' = 36(1 - \frac{16}{3}u + 5u^2)$
Velocity	3	$t_s^2 h' = 120(1 - 10u + \frac{45}{2}u^2 - 14u^3)$
Acceleration	2	$t_s^3 h'' = 60(1 - 6u + 6u^2)$
Acceleration	3	$t_s^3 h'' = 480(1 - \frac{45}{4}u + 27u^2 - \frac{35}{2}u^3)$

Equations for optimum smoothing times and minimum rms error are given in Table C.5, in terms of noise density W_n (single-sided) and magnitude of the time derivatives of x. As an example, assume that range measurements are made with a noise error of 3 m rms in a bandwidth of 10 Hz. The target has a radial acceleration of 2 g, or 20 m/sec², and a fourth derivative of 0.2 g/sec², or 2 m/sec⁴.

$$W_n = \frac{\sigma_r^2}{\beta_n} = \frac{9}{10} \text{ m}^2/\text{Hz.}$$

Using a second- or third-degree midpoint estimator,

$$\text{Optimum } t_s = 5.2(0.9/4)^{1/9} = 4.4 \text{ sec,}$$

$$\text{Minimum } \sigma_r = 0.491(0.9^4 \times 2)^{1/5} = 0.5 \text{ m.}$$

If the data were fitted to a first-degree polynomial, the optimum smoothing interval would be

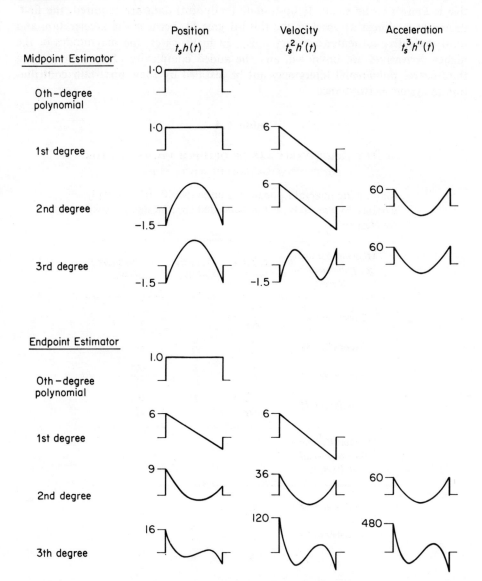

Fig. C.7 Weighting functions for polynomial signal (not to scale).

$$t_s = 2.36(0.9/400)^{1/5} = 0.71 \text{ sec,}$$

$$\sigma_r = 0.9 \text{ m.}$$

Thus, an improvement factor of about two in rms error can be achieved by use of a high-order polynomial smoothing process in this case, because the fourth derivative is known to be small. If up-to-date (endpoint) data are required, the first-degree fit is almost as good as the third-degree fit, because the acceleration and third derivative estimates are very noisy. In many cases, the magnitudes of the higher derivatives are unknown, and the added complexity of the second- and third-degree polynomial filters may not be justified by their uncertain contribution to system performance.

Table C.4

OUTPUT ERROR VARIANCES OF OPTIMUM ESTIMATORS FOR
POLYNOMIAL SIGNALS IN WHITE NOISE

(Smoothing interval t_s, one-sided noise density W_n; multiply variance by two if W_n is the two-sided spectral density used by Nesline.)

Midpoint Estimator for Polynomial of Degree:	*Position* σ_x^2/W_n	*Velocity* $\sigma_{\dot{x}}^2/W_n$	*Acceleration* $\sigma_{\ddot{x}}^2/W_n$
0 (constant x)	$\dfrac{1}{2t_s}$	–	–
1 (constant $\dot{x}$)	$\dfrac{1}{2t_s}$	$\dfrac{6}{t_s^3}$	–
2 (constant $\ddot{x}$)	$\dfrac{9}{8t_s}$	$\dfrac{6}{t_s^3}$	$\dfrac{360}{t_s^5}$
3 (constant $\dddot{x}$)	$\dfrac{9}{8t_s}$	$\dfrac{75}{2t_s^3}$	$\dfrac{360}{t_s^5}$

Endpoint Estimator for Polynomial of Degree:			
0 (constant x)	$\dfrac{1}{2t_s}$	–	–
1 (constant $\dot{x}$)	$\dfrac{2}{t_s}$	$\dfrac{6}{t_s^3}$	–
2 (constant $\ddot{x}$)	$\dfrac{9}{2t_s}$	$\dfrac{96}{t_s^3}$	$\dfrac{360}{t_s^5}$
3 (constant $\dddot{x}$)	$\dfrac{8}{t_s}$	$\dfrac{600}{t_s^3}$	$\dfrac{12960}{t_s^5}$

Table C.5

OPTIMUM SMOOTHING TIMES AND MINIMUM ERRORS FOR POLYNOMIAL SIGNALS IN WHITE NOISE

(from Nesline)

Quantity Estimated	Degree of Polynomial	Optimum t_s	Minimum Rms Error
Midpoint Estimates			
Position	0, 1	$2.36(W_n/\ddot{x}^2)^{1/5}$	$0.515(W_n^2\ddot{x})^{1/5}$
	2, 3	$5.2(W_n/\overline{x}^2)^{1/9}$	$0.491(W_n^4\overline{x})^{1/9}$
Velocity	1, 2	$4.4(W_n/\ddot{x}^2)^{1/7}$	$0.486(W_n^2\ddot{x}^3)^{1/7}$
Acceleration	2, 3	$4.83(W_n/\overline{x}^2)^{1/9}$	$0.557(W_n^2\overline{x}^5)^{1/9}$
Endpoint Estimates			
Position	0	$(W_n/\dot{x}^2)^{1/3}$	$0.855(W_n\dot{x})^{1/3}$
	1	$2.36(W_n/\ddot{x}^2)^{1/5}$	$1.02(W_n^2\ddot{x})^{1/5}$
	2	$4.74(W_n/\ddot{x}^2)^{1/7}$	$1.06(W_n^3\ddot{x})^{1/7}$
	3	$5.2(W_n/\overline{x}^2)^{1/9}$	$1.32(W_n^4\overline{x})^{1/9}$
Velocity	1	$2.05(W_n/\ddot{x}^2)^{1/5}$	$1.32(W_n\ddot{x}^3)^{1/5}$
	2	$4.4(W_n/\ddot{x}^2)^{1/7}$	$1.93(W_n^2\ddot{x}^3)^{1/7}$
	3	$5.06(W_n/\overline{x}^2)^{1/9}$	$2.65(W_n^3\overline{x}^3)^{1/9}$
Acceleration	2	$3.96(W_n/\ddot{x}^2)^{1/7}$	$1.92(W_n\ddot{x}^5)^{1/7}$
	3	$4.83(W_n/\overline{x}^2)^{1/9}$	$3.33(W_n^2\overline{x}^5)^{1/9}$

Appendix D

Atmospheric Propagation Errors

The previous sections of this handbook have discussed errors which depend on equipment design and target characteristics. In this appendix, we summarize the effects of the earth's atmosphere on measurements of range, angle, and velocity. The regular (and largely predictable) effects of the troposphere are described first, on the basis of the "Exponential Reference Atmosphere" put forth as a model by the National Bureau of Standards (Bean and Thayer, 1959). The second section discusses tropospheric fluctuations in some detail, using the "frozen atmosphere" model, in which the fluctuations are assumed to result from a random pattern of refractivity variations fixed in the air mass and drifting across the measurement path with the average wind velocity. While not an accurate description of the troposphere, this model has been found to yield predictions of error which are far better than the "exponential correlation" model (Muchmore and Wheelon, 1955). Finally, the influence of the ionosphere is summarized, on the basis of rather scanty knowledge of its average densities and variations with height and sunspot season.

D.1 Tropospheric Refraction

Exponential Reference Atmosphere

The refractive index of the troposphere, for frequencies below about 20 GHz, can be expressed by using the "Smith-Weintraub constants" in the equation

$$N = (n - 1) \times 10^6 = \frac{77.6}{T}\left(P + \frac{4810p}{T}\right), \qquad (D.1)$$

where T is the temperature in degrees Kelvin, P is the total pressure in millibars, p is the partial pressure of the water-vapor component, n is the refractive index, and N is a scaled-up value known as the "refractivity." At sea level, N is typically between 300 and 350, with mean and extreme height profiles as shown in Fig. D.1. The approximate straight-line relationship of this data on the log-linear plot suggests the use of an exponential form for a mathematical model, and the following has been used for mean conditions:

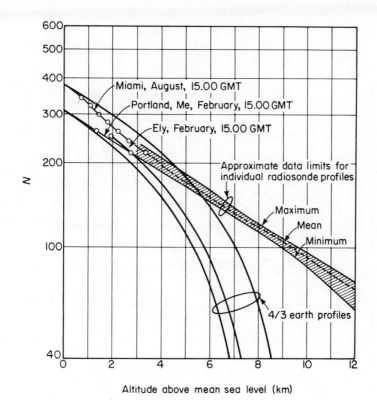

Fig. D.1 Refractivity vs. altitude for mean and extreme conditions, compared to 4/3 earth radius profiles (Bean and Thayer, 1959).

$$N(h) = 313.0 \exp(0.14386h), \tag{D.2}$$

with h in kilometers above mean sea level. The mean value $N(0) = N_o = 313.0$ will be used in the following discussions of error, and all error values may be scaled upwards or downwards in accordance with actual surface refractivity.

Range and Elevation Bias Errors

Refraction of the radar wave in the troposphere causes an extra time delay in transmission of the signal, and an increase in the elevation angle measured by the antenna system. The ray actually follows the path of minimum delay to reach the target and return, and this carries it above the straight geometrical line between radar and target. Figures D.2 and D.3 show the range and elevation angle errors for the exponential reference atmosphere with sea level refractivity $N_o = 313$. The worst error in range is about 100 m for a horizontal ray. The error in elevation angle is slightly less than 1 deg for a ray leaving the radar horizontally and passing through the entire atmosphere. In all cases, the errors are directly proportional

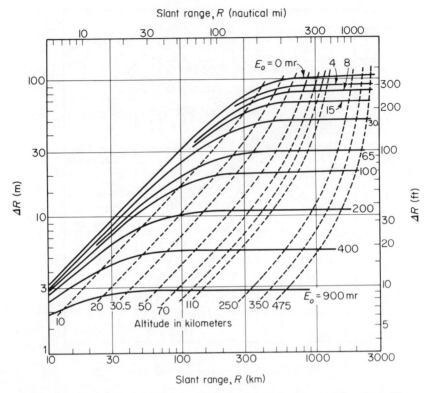

Fig. D.2 Range bias vs. range for CRPL exponential reference atmosphere ($N_o =$ 313).

to surface refractivity N_s, whether this changes as a result of local weather or operation of the radar above sea level.

For targets which lie well outside the troposphere, the errors in range and elevation angle are seen to level off at values which depend upon elevation angle but not on range. These errors can be approximated, above $E = 5$ deg, by the following simple expressions:

$$\Delta R = 0.007 \, N_s \, \text{csc} \, E_o \quad \text{(meters)}, \tag{D.3}$$

$$\Delta E_o = N_s \cot E_o \quad (\mu\text{rad}). \tag{D.4}$$

Equation (D.4) can be modified for lower angles, using the constants a and b shown in Fig. D.4 as in the following equation:

$$\Delta E_o = b N_s + a \quad (\mu\text{rad}). \tag{D.5}$$

The actual refraction error will vary from these average values by a few percent,

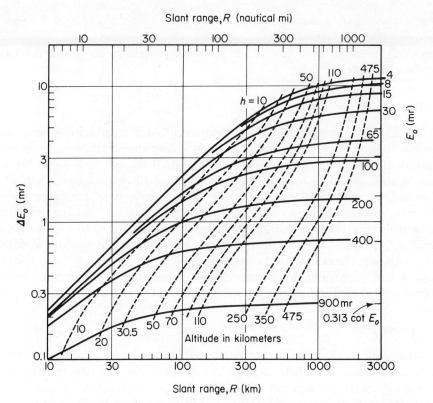

Fig. D.3 Tracker elevation angle bias vs. range for CRPL exponential reference atmosphere ($N_o = 313$).

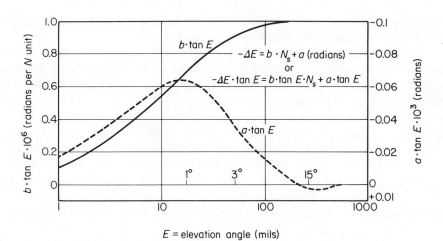

Fig. D.4 Refraction correction constants for radio astronomy case (after Bean and Cahoon, 1957).

depending upon the departure of the tropospheric refractivity profile from the exponential model.

Variations in Refractivity

Results of a large number of observations of surface refractivity have been published (Bean, Horn, and Ozanich, 1960). These show variations in N_s as a function of time of day and of season, with diurnal changes of 20 to 40 N-units peak-to-peak, added to seasonal changes of about 10 N-units rms. Thus, at a given site, the refractivity may change by 100 N-units or more during the course of a year, affecting the scale of the errors plotted in Figs. D.2 and D.3. At angles below 0.5 deg, extreme refraction effects are sometimes observed as a result of "ducting," in which a temperature inversion causes a reversal in the slope of refractivity vs height. Rays leaving the surface below some critical angle can then be trapped and propagated for considerable distances around the earth, leading to large and unpredictable errors in measurement of low-altitude target position.

Velocity Error

Measurements of target velocity are affected by the average tropospheric refractivity in two ways. First, those targets which lie within the troposphere will exhibit a change in apparent range (time delay) which is proportional to the local refractive index n_t at the target. The Doppler shift is increased by this same factor:

$$\left. \begin{array}{l} f_d = -\dfrac{2f_o\dot{R}}{c_t} = -\dfrac{2\dot{R}n_t}{\lambda} \\[2mm] \Delta f_d = f_d(n_t - 1) = f_d N_t \times 10^{-6} \end{array} \right\} \quad \text{(first-order term).} \qquad (D.6)$$

Here, $c_t = c_o/n_t$ is the velocity of light in the medium which surrounds the target, and λ is the vacuum wavelength of the transmission.

A second effect, often larger than the first, was described by Millman (1958). Because of ray bending, the radar signal arrives at the target from a direction slightly above the straight-line (geometrical) path, as shown in Fig. D.5. The signal becomes sensitive to an apparent velocity vector of the target, equal in amplitude to the true vector but rotated by an angle $\Delta\alpha_t$ with respect to the true vector. Since $\Delta\alpha_t$ is small, the contribution of the true radial component of velocity is essentially unchanged, but error is added by the presence of a tangential component v_t. The rotation of apparent velocity through the angle $\Delta\alpha_t$ changes the observed radial velocity by an amount $\Delta v = v_t \sin \Delta\alpha_t \cong v_t \Delta\alpha_t$. Average values of $\Delta\alpha_t$ may reach 2 to 4 mr for observations near zero elevation, and hence this error may be appreciable there. An approximation for $\Delta\alpha_t$ can be found from the tropospheric bias error ΔE_o, given in Fig. D.3:

$$\Delta\alpha_t \cong \Delta E_o \frac{L}{R}, \qquad (D.7)$$

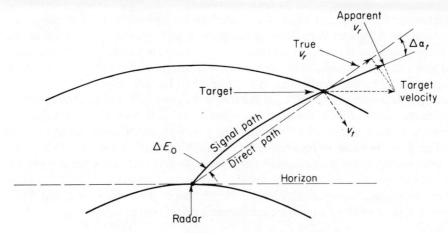

Fig. D.5   Effect of ray bending on Doppler measurement.

where L is the length of the tropospheric portion of the path (below approximately 5 km, see Fig. D.8 below). From this,

$$\sigma_\alpha \cong \sigma_E \frac{L}{R},$$

$$\sigma_{vE} \cong v_t \sigma_E \frac{L}{R} = \dot{E}_t L \sigma_E, \tag{D.8}$$

where σ_E is the unknown or uncorrected portion of ΔE_o and $\dot{E}_t$ is the elevation rate of the target. This is an error in radial (Doppler) velocity measurement, which applies even to those systems which attempt to ignore angular refraction by combining range and Doppler measurements from multiple sites. Even when accurate corrections have reduced σ_E to 0.1 mr, the Doppler error can be significant for high-velocity targets.

An expression similar to Eq. (D.8) can be written for the azimuth coordinate, in which an error σ_A describes the tropospheric bias error in azimuth and $\dot{A}$ is azimuth rate:

$$\sigma_{vA} \cong \dot{A} L \sigma_A. \tag{D.9}$$

Azimuth bias in the order of 0.1 mr can result from gradients in the horizontal structure of refractive index, which are seldom predictable or correctable.

D.2 Tropospheric Fluctuations

Range Fluctuation Spectra

Radar measurements are affected by the irregular refractivity fluctuations in the troposphere, as well as by the bias errors discussed above. Of the many ways in which tropospheric fluctuations can be described, the model which is most useful for prediction of errors is based on the power spectrum of range fluctuations. Such spectra have been measured over prolonged periods by Thompson and his associates at the National Bureau of Standards in Colorado. Both continental and maritime atmospheres were measured, using radio paths from the lowest local surfaces to high mountains in Colorado and Hawaii. Figure D.6 shows the resulting data over nine decades of the frequency spectrum. Below about 10^{-4} Hz, the range spectra follow the spectra of surface refractive index fluctuations, and these have been measured at various sites over periods of many years (N_s data in Fig. D.6). In order to equate the range and N_s spectral densities in this region, we express the range fluctuation in parts per million (ppm) of the pathlength, and the power spectral density in (ppm)^2/Hz. The scale at the right side of the figure gives the corresponding density in cm^2/Hz for tropospheric paths of about 30-km length, typical of those over which the measurements were made.

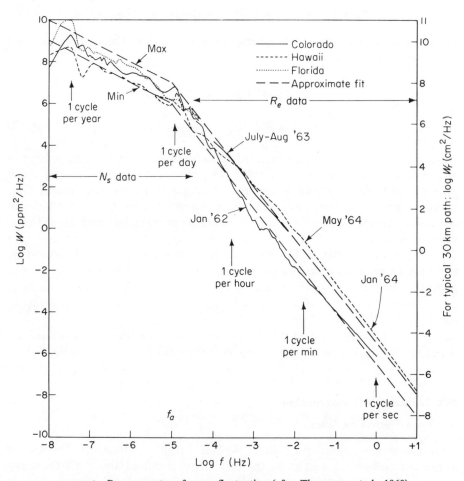

Fig. D.6   Power spectra of range fluctuation (after Thompson et al., 1960).

The data shown in Fig. D.6 were taken at elevation angles of three and seven degrees, on paths which extended from the surface to altitudes of two and three kilometers. The great variability of the atmospheric conditions tends to mask any effects of elevation angle or altitude above sea level, but the lines labeled "Max" and "Min" can be taken as representative of the spread of conditions encountered over most paths anywhere in the world. Unusually "quiet" conditions might be expected to fall another factor of 10 in power density below the "Min" curve, and some disturbances may also appear ten times as intense as the "Max" curve shown.

In the computations to follow, we will use a mean spectrum midway between the "Min" and "Max." For the two experimental paths, this corresponds to

$$\overline{W}(f) = 32f^{-1} \quad (10^{-8} < f < 10^{-5} \text{ Hz}),$$
$$\overline{W}(f) = 10^{-6} C_1 f^{-2.5} \quad (f > 10^{-5} \text{ Hz}). \tag{D.10}$$

Here, C_1 is a dimensional constant equal to 1.0 $(\text{ppm})^2\text{-Hz}^{1.5}$, included to express W in $(\text{ppm})^2/\text{Hz}$. In the region below 10^{-8} Hz, the spectral densities tend to fall off, with negligible contribution to total range fluctuation. The total fluctuation under the mean spectrum is

$$\frac{\overline{\sigma_r}}{L} \times 10^6 = \left[\int_{10^{-8}}^{\infty} \overline{W}(f) \, df\right]^{1/2} = 15 \text{ ppm}. \tag{D.11}$$

For $L = 30$ km, this gives $\overline{\sigma_r} = 0.45$ m over a period of many months. Most of the power lies below the break frequency $f_a = 10^{-5}$ Hz in the spectrum (roughly one cycle per day).

Effect of Tropospheric Pathlength

Spectral density in the region below 3×10^{-5} Hz is independent of pathlength, when expressed in $(\text{ppm})^2/\text{Hz}$. This is because these long-term components are governed by diurnal and seasonal effects on N_s which apply simultaneously along the entire path, adding directly to cause a fluctuation amplitude proportional to L. Above this frequency, the spectral density will vary with the pathlength as shown in Fig. D.7. Here we have made a rather arbitrary separation of the spectrum into long-term and drift components, the latter describing refractivity variations which are assumed "frozen" into the air mass, which drifts across the path with a mean velocity v_a. The top scale indicates the spatial wavelength of these fluctuations, based on an assumed velocity $v_a = 3$ m/sec for the air mass during the measurement programs. Although the drift component contains wavelengths as long as 10^6 m, the waves beyond 10^5 m tend to be masked by diurnal effects. The drift component of refractivity is well described by a spectral density

$$W_N(f) = \frac{\sqrt{3v_a}}{300} f^{-1.5} \quad (\text{ppm})^2/\text{Hz} \quad (f > 10^{-5} \text{ Hz}) \tag{D.12}$$

$$= 10^{-2}f^{-1.5} \quad (v_a = 3 \text{ m/sec}, \ f > 10^{-5} \text{ Hz}). \tag{D.13}$$

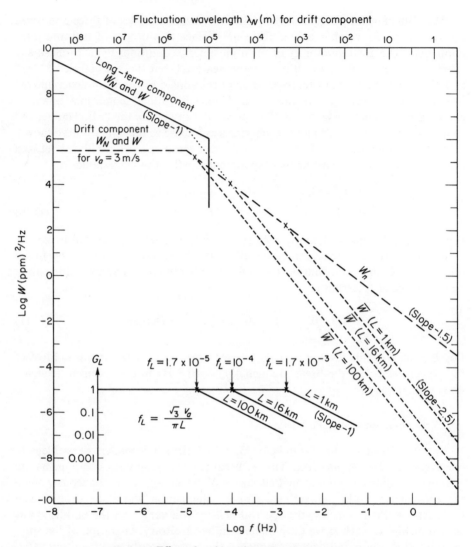

Fig. D.7 Effect of pathlength on fluctuation spectrum.

(Dimensional constants are omitted for simplicity.) The total power in the drift component corresponds to an rms fluctuation in refractivity of 6 ppm.

Refractivity fluctuations whose wavelengths are shorter than about two pathlengths will not affect the entire path simultaneously, so the corresponding range fluctuations will be smaller. The function $G_L(f)$ which relates refractivity to range spectral density is plotted in Fig. D.7:

$$G_L(f) = 1 \qquad \left(f < f_L = \frac{\sqrt{3}\, v_a}{\pi L} \right),$$

$$G_L(f) = \frac{f_L}{f} \qquad (f > f_L), \tag{D.14}$$

$$\bar{W}(f) = W_N(f)G_L(f) \qquad (\text{ppm})^2/\text{Hz}. \tag{D.15}$$

Thus, in the region above the pathlength crossover frequency f_L,

$$\bar{W}(f) = \frac{v_a^{1.5}}{100\pi L f^{2.5}} \qquad (\text{ppm})^2/\text{Hz}, \tag{D.16}$$

where a dimensional constant is again omitted. This spectrum will find later use in deriving angular error, and will also be applied directly to estimation of the magnitude and correlation distance of phase errors across an aperture.

Range fluctuation spectra for pathlengths of 1, 16, and 100 km are shown in Fig. D.7. At pathlengths less than 16 km, the spectrum will have three distinct components:

(a) the long-term component below 3×10^{-5} Hz, where W is independent of drift velocity and pathlength;

(b) the region from 3×10^{-5} Hz to f_L, where $W = W_N$, independent of pathlength but dependent on drift velocity;

(c) the region above f_L, where Eq. (D.16) applies. Considering the pathlengths over which data have been taken (Fig. D.6), the difference between 16 km and

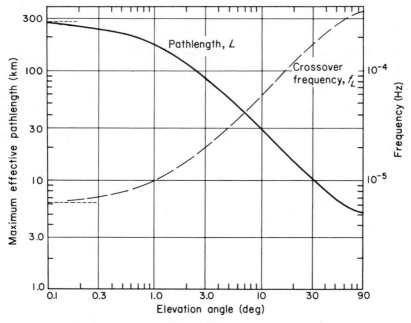

Fig. D.8 Pathlength and crossover frequency vs. elevation for paths extending above $h = 5$ km.

25 km is so small that the inherent variability of tropospheric conditions prevents our recognizing a change in spectral shape. For both these paths, the long-term and drift components fit together to form the spectrum expressed by Eq. (D.10). On paths which are much longer or shorter, however, the division into three segments of different slopes should become apparent.

To estimate the effective length of the tropospheric part of a radar path, we may consider that portion of the path which lies below 5-km altitude. Figure D.8 shows the maximum value of this length as a function of elevation angle, and the corresponding frequency f_L. This curve applies to targets which lie above 5-km altitude. For lower targets, the full range to the target should be used. This low-altitude part of the path should also be used to estimate the drift velocity v_a, because most of the refractivity variations are in the dense, moist portions of the air mass.

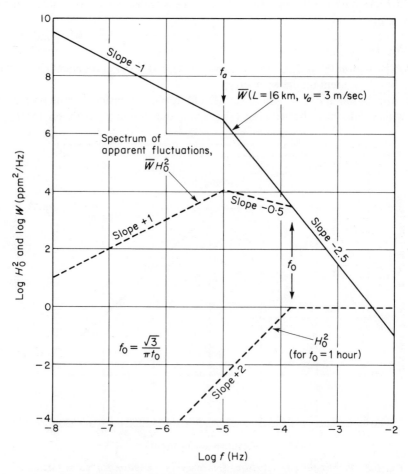

Fig. D.9 Spectra showing effect of restricted observation time.

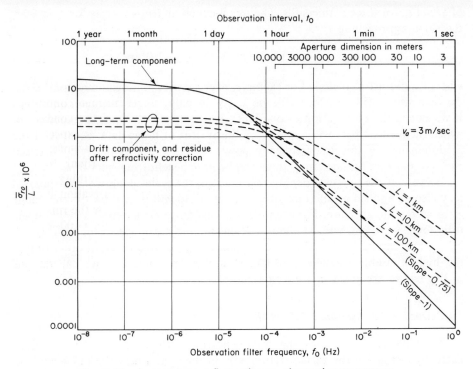

Fig. D.10 Apparent range fluctuation vs. observation parameters.

Apparent Range Fluctuation

When the observation period, t_o, is restricted, the full magnitude of long-period fluctuations cannot be seen. The effect on observed data can be described by applying to the fluctuation spectrum a high-pass filter function, $H_o(f)$:

$$H_o(f) = \frac{f}{f_o} \qquad \left(f < f_o = \frac{\sqrt{3}}{\pi t_o}\right),$$
$$H_o(f) = 1 \qquad (f > f_o). \tag{D.17}$$

The apparent range fluctuation $\overline{\sigma_{ro}}$ is then found by integrating the product of spectral density times this filter function squared:

$$\frac{\overline{\sigma_{ro}}}{L} \times 10^6 = \left[\int_{10^{-8}}^{\infty} \overline{W}(f) H_o^2(f)\, df\right]^{1/2} \text{ppm}. \tag{D.18}$$

Plots of the filter function and of the resulting spectrum of apparent range fluctuations are shown in Fig. D.9, calculated for a medium path with $t_o = 1$ hr. The magnitude of apparent fluctuation in ppm is shown in Fig. D.10 as a function of f_o and t_o, for three pathlengths. The rms fluctuation drops sharply for $t_o < 3$ hr, the slope approaching $t_o^{3/4}$.

$$\overline{\sigma_{ro}} \cong 1.4 \times 10^{-7} L^{0.5} t_o^{0.75} \qquad (t_o < 10^4 \text{ sec}) \qquad \text{(D.19)}$$

where lengths are in meters. For example, over a period of one minute, only 0.07-cm fluctuation will be observable on a 10-km path, under average conditions. If we extend the observation to one hour, 1.6-cm fluctuation can be expected, of which the greatest part will appear as a linear drift of range data. Because of the predominance of this low-frequency power in the range error (lying between 10^{-5} Hz and f_o in the spectrum) there can be little benefit from smoothing.

Because of the time-wavelength relationship, the scale of observation interval may also be applied to aperture size, and Fig. D.10 will then show the rms range error across the aperture. For example, the rms range error across a 100-m aperture, for $L = 10$ km, will be about 0.05 cm for the mean spectrum, with $v_a = 3$ m/sec. This can be converted to phase angle at a given frequency, in order to find the highest usable frequency or largest usable aperture. Further calculations of this nature are given in a later paragraph.

Correction for Measured Refractivity

The NBS work has shown that a large portion of the range error can be eliminated by measuring the surface refractivity in the region near the radar site and applying a correction in the general form of the plots shown in Fig. D.2. Because the range and refractivity spectra diverge at frequencies above about 10^{-4} Hz, the correction must be averaged over a period of about three hours to achieve its best effect (Norton, 1963). The residual error spectrum will be as shown in Fig. D.11, where curves are given for two different accuracies in reading of N_s. The correction takes the form of a filter transfer function $H_c(f)$, which reduces the low-frequency components to a level set by the accuracy of N_s measurement. The filter attenuation is reduced as frequency increases, and components above 10^{-4} Hz are passed unchanged. Although the diurnal component and lower frequencies are greatly reduced, the residual fluctuation error σ_{rc} cannot be reduced below about ten percent of σ_r (see drift component, Fig. D.10). This is because the error power near 10^{-4} is not affected by the correction, and this contributes an error very near ten percent of the total observed in the uncorrected data. Thus, attempts to use very accurate refractometers for range correction do not appear worthwhile, except perhaps on ground-to-ground paths where readings may be taken at many points along the path and averaged over space instead of in time.

Angular Fluctuation in Azimuth Measurements

The fluctuation in apparent angle of arrival of a wavefront can be found from the spatial equivalent of the time fluctuation spectrum in range data. A reasonable

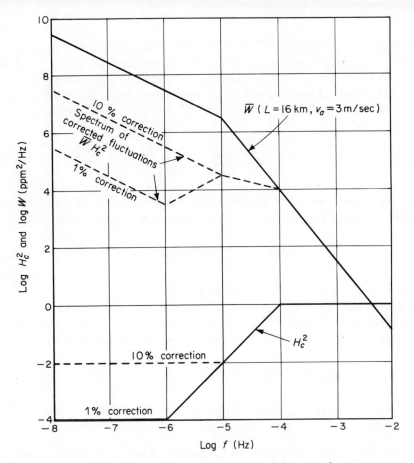

Fig. D.11 Spectra showing effect of refractivity correction.

approximation of this spatial spectrum is obtained as described above, by assuming the time fluctuations to be the result of the drift of a "frozen" pattern of irregularities past the radar line of sight. This "frozen" model, which is admittedly a rough approximation of a more complex process, can be used to predict with acceptable accuracy the magnitude and spectral characteristics of angular fluctuation errors caused by the troposphere in radar and interferometer systems. The spatial spectrum of phase error, derived from range, also permits evaluation of beam broadening, lost efficiency, and sidelobe increase in large antennas and arrays.

The process of angular measurement across a baseline or aperture can be related to differentiation, or slope measurement, in the time domain. The coordinate x measured along the aperture is related to time by $x = v_a t$, and the aperture width is related to the total smoothing interval by $w = v_a t_s$, where v_a is the drift velocity. The slope measurement is described by a differentiator transfer function (see App. C):

$$H'(f) = 2\pi f \quad (f < f_w).$$

The shape of the function H' above the differentiator passband is indicated in Fig. C.3 for two-point measurement (the interferometer case) and in Fig. C.4 for continuous measurement (use of a linear-odd difference illumination over an aperture). The measured angle of arrival is equivalent to the slope of the phasefront. For purposes of evaluating error caused by drifting tropospheric irregularities, we may write

$$\theta = \frac{dR}{dx} = \frac{1}{v_a} \frac{dR}{dt}, \tag{D.20}$$

where the slope is smoothed across the aperture (or baseline) in space, and irregularities are smoothed over $t_s = w/v_a$ sec in time. The angular error is then

$$\bar{\sigma}_\theta = \frac{1}{v_a} \bar{\sigma}_i = \frac{L}{v_a} \left[\int_0^\infty H'^2(f) \bar{W}(f) \, df \right]^{1/2}. \tag{D.21}$$

When $\bar{W}$ is in $(\text{ppm})^2/\text{Hz}$, σ_θ will be in μrad.

Consider first the optimum (linear-odd) aperture illumination function for measurement [Fig. C.4(a)]. When computing rms error for a continuous spectrum, the frequency response (transfer function) may be approximated by

$$H'(f) = 2\pi f \quad \left(f \leq f_w = \frac{\sqrt{3} \, v_a}{\sqrt[4]{2} \, \pi w} = \frac{0.44 v_a}{w} \right),$$

$$H'(f) = \frac{2\pi f_w^2}{f} \quad (f > f_w). \tag{D.22}$$

The spectrum of angular fluctuations will appear as shown in Fig. D.12, with four segments of slopes $+2$ (for $f < 10^{-5}$), $+0.5$ (for $10^{-5} < f < f_L$), -0.5 (for $f_L < f < f_w$), and -4.5 (for $f_w < f$). Evaluation of the area under the curve, corresponding to the integral in Eq. (D.21), yields

$$\bar{\sigma}_\theta \cong 0.44 \sqrt{\frac{L}{\sqrt{w}}} \, \mu\text{rad} \quad (w \ll L, \text{ both in meters}). \tag{D.23}$$

Here, the overbar indicates that the rms error has been computed for the mean fluctuation spectrum described earlier. The resulting angular error is independent of the drift velocity. Most of the power lies in the region just below f_w in the spectrum.

For an interferometer, the transfer function is also that of an ideal differentiator in the low-frequency region, but the high-frequency response is

$$H'(f) = 2\pi f_b \quad \left(f > f_b = \frac{v_a}{\sqrt{2} \, \pi b} \right), \tag{D.24}$$

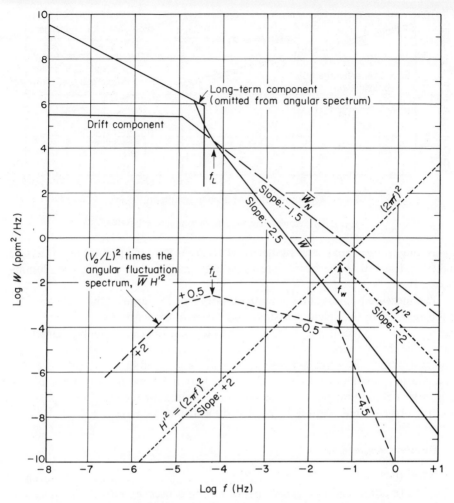

Fig. D.12 Transfer function and fluctuation spectrum of angle measurement. Drawn for linear-odd illumination, where $W = 30$ m, $L = 30$ km, $v_a = 3$ m/sec.

where b is the interferometer baseline length. This type of response is plotted in Fig. C.3(a). Evaluation of the integral for this function gives (for b normal to the line of sight)

$$\bar{\sigma}_\theta \cong 0.4\left[\sqrt{L}\left(\sqrt{\frac{L}{b}} - 0.74\right)\right]^{1/2} \qquad (b \leq L) \qquad (D.25)$$

$$\cong 0.4\sqrt{\frac{L}{\sqrt{b}}}\,\mu\text{rad} \qquad (b \ll L). \qquad (D.26)$$

Results for both aperture antennas and interferometers are plotted in Fig. D.13, along with experimental points taken from NBS measurements (Thompson,

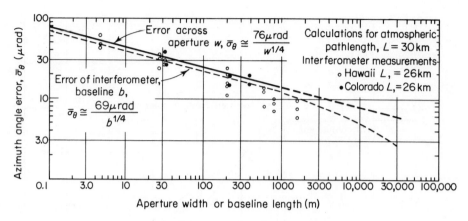

Fig. D.13   Angle error vs. size of antenna or baseline.

Janes and Grant, 1965). The similarity of error curves for aperture and interferometer indicates that the results are essentially independent of aperture illumination.

The measured points show a reduction of error for large b which is somewhat
more rapid than the calculated slope: $\bar{\sigma}_\theta$ varies more nearly as $b^{-1/3}$ than as the
calculated $b^{-1/4}$. This is due, in part, to the fact that the experimental "target" lay
within the troposphere, so that the two paths converged as they neared the target.
The range-difference error between such paths is reduced in a way similar to that
of a shorter pathlength L in Eq. (D.25), so that the slope is steeper for $b \rightarrow L$.
For targets at long range, the paths remain separated throughout the troposphere,
and the error should be found closer to the calculated curves.

Elevation Angle Error

When the baseline is not normal to the line of sight, as in elevation measurement
(Fig. D.14), additional fluctuation errors will be caused by refractivity variations
in the short segment z of the path which lies between the more distant receiving

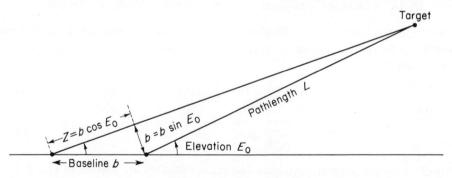

Fig. D.14   Elevation measurement with interferometer.

point and the projected baseline b'. This effect is of dominant importance when elevation angles below about 30 deg are measured with a horizontal interferometer. The procedure for calculating error in this case is as follows:

1. The normal component of error $\bar{\sigma}_\theta$ is computed, or read from Fig. D.13, for the projected baseline $b' = b \sin E_o$.

2. The added range fluctuation variance $\bar{\sigma}_z^2$ for the path segment $z = b \cos E$ is computed by substituting z for L in Eq. (D.15) and integrating $W(f)$.

3. The two variances are added to find total elevation error:

$$\bar{\sigma}_E^2 = \bar{\sigma}_\theta^2 + \left(\frac{\bar{\sigma}_z}{b'}\right)^2. \tag{D.27}$$

Since, from Fig. D.10, we find $\bar{\sigma}_z = 16 \times 10^{-6} z$ for all paths (when the long-term component is included), the error will be

$$\bar{\sigma}_E^2 = \bar{\sigma}_\theta^2 + (16 \cot E_o)^2 \qquad (\mu \, \text{rad})^2. \tag{D.28}$$

The result, for $E_o = 6$ deg, agrees well with the trend of experimental data (Fig. D.15). Although the vertical structure of refractivity variations must necessarily differ from the horizontal structure at long wavelengths, this effect is not evident

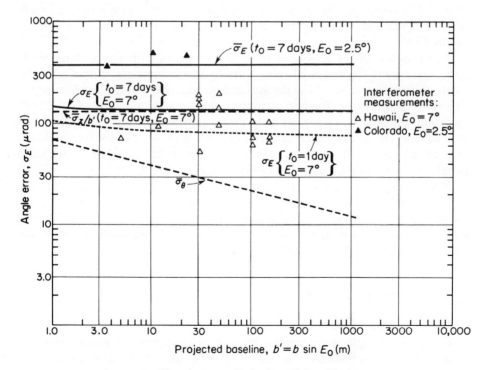

Fig. D.15 Elevation error for horizontal interferometer.

for baselines as great as 1500 m, as long as the long-term error components are included.

In tracking a target at long range, where the two paths traverse the troposphere in parallel lines, the long-term components will be correlated and will produce no angular error. (This will also be true of the exponential altitude gradient, which produces no bias error in a horizontal interferometer for cases where a flat-earth assumption is warranted.) The drift component of refractivity variation will still introduce an error $\bar{\sigma}_z = 2 \times 10^{-6} z$, so that

$$\bar{\sigma}_E^2 = \bar{\sigma}_\theta^2 + (2 \cot E_o)^2 \qquad (\mu \text{ rad})^2. \tag{D.29}$$

This is also the error approached in short-range tracking when the long-term component is corrected on the basis of accurate measurements of refractivity.

In the preceding discussions, it has been assumed that observations are carried out over periods longer than b/v_a, so that the fluctuation power below f_b is evident at the output. Some of the early measurements (Norton, et al., 1961) suffered from a restriction of the observation period to 1000 sec, and from the further reduction of the linear drift which resulted from the long-term "trends" in the data. The result was to produce an *apparent* error which varied inversely with b, rather than with the fourth root of b, for $b > 1000$ m. The more recent measurements (Fig. D.13), taken over longer periods, confirm the lesser dependence of $\overline{\sigma_\theta}$ on b, in close agreement with the error model which uses the "frozen atmosphere" assumption.

The presence of the long-term error component in $\overline{\sigma_z}$ makes it desirable to apply corrections for measured refractivity to interferometer elevation measurements on targets near or within the atmosphere, regardless of the baseline length. This is in contrast to the situation in azimuth measurement. Figures D.11 and D.12 show that the fluctuation power from the drift component (near and above 10^{-4} Hz) will govern the performance in azimuth, for all but the longest baselines ($b > 10^4$ m). As a result, corrections for measured refractivity will be of little value in reducing azimuth error on most interferometers.

When elevation measurements are made with an aperture which tracks the target, there will be no component $\overline{\sigma_z}$. The azimuth error analysis given earlier can be applied directly to all tracking measurements of angle. Fixed array antennas will be subject to the $\overline{\sigma_z}$ component, but this will usually be small in comparison with other errors in this type of system.

Maximum Usable Aperture

The maximum aperture dimension which can be employed over a given path at a given wavelength is determined by angular fluctuation or beam spreading, caused by phase fluctuations. The beamwidth of an antenna is in the order of λ/w radians, and the gain will begin to deteriorate when the rms pointing error reaches about one-tenth of this value. For a system which does not track the local apparent angle of arrival, Eq. (D.23) leads to the following limit:

$$\theta_3 \cong \frac{\lambda}{w} \geq 10\overline{\sigma}_\theta \cong 0.44 \sqrt{\frac{L}{\sqrt{w}}} \times 10^{-5},$$

$$w_{max} \cong 1.3 \times 10^7 \lambda^{4/3} L^{-2/3} \text{ meters.} \qquad (D.30)$$

For example, with $\lambda = 0.03$ m (X-band), $L = 180$ km (elevation 1.0 deg),

$$w_{max} \cong \frac{1.3 \times 10^7 \times 0.01}{3200} = 40 \text{ m,}$$

$$\theta_3 \cong 0.75 \text{ mr}; \qquad \overline{\sigma}_\theta \cong 0.075 \text{ mr.}$$

When the apparent angle of arrival is tracked by the antenna, the limitation may be caused by spreading of the main beam (resulting from nonlinear phase fluctuations over the aperture). This can be traced to fluctuations in the octave just above f_w in the W spectrum, corresponding to fluctuation wavelengths between w and $2w$. The phase error is

$$\sigma_\phi^2 = \left[\frac{2\pi L}{\lambda}\right]^2 \int_{f_w}^{2f_w} \overline{W}(f)\, df \times 10^{-12} \cong 0.18 \times 10^{-12} \frac{L w^{3/2}}{\lambda^2}. \qquad (D.31)$$

The numerical constants here apply to average conditions, as noted earlier in the discussion of the refractivity spectrum. For a loss of 1 db in antenna gain, we can allow $\sigma_\phi = 0.5$ rad, giving the limitation in size as

$$w_{max} \cong \lambda^{4/3} L^{-2/3} \times 10^8 \qquad \text{(all dimensions in meters).} \qquad (D.32)$$

In the example ($\lambda = 0.03$ m, $L = 180$ km), we find $w_{max} \cong 300$ m for 1-db loss at X-band.

Velocity Error

Target velocity is found by differentiation of range and angle data, sometimes supplemented by Doppler measurement of range rate. As far as tropospheric fluctuations are concerned, the velocity error is the same for differentiated position data as for Doppler data. The velocity error power spectrum is just $(2\pi f)^2$ times the corresponding position spectrum, and the rms velocity error is strongly dependent upon the high-frequency portion of this spectrum. As a result, the frequencies below about 10^{-4} Hz can be neglected, and only the high-frequency segment of the fluctuation spectrum plots in Fig. D.7 need be considered. These are described by Eq. (D.16). Because some form of smoothing is always used in a practical differentiator, we include an additional transfer function of the form $H_s(f) = f_d/f$ at the upper end of the spectrum, to account for averaging across an aperture or in the differentiating circuits. The range-rate error is then

$$\overline{\sigma_{\dot{r}}^2} \cong L^2 \int_{10^{-4}}^{\infty} (2\pi f)^2 H_s^2(f) \overline{W}(f) \, df \times 10^{-12}$$

$$\cong 0.33 \times 10^{-12} v_a^{3/2} f_d^{1/2} L \qquad (\text{m/sec})^2. \tag{D.33}$$

For example, on a path penetrating the troposphere at 2.5-deg elevation, $L = 100$ km, and for a drift velocity $v_a = 3$ m/sec with smoothing above $f_d = 1$ Hz, we find $\overline{\sigma_{\dot{r}}} \cong 0.04$ cm/sec, with most of the power just below the 1-Hz upper limit.

The angular velocity error is found by applying this time differentiation to that performed by the aperture or baseline, in Eq. (D.21).

$$\overline{\sigma_{\dot{\theta}}} \cong \frac{L}{v_a} \left[\int_{10^{-4}}^{\infty} (2\pi f)^2 H_s^2(f) H'^2(f) \overline{W}(f) \, df \right]^{1/2} \mu\text{rad/sec.} \tag{D.34}$$

For an aperture system, this gives

$$\overline{\sigma_{\dot{\theta}}} \cong 0.45 v_a L^{1/2} w^{-5/4} \quad \mu\text{rad/sec} \qquad (f_d > 0.44 \, v_a/w) \tag{D.35}$$

$$\cong 2.5 f_d L^{1/2} w^{-1/4} \quad \mu\text{rad/sec} \qquad (f_d < 0.44 \, v_a/w). \tag{D.36}$$

For an interferometer,

$$\overline{\sigma_{\dot{\theta}}} \cong 0.8 v_a^{3/4} f_d^{1/2} L^{1/2} b^{-1} \qquad (f_d > 0.22 \, v_a/b) \tag{D.37}$$

$$\cong 0.3 f_d^{5/4} L^{1/2} v_a^{-1/4} \qquad (f_d < 0.22 \, v_a/b). \tag{D.38}$$

All lengths are expressed in meters. Figure D.16 shows the error vs aperture or baseline length for a 30-km path with $f_d = 0.1$ Hz and two values of drift velocity. Note that the errors are roughly proportional to drift velocity and inversely proportional to aperture or baseline length for large systems, but become dependent primarily on smoothing time for small systems or high drift velocities.

The dependence of rate error on drift velocity leads to a dependence on angular tracking rate. The beams move through the troposphere as a high-velocity target is tracked, and the effect is similar to an increase in drift velocity of the troposphere. Although the component caused by beam rotation varies linearly with range from the antenna, a good approximation is provided by using the value at the midpoint of the 5-km effective height of the troposphere:

$$v_b \cong \frac{2500\omega}{\sin E_t} \qquad \text{m/sec,} \tag{D.39}$$

where ω is the angular tracking rate in rad/sec, E_t is the elevation angle, and the target is assumed to lie above 5-km altitude. When this beam velocity v_b rises above the drift velocity v_a, the tracking motion is dominant and v_b should be substituted

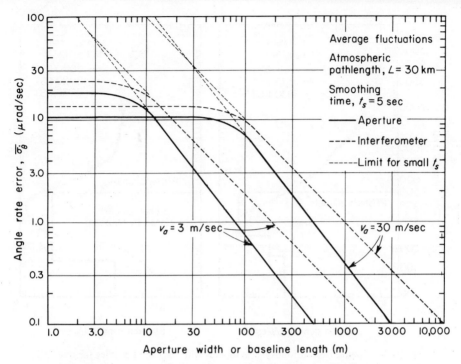

Fig. D.16   Angle rate error vs. size of aperture or baseline.

for v_a in Eqs. (D.33) through (D.38). For example, the curves in Fig. D.16 for $v_a = 30$ m/sec will apply to a target moving at 20 mr/sec at 10 deg elevation.

D.3 Ionospheric Effects

When radar measurements are made on targets above about 100 km, the effects of the ionospheric layers must be considered. These effects are all dependent upon operating frequency, varying directly with the square of the wavelength in the commonly used radar bands. A brief summary of the errors produced by the ionosphere will be given, based on simplified models of the ionosphere for day and night conditions.

Ionospheric Profiles

Figure D.17 shows typical day and night profiles of electron density vs altitude, based on backscattering data (Bowles, 1961), as compared to the "Chapman distribution" and with the simple, rectangular distributions used in an Air Force study

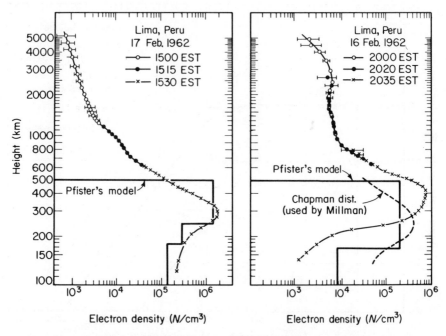

Fig. D.17 Comparison of ionospheric models with measured profiles for day and night conditions.

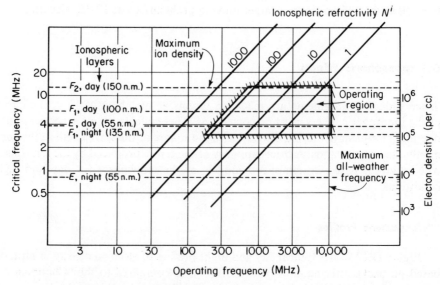

Fig. D.18 Ionospheric refractivity vs. operating and critical frequencies and electron density.

of ionospheric effects (Pfister and Keneshea, 1956). These profiles may be used to calculate the refractivity of the ionosphere as a function of frequency:

$$N_i = (n - 1) \times 10^6 \cong \frac{40N_e}{f^2} = \frac{1}{2}\left[\frac{f_c}{f}\right]^2 \tag{D.40}$$

where N_e is the electron density per m³, f is in Hz, and f_c is the "critical frequency" ($f_c \cong 9\sqrt{N_e}$). When N_e is expressed in electrons per cm³, the same expressions can be used with f and f_c in kHz. These relationships are plotted in Fig. D.18, which shows the operating region for accurate measurement systems (better than one part in 10^4).

Ionospheric Errors in Range and Angle

Plots of range and angular refraction errors, based on the approximate rectangular profiles of the daytime ionosphere, are shown in Figs. D.19 and D.20.

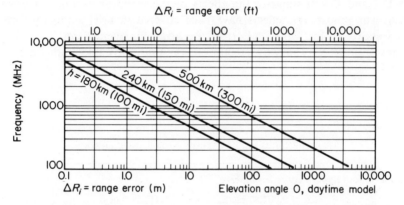

Fig. D.19 Ionospheric range error vs. frequency (after Pfister and Keneshea).

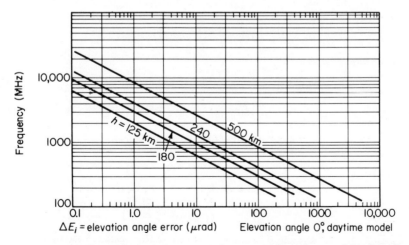

Fig. D.20 Ionospheric angle error vs. frequency (after Pfister and Keneshea).

These may be reduced by a factor of about 3 for night operation, and increased by a similar factor for operation during periods of extreme ionospheric disturbance. The expected value of range fluctuation is shown in Fig. D.21, for normal and disturbed conditions. The assumptions used in this figure are that the scale length of the ionospheric irregularities is 5 km, and the rms refractivity fluctuation is one percent of the mean value. It is estimated that the correlation period of the range fluctuations is about 7 sec for normal conditions and 50 sec for disturbed conditions, when the radar beam is moving at a rate determined by rotation of the earth (Smith, 1967).

D.4 Summary of Propagation Errors

Tables D.1 and D.2 summarize the effects of the troposphere on measurement accuracy. The ionospheric effects have been covered in brief summary form in Sec. D.3 above. In the tables, all errors are given for average conditions in the troposphere. Errors may vary by a factor of 3 (and rarely by 10) in either direction.

Table D.1

TROPOSPHERIC REFRACTIVITY AND RANGE ERROR

Range error on path through entire troposphere

$$\Delta R = 0.007 \, N_s \csc E_o \quad \text{(m)} \quad (E_o > 5 \text{ deg}) \tag{D.3}$$

$$(\Delta R)_{max} \cong 100 \text{ m} \quad (E_o \cong 0 \text{ deg}) \tag{Fig. D.2}$$

Refractivity at sea level

$$\overline{N_o} = 313 \quad \text{(world-wide annual average)}$$

Average refractivity spectrum

$$\overline{W_N}(f) = 32f^{-1} \quad (10^{-8} < f < 10^{-5} \text{ Hz}) \tag{D.10}$$

$$= \frac{\sqrt{3} \, v_a}{300 f^{1.5}} \quad (f > 10^{-5} \text{ Hz}) \tag{D.13}$$

$$\overline{\sigma_N} = \left[\int_{10^{-8}}^{\infty} \overline{W_N}(f) \, df \right]^{1/2} \cong 15 \text{ N-units}$$

Effect of pathlength

$$\left. \begin{array}{ll} G_L(f) = 1 & \left(f < f_L = \dfrac{\sqrt{3} \, v_a}{\pi L} \right) \\[2mm] = f_L/f & (f > f_L) \end{array} \right\} \tag{D.14}$$

Table D.1—*Cont.*

Average range fluctuation spectrum

$$\overline{W}(f) = \overline{W}_N(f)G_L(f) \qquad (\text{ppm})^2/\text{Hz} \tag{D.15}$$

$$= 32f^{-1} \qquad (10^{-8} < f < 10^{-5} \text{ Hz}) \tag{D.10}$$

$$= \frac{v_a^{1.5}}{100\pi L f^{2.5}} \qquad (f > 10^{-5} \text{ Hz}) \tag{D.16}$$

$$\frac{\overline{\sigma}_r}{L} \times 10^6 = \left[\int_{10^{-8}}^{\infty} \overline{W}(f)\, df\right]^{1/2} \cong 15 \text{ ppm} \tag{D.11}$$

Apparent range error in restricted observation period, t_o

$$\left.\begin{array}{l} H_o(f) = \dfrac{f}{f_o} \qquad \left(f < f_o = \dfrac{\sqrt{3}}{\pi t_o}\right) \\[2mm] = 1 \qquad (f > f_o) \end{array}\right\} \tag{D.17}$$

$$\frac{\overline{\sigma}_{ro}}{L} \times 10^6 = \left[\int_{10^{-8}}^{\infty} \overline{W}(f)H_o^2(f)\, df\right]^{1/2} \qquad (\text{ppm}) \tag{D.18}$$

$$\overline{\sigma}_{ro} \cong 1.4 \times 10^{-7} L^{0.5} t_o^{0.75} \qquad (t_o < 10^4 \text{ sec}) \tag{D.19}$$

Range error after refractivity correction

(For correction filter function, H_c, see Fig. D.11.)

$$\frac{\overline{\sigma}_{rc}}{L} \times 10^6 = \left[\int_{10^{-8}}^{\infty} \overline{W}(f)H_c^2(f)\, df\right]^{1/2}$$

$$\cong 0.1 \frac{\overline{\sigma}_r}{L} \times 10^6$$

Brief definitions of symbols

ΔR = Range bias error

N_s = Surface refractivity = $(n - 1) \times 10^6$

E_o = Ray elevation angle at surface

N_o = Average sea-level refractivity

$\overline{W}_N(f)$ = Refractivity spectrum for average conditions, in $(N\text{-units})^2/\text{Hz}$

v_a = Atmospheric drift speed (average wind speed)

$G_L(f)$ = Factor relating refractivity to range spectrum

f_L = Break frequency in G_L function

$\overline{\sigma}_r$ = Range fluctuation for average conditions

L = Pathlength within troposphere

$H_o(f)$ = High-pass filter function for restricted observation interval

f_o = Break frequency in H_o function

$\overline{\sigma}_{ro}$ = Apparent range fluctuation observed in restricted interval

$H_c(f)$ = Correction filter function

$\overline{\sigma}_{rc}$ = Range fluctuation after refractivity correction

Table D.2

TROPOSPHERIC ANGLE ERROR

Elevation error on path through entire troposphere

$$\Delta E_o = N_s \cot E_o \qquad (E_o > 5 \text{ deg}) \qquad\qquad\qquad \text{(D.4)}$$

$$(\Delta E_o)_{max} \cong 15 \text{ mr} \qquad (E_o \cong 0 \text{ deg}) \qquad\qquad \text{(Fig. D.3)}$$

Angular error for "frozen atmosphere" assumption

$$\overline{\sigma}_\theta = \frac{\overline{\sigma}_t}{v_a} = \frac{L}{v_a}\left[\int_0^\infty H'^2(f)\overline{W}(f)\,df\right]^{1/2} \qquad \mu\text{rad} \qquad\qquad \text{(D.21)}$$

Angle error for aperture width w

$$\overline{\sigma}_\theta \cong 0.44\sqrt{\frac{L}{\sqrt{w}}} \qquad \mu\text{rad} \qquad\qquad\qquad \text{(D.23)}$$

Azimuth error for baseline length b

$$\overline{\sigma}_\theta \cong 0.4\sqrt{\frac{L}{\sqrt{b}}} \qquad \mu\text{rad} \qquad (b \ll L) \qquad\qquad \text{(D.26)}$$

Elevation error for baseline system $(R \cong L)$

$$\overline{\sigma}_E^{\,2} = \overline{\sigma}_\theta^{\,2} + (16 \cot E_o)^2 \qquad (\mu\text{rad})^2 \qquad\qquad \text{(D.28)}$$

Elevation for long-range target $(R \gg L)$, or for system corrected for measured refractivity

$$\overline{\sigma}_E^{\,2} = \overline{\sigma}_\theta^{\,2} + (2 \cot E_o)^2 \qquad (\mu\text{rad})^2 \qquad\qquad \text{(D.29)}$$

Further definitions

ΔE_o = elevation bias error

$\overline{\sigma}_\theta$ = angular error under average conditions

w = aperture width

b = baseline length

$\overline{\sigma}_E$ = elevation error from baseline system

$H'^2(f)$ = differentiator function [see Eqs. (D.22), (D.24)]

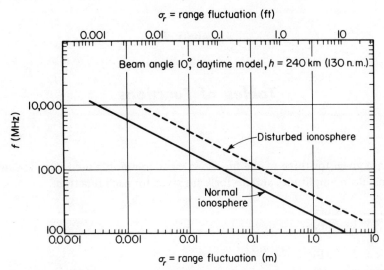

Fig. D.21 Ionospheric range fluctuation vs. frequency (scale length 5 km, $\overline{\Delta N}^2$ $= 3 \times 10^{-4} N^2$).

Appendix E

Tables of Functions

Tables are given for three of the functions most commonly used in radar analysis problems. Both amplitudes and decibels are given for each function.

Table E.1 $\dfrac{\sin(\pi u)}{(\pi u)}$

Table E.2 $20 \log_{10}\left[\dfrac{\sin(\pi u)}{(\pi u)}\right]$

Table E.3 $\dfrac{\sin(2.7831\ u)}{2.7831\ u}$

Table E.4 $20 \log_{10}\left[\dfrac{\sin(2.7831\ u)}{2.7831\ u}\right]$

Table E.5 $\exp(-1.3863\ u^2)$

Table E.6 $20 \log_{10}[\exp(-1.3863\ u^2)]$

Table E.1

VALUES OF $\sin(\pi u)/(\pi u)$

(Intervals of 0.01 from $u = 0$ to $u = 1.0$, intervals of 0.1 from $u = 1.0$ to $u = 10.9$)

u	0	1	2	3	4	5	6	7	8	9
0.0		0.9998	0.9993	0.9985	0.9973	0.9958	0.9940	0.9919	0.9895	0.9867
0.1	0.9836	0.9802	0.9764	0.9724	0.9680	0.9633	0.9584	0.9531	0.9475	0.9416
0.2	0.9354	0.9290	0.9222	0.9152	0.9079	0.9003	0.8924	0.8843	0.8759	0.8672
0.3	0.8583	0.8492	0.8398	0.8302	0.8204	0.8103	0.8000	0.7895	0.7788	0.7679
0.4	0.7568	0.7455	0.7340	0.7224	0.7106	0.6986	0.6865	0.6742	0.6618	0.6492
0.5	0.6366	0.6238	0.6109	0.5979	0.5848	0.5716	0.5583	0.5449	0.5315	0.5180
0.6	0.5045	0.4909	0.4773	0.4636	0.4500	0.4363	0.4226	0.4089	0.3952	0.3815
0.7	0.3678	0.3542	0.3406	0.3270	0.3135	0.3001	0.2867	0.2733	0.2601	0.2469
0.8	0.2338	0.2208	0.2079	0.1952	0.1825	0.1700	0.1575	0.1453	0.1331	0.1211
0.9	0.1092	0.0975	0.0860	0.0746	0.0634	0.0524	0.0415	0.0308	0.0203	0.0101
1.0	0.0000	−0.0894	−0.1559	−0.1980	−0.2162	−0.2122	−0.1892	−0.1514	−0.1039	−0.0517
2.0	0.0000	0.0468	0.0850	0.1119	0.1261	0.1273	0.1164	0.0953	0.0668	0.0339
3.0	0.0000	−0.0317	−0.0584	−0.0780	−0.0890	−0.0909	−0.0840	−0.0696	−0.0492	−0.0252
4.0	0.0000	0.0239	0.0445	0.0598	0.0688	0.0707	0.0658	0.0547	0.0389	0.0200
5.0	0.0000	−0.0192	−0.0359	−0.0485	−0.0560	−0.0578	−0.0540	−0.0451	−0.0322	−0.0166
6.0	0.0000	0.0161	0.0301	0.0408	0.0473	0.0489	0.0458	0.0384	0.0275	0.0142
7.0	0.0000	−0.0138	−0.0259	−0.0352	−0.0409	−0.0424	−0.0398	−0.0334	−0.0239	−0.0124
8.0	0.0000	0.0121	0.0228	0.0310	0.0360	0.0374	0.0352	0.0296	0.0212	0.0110
9.0	0.0000	−0.0108	−0.0203	−0.0276	−0.0322	−0.0335	−0.0315	−0.0265	−0.0190	−0.0099
10.0	0.0000	0.0097	0.0183	0.0250	0.0291	0.0303	0.0285	0.0240	0.0173	0.0090

Table E.2

VALUES OF $20\log_{10}[\sin(\pi u)/(\pi u)]$ (decibels)

(Intervals of 0.01 from $u = 0$ to $u = 1.0$, intervals of 0.1 from $u = 1.0$ to $u = 10.9$)

u	0	1	2	3	4	5	6	7	8	9
0.0	0.00	0.00	0.00	-0.01	-0.02	-0.03	-0.05	-0.07	-0.09	-0.11
0.1	-0.14	-0.17	-0.20	-0.24	-0.28	-0.32	-0.36	-0.41	-0.46	-0.52
0.2	-0.57	-0.63	-0.70	-0.76	-0.83	-0.91	-0.98	-1.06	-1.15	-1.23
0.3	-1.32	-1.41	-1.51	-1.61	-1.71	-1.82	-1.93	-2.05	-2.17	-2.29
0.4	-2.42	-2.55	-2.68	-2.82	-2.96	-3.11	-3.26	-3.42	-3.58	-3.75
0.5	-3.92	-4.09	-4.28	-4.46	-4.65	-4.85	-5.06	-5.27	-5.48	-5.71
0.6	-5.94	-6.17	-6.42	-6.67	-6.93	-7.20	-7.48	-7.76	-8.06	-8.36
0.7	-8.68	-9.01	-9.35	-9.70	-10.07	-10.45	-10.85	-11.26	-11.69	-12.14
0.8	-12.62	-13.11	-13.63	-14.18	-14.77	-15.39	-16.04	-16.75	-17.51	-18.33
0.9	-19.22	-20.21	-21.30	-22.53	-23.95	-25.61	-27.62	-30.20	-33.80	-39.91
1.0	$-\infty$	-20.97	-16.14	-14.06	-13.30	-13.46	-14.46	-16.39	-19.66	-25.71
2.0	$-\infty$	-26.58	-21.40	-19.01	-17.98	-17.90	-18.67	-20.41	-23.50	-29.39
3.0	$-\infty$	-29.97	-24.66	-22.15	-21.00	-20.82	-21.50	-23.14	-26.15	-31.96
4.0	$-\infty$	-32.39	-27.02	-24.45	-23.24	-23.00	-23.63	-25.22	-28.18	-33.94
5.0	$-\infty$	-34.29	-28.87	-26.26	-25.02	-24.75	-25.34	-26.90	-29.82	-35.55
6.0	$-\infty$	-35.85	-30.40	-27.77	-26.50	-26.20	-26.76	-28.30	-31.20	-36.91
7.0	$-\infty$	-37.16	-31.70	-29.05	-27.76	-27.44	-27.99	-29.51	-32.40	-38.09
8.0	$-\infty$	-38.31	-32.83	-30.16	-28.86	-28.53	-29.06	-30.57	-33.44	-39.13
9.0	$-\infty$	-39.32	-33.83	-31.15	-29.84	-29.49	-30.02	-31.51	-34.38	-40.05
10.0	$-\infty$	-40.23	-34.73	-32.04	-30.71	-30.36	-30.88	-32.37	-35.22	-40.89

Table E.3

VALUES OF $\sin(2.7831\,u)/2.7831\,u$

(Intervals of 0.01 from $u = 0$ to $u = 1.0$, intervals of 0.1 from $u = 1.0$ to $u = 10.9$)

u	0	1	2	3	4	5	6	7	8	9
0.0	1.0000	0.9998	0.9994	0.9988	0.9979	0.9967	0.9953	0.9936	0.9917	0.9895
0.1	0.9871	0.9844	0.9815	0.9783	0.9748	0.9712	0.9672	0.9631	0.9586	0.9540
0.2	0.9491	0.9440	0.9386	0.9330	0.9272	0.9212	0.9149	0.9085	0.9018	0.8949
0.3	0.8877	0.8804	0.8729	0.8652	0.8573	0.8491	0.8408	0.8324	0.8237	0.8148
0.4	0.8058	0.7966	0.7873	0.7778	0.7681	0.7583	0.7483	0.7382	0.7280	0.7176
0.5	0.7071	0.6964	0.6857	0.6748	0.6638	0.6527	0.6415	0.6302	0.6189	0.6074
0.6	0.5959	0.5842	0.5726	0.5608	0.5490	0.5371	0.5252	0.5132	0.5012	0.4892
0.7	0.4771	0.4650	0.4529	0.4408	0.4287	0.4165	0.4044	0.3923	0.3801	0.3680
0.8	0.3560	0.3439	0.3319	0.3199	0.3079	0.2960	0.2842	0.2724	0.2606	0.2490
0.9	0.2373	0.2258	0.2144	0.2030	0.1917	0.1805	0.1694	0.1584	0.1475	0.1367
1.0	0.1260	0.0261	-0.0589	-0.1267	-0.1758	-0.2057	-0.2170	-0.2113	-0.1908	-0.1586
2.0	-0.1180	-0.0726	-0.0260	0.0183	0.0577	0.0897	0.1126	0.1254	0.1280	0.1209
3.0	0.1053	0.0829	0.0556	0.0259	-0.0039	-0.0319	-0.0558	-0.0743	-0.0863	-0.0912
4.0	-0.0889	-0.0801	-0.0657	-0.0471	-0.0257	-0.0033	0.0182	0.0376	0.0533	0.0643
5.0	0.0701	0.0703	0.0652	0.0554	0.0417	0.0254	0.0078	-0.0097	-0.0260	-0.0398
6.0	-0.0500	-0.0562	-0.0579	-0.0551	-0.0483	-0.0380	-0.0251	-0.0108	0.0039	0.0180
7.0	0.0303	0.0399	0.0463	0.0489	0.0478	0.0430	0.0351	0.0248	0.0128	0.0002
8.0	-0.0121	-0.0232	-0.0323	-0.0387	-0.0420	-0.0420	-0.0389	-0.0328	-0.2440	-0.0143
9.0	-0.0033	0.0075	0.0177	0.0263	0.0327	0.0365	0.0374	0.0354	0.0308	0.0239
10.0	0.0154	0.0058	-0.0039	-0.0133	-0.0214	-0.0277	-0.0319	-0.0335	-0.0325	-0.0290

Table E.4

VALUES OF $20 \log_{10}[\sin(2.7831 u)/2.7831 u]$

(Intervals of 0.01 from $u = 0$ to $u = 1.0$, intervals of 0.1 from $u = 1.0$ to $u = 10.9$)

u	0	1	2	3	4	5	6	7	8	9
0.0	0.00	0.00	0.00	−0.01	−0.01	−0.02	−0.04	−0.05	−0.07	−0.09
0.1	−0.11	−0.13	−0.16	−0.19	−0.22	−0.25	−0.28	−0.32	−0.36	−0.40
0.2	−0.45	−0.50	−0.54	−0.60	−0.65	−0.71	−0.77	−0.83	−0.89	−0.96
0.3	−1.03	−1.10	−1.18	−1.25	−1.33	−1.41	−1.50	−1.59	−1.68	−1.77
0.4	−1.87	−1.97	−2.07	−2.18	−2.29	−2.40	−2.51	−2.63	−2.75	−2.88
0.5	−3.01	−3.14	−3.27	−3.41	−3.55	−3.70	−3.85	−4.00	−4.16	−4.32
0.6	−4.49	−4.66	−4.84	−5.02	−5.20	−5.39	−5.59	−5.79	−5.99	−6.20
0.7	−6.42	−6.64	−6.87	−7.11	−7.35	−7.60	−7.86	−8.12	−8.39	−8.68
0.8	−8.97	−9.27	−9.57	−9.89	−10.22	−10.57	−10.92	−11.29	−11.67	−12.07
0.9	−12.49	−12.92	−13.37	−13.84	−14.34	−14.86	−15.41	−16.00	−16.62	−17.28
1.0	−17.98	−31.64	−24.59	−17.94	−15.09	−13.73	−13.26	−13.50	−14.38	−15.99
2.0	−18.55	−22.77	−31.67	−34.71	−24.76	−20.93	−18.96	−18.02	−17.85	−18.34
3.0	−19.54	−21.62	−25.08	−31.72	−47.98	−29.92	−25.05	−22.56	−21.27	−20.79
4.0	−21.01	−21.91	−23.63	−26.53	−31.78	−49.40	−34.77	−28.49	−25.46	−23.82
5.0	−23.08	−23.05	−23.70	−25.12	−27.57	−31.87	−42.10	−40.19	−31.68	−28.00
6.0	−26.00	−24.99	−24.74	−25.16	−26.31	−28.39	−31.97	−39.33	−47.98	−34.87
7.0	−30.36	−27.96	−26.68	−26.20	−26.40	−27.31	−29.07	−32.09	−37.81	−73.47
8.0	−38.31	−32.66	−29.80	−28.23	−27.52	−27.51	−28.19	−29.66	−32.24	−36.86
9.0	−49.41	−42.39	−35.01	−31.58	−29.69	−28.75	−28.53	−29.00	−30.21	−32.40
10.0	−36.24	−44.66	−47.99	−37.51	−33.37	−31.11	−29.92	−29.49	−29.75	−30.72

Table E.5

VALUES OF $\exp(-1.3863\,u^2)$

(Intervals of 0.01 from $u = 0$ to $u = 1.0$, intervals of 0.1 from $u = 1.0$ to $u = 10.9$)

u	0	1	2	3	4	5	6	7	8	9
0.0	1.0000	0.9998	0.9994	0.9987	0.9977	0.9965	0.9950	0.9932	0.9911	0.9888
0.1	0.9862	0.9833	0.9802	0.9768	0.9731	0.9692	0.9651	0.9607	0.9560	0.9511
0.2	0.9460	0.9406	0.9351	0.9292	0.9232	0.9170	0.9105	0.9038	0.8970	0.8899
0.3	0.8827	0.8752	0.8676	0.8598	0.8519	0.8438	0.8355	0.8271	0.8185	0.8098
0.4	0.8010	0.7921	0.7830	0.7738	0.7646	0.7552	0.7457	0.7362	0.7265	0.7168
0.5	0.7071	0.6972	0.6873	0.6774	0.6674	0.6574	0.6474	0.6373	0.6272	0.6171
0.6	0.6070	0.5969	0.5869	0.5768	0.5667	0.5567	0.5466	0.5367	0.5267	0.5168
0.7	0.5069	0.4971	0.4874	0.4777	0.4680	0.4585	0.4490	0.4395	0.4302	0.4209
0.8	0.4117	0.4027	0.3937	0.3848	0.3759	0.3672	0.3586	0.3501	0.3417	0.3335
0.9	0.3253	0.3172	0.3093	0.3014	0.2937	0.2861	0.2787	0.2713	0.2641	0.2569
1.0	0.2499	0.1868	0.1358	0.0960	0.0660	0.0441	0.0287	0.0181	0.0112	0.0067
2.0	0.0039	0.0022	0.0012	0.0006	0.0003	0.0001	0.0000	0.0000	0.0000	0.0000
3.0	0.0000	0.0000	0.0000	0.0000	0.0000	0.0000	0.0000	0.0000	0.0000	0.0000
4.0	0.0000	0.0000	0.0000	0.0000	0.0000	0.0000	0.0000	0.0000	0.0000	0.0000
5.0	0.0000	0.0000	0.0000	0.0000	0.0000	0.0000	0.0000	0.0000	0.0000	0.0000
6.0	0.0000	0.0000	0.0000	0.0000	0.0000	0.0000	0.0000	0.0000	0.0000	0.0000
7.0	0.0000	0.0000	0.0000	0.0000	0.0000	0.0000	0.0000	0.0000	0.0000	0.0000
8.0	0.0000	0.0000	0.0000	0.0000	0.0000	0.0000	0.0000	0.0000	0.0000	0.0000
9.0	0.0000	0.0000	0.0000	0.0000	0.0000	0.0000	0.0000	0.0000	0.0000	0.0000
10.0	0.0000	0.0000	0.0000	0.0000	0.0000	0.0000	0.0000	0.0000	0.0000	0.0000

Table E.6

VALUES OF $20 \log_{10}[\exp(-1.3863\ u^2)]$

(Intervals of 0.01 from $u = 0$ to $u = 1.0$, intervals of 0.1 from $u = 1.0$ to $u = 8.9$)

u	0	1	2	3	4	5	6	7	8	9
0.0	0.00	0.00	0.00	−0.01	−0.01	−0.03	−0.04	−0.05	−0.07	−0.09
0.1	−0.12	−0.14	−0.17	−0.20	−0.23	−0.27	−0.30	−0.34	−0.39	−0.43
0.2	−0.48	−0.53	−0.58	−0.63	−0.69	−0.75	−0.81	−0.87	−0.94	−1.01
0.3	−1.08	−1.15	−1.23	−1.31	−1.39	−1.47	−1.56	−1.64	−1.73	−1.83
0.4	−1.92	−2.02	−2.12	−2.22	−2.33	−2.43	−2.54	−2.65	−2.77	−2.89
0.5	−3.01	−3.13	−3.25	−3.38	−3.51	−3.64	−3.77	−3.91	−4.05	−4.19
0.6	−4.33	−4.48	−4.62	−4.77	−4.93	−5.08	−5.24	−5.40	−5.56	−5.73
0.7	−5.90	−6.06	−6.24	−6.41	−6.59	−6.77	−6.95	−7.13	−7.32	−7.51
0.8	−7.70	−7.90	−8.09	−8.29	−8.49	−8.69	−8.90	−9.11	−9.32	−9.53
0.9	−9.75	−9.97	−10.19	−10.41	−10.63	−10.86	−11.09	−11.32	−11.56	−11.80
1.0	−12.04	−14.56	−17.33	−20.34	−23.60	−27.09	−30.82	−34.79	−39.01	−43.46
2.0	−48.16	−53.10	−58.27	−63.69	−69.35	−75.25	−81.39	−87.78	−94.40	−101.26
3.0	−108.37	−115.71	−123.30	−131.12	−139.19	−147.50	−156.05	−164.84	−173.87	−183.14
4.0	−192.65	−202.41	−212.40	−222.64	−233.11	−243.83	−254.79	−265.99	−277.43	−289.11
5.0	−301.03	−313.19	−325.59	−338.23	−351.12	−364.24	−377.61	−391.22	−405.06	−419.15
6.0	−433.48	−448.05	−462.86	−477.91	−493.20	−508.74	−524.51	−540.53	−556.78	−573.28
7.0	−590.02	−606.99	−624.21	−641.67	−659.37	−677.32	−695.50	−713.92	−732.58	−751.49
8.0	−770.63	−790.02	−809.65	−829.52	−849.63	−869.98	−890.57	−911.40	−932.47	−953.78

Appendix F

List of Symbols

Symbol	*Meaning*	*Section Number*
A	Area, physical aperture, amplitude, azimuth angle	
$A(f)$	Signal voltage spectrum	1.1
$A_1(f)$	Envelope of $A(f)$	4.1
A_c	Correlator output spectrum	4.2
A_d	Difference-channel spectrum	3.1
A_i	Signal voltage sample	1.1
A_m	Maximum signal spectral density	3.1
A_{mc}	Maximum correlator output spectral density	4.2
A_o	Maximum of A_{mc} for matched filter	4.2
A_r	Effective receiving aperture	1.4
A_x	Filter output spectrum	3.1
a	Width of feed horn	2.4
$a(t)$	Signal voltage waveform	1.2
$a_1(t)$	Waveform of single pulse	4.1
a_c	Correlator output waveform	4.2
a_d	Difference-channel waveform	3.1
a_i	Interfering signal voltage	5.2
a_m	Maximum signal voltage	3.1
a_{mx}	Maximum filter output voltage	3.1
a_o	Maximum output voltage of matched filter	3.1
a_s	Sampled signal voltage	7.1
a_{sq}	Quantized sample of a_s	7.2
a_t	Target acceleration	8.3
a_x	Filter output waveform	3.1
a_1, a_2	Monopulse comparator errors	8.3
B	Total bandwidth of filter or signal	App. B
B_a	Total signal bandwidth	3.1
B_e	Width of error spectrum	5.2
B_h	Total width of filter passband	3.1

Symbol	*Meaning*	*Section Number*
B_n	Receiver noise bandwidth (see B_{nh})	1.4
B_{na}	Signal noise bandwidth	3.1
B_{nh}	Filter noise bandwidth	3.1
B_{ns}	Video spectrum noise bandwidth	6.2
B_o	Output bandwidth of matched correlator	4.2
B_s	Upper limit of data filter response	App. C
B_v	Noise bandwidth of video filter	2.6
B_{vg}	Equivalent video bandwidth of range gate	2.6
B_x	Bandwidth at filter output	3.1
B_3	Half-power (3-db) bandwidth	App. B
B_{3a}	Half-power bandwidth of signal	3.1
B_{3a1}	Half-power bandwidth of spectral envelope	4.2
B_{3c}	Half-power width of correlator output spectrum	4.2
B_{3h}	Half-power width of filter passband	3.1
B_{3x}	Half-power width of filter output spectrum	3.1
b	Length of interferometer baseline	App. D
b'	Projected length of baseline	App. D
C	Dimensional constant	1.1
	Integral of squared antenna pattern	App. A
C_f	Scanning coefficient for frequency scan	8.3
C_v	Velocity error ratio	App. C
C_1	Dimensional constant	App. D
c	Speed of light	3.1
c_o	Speed of light in vacuum	App. D
c_t	Speed of light in medium at target	App. D
D	Diameter of circular antenna	App. A
	Dispersion factor in pulse compression	3.3
D_u	Duty factor, τf_r	4.6
d	Fractional bandwidth	4.6
	Array element spacing	7.1
d_c	Correlation distance of surface features	5.4
d_1, d_2	Monopulse comparator errors	8.3
E	Signal energy, elevation angle, signal voltage	
E_o	Output angle of tracker	5.4
E_r	Reference signal voltage	5.4
E_t	Target elevation angle	5.4
E_1	Single-pulse signal energy	3.1
	Transition elevation angle	8.2
E_2	Multipath reflection voltage	5.4
	Transition elevation angle	8.2
$F(\theta)$	Voltage gain (pattern) function	2.1

Symbol	Meaning	Section Number
$F_d(\theta)$	Difference pattern function	2.1
F_m	Maximum voltage gain	App. A
f	Frequency	1.1
f_a	Half-power width of Markoffian spectrum	8.1
f_b	Break frequency in interferometer response	App. D
f_c	IF carrier frequency	4.5
	Correlation frequency interval	6.2
	Frequency of sinusoidal error	8.1
	Critical frequency of ionosphere	App. D
f'_c	Correlation frequency interval of video signal	6.2
f_d	Doppler frequency shift	1.1
	Upper break frequency in fluctuation spectrum	App. D
f_L	Pathlength break point in fluctuation spectrum	App. D
f_m	Frequency to which receiver is tuned	4.2
f_o	Carrier frequency	3.1
f_r	Pulse repetition frequency	1.4
f_s	Conical scan frequency	6.3
f_w	Aperture break frequency in fluctuation spectrum	App. D
f_1	Upper frequency of "bias" spectrum	8.1
$G(\theta)$	Antenna (power) gain function	2.1
G_b	Beacon antenna gain	1.4
G_d	Difference-channel gain	2.1
G_i	Antenna gain on ith sidelobe peak	7.2
G_L	Ratio of range to refraction spectral density	App. D
G_m	Directive gain for weighted illumination	2.1
G_n	Null depth of monopulse antenna	8.3
G_o	Directive gain for uniform illumination	2.1
G_r	Receive antenna gain	1.4
G_{se}	Difference-channel sidelobe ratio	2.4
G_{sr}	Sum-channel sidelobe ratio	2.4
G_{st}	Transmitting sidelobe ratio	5.2
G_t	Transmitting antenna gain	1.4
$G_1, G_2,$ G_3, G_4	Gain values on sample patterns	5.4
$g(x, y)$	Aperture illumination function	2.1
$g_d(x, y)$	Difference-channel illumination function	2.1
$H(f)$	Filter transfer function	1.1
$H'(f)$	Differentiator transfer function	3.5
H_c	Correction filter function	App. D
H_d	Discriminator transfer function	4.2

Symbol	*Meaning*	*Section Number*
H_i	Sample weighting factor	1.1
H_m	Maximum of transfer function	3.1
H_s	Aperture averaging transfer function	App. D
H_1	IF filter transfer function	3.7
H_2	Video filter transfer function	3.7
	Discriminator transfer function	4.5
h	Antenna height (along y-axis)	2.1
	Altitude above sea level	App. D
$h(t)$	Filter weighting function, or impulse response	1.1
$h'(t)$	Differentiator weighting function	3.5
h_d	Discriminator weighting function	4.2
h_m	Maximum of filter weighting function	3.1
h_r	Effective antenna height	2.1
h_1	Sum-channel filter weighting function	4.5
h_2	Discriminator weighting function	4.5
I	Interference power in sum channel	5.1
I_Δ	Difference-channel interference power	5.1
i	Integer identifying the ith pulse	6.2
K	Difference slope	2.1
K_a	Acceleration error constant	8.3
K_c	Pulse compression ratio	3.3
K_f	Slope of frequency difference channel	4.2
K_{f1}	Single-pulse frequency error slope	4.7
K_h	Pulse broadening factor $= \tau_{3x}/\tau_{3a}$	3.3
K_o	Maximum possible value of K	2.1
K_{of}	Maximum possible value of K_f	4.2
K_r	Difference slope ratio $= K/K_o$	2.1
K_{rf}	Difference slope ratio $= K_f/K_{of}$	4.2
K_s	Spectrum broadening factor $= B_{3c}/B_{3a}$	4.4
K_v	Velocity error constant	8.3
K_z	Difference slope in general coordinate z	5.1
K_θ	Sum-channel beamwidth ratio	2.1
k	Cross-section ratio for two-point target	6.1
	Boltzmann's constant	1.4
k_m	Normalized monopulse slope	2.1
k'_m	Monopulse slope with gain and phase errors	8.3
k_p	Slope factor for scanning antenna	2.5
k_s	Conical-scan error slope	2.1
L	Loss factor, length	
	Length (span) of target	6.1
	Length of tropospheric path	App. D

Symbol	*Meaning*	*Section Number*
L	Total radar loss factor	8.3
L_c	Collapsing loss	2.6
L_h	Loss caused by detuning	4.7
L_k	Crossover loss in conical scan	2.5
L_m	Filter matching loss	1.4
L_{mf}	Matching loss of fine-line filter	4.6
L_{mv}	Matching loss of video filter	2.6
L_{m1}	Matching loss of receiver to single pulse	4.6
L_{nr}	Equivalent radial target span	6.2
L_{nx}	Equivalent cross-range target span	6.2
L_p	Beamshape loss in scanning case	2.5
L_q	Quantizing loss in array antenna	7.2
L_r	Radial target span	6.1
L_x	Detector loss	2.6
	Cross-range target span	6.1
L_1	Loss in received energy	1.4
L_θ	Off-axis loss $= G_m/G(\theta)$	2.6
$\mathscr{L}$	Rms aperture width	1.3
$\mathscr{L}_o$	Rms aperture width unweighted	2.1
$\mathscr{L}_s$	Rms aperture illumination power width	2.1
$\mathscr{L}_{s2}$	Rms width of aperture convolved with itself	2.5
$\mathscr{L}_\theta$	Rms aperture illumination voltage width	2.1
m	Number of bits of array phase quantization	7.2
	Number of interference pulses	5.5
N	Tropospheric refractivity	App. D
	Noise power	1.4
$N^2(f)$	Noise power density	1.1
N_b	Number of bits in quantized signal	7.2
N_d	Noise power in difference channel	4.2
N_e	Ionospheric electron density	App. D
$\overline{NF_o}$	Operating noise figure $= T_i/T_o$	1.4
N_i	Ionospheric refractivity	App. D
	Individual noise sample	1.1
N_o	Density of uniform noise	1.4
	World-wide average of sea-level refractivity	App. D
N_s	Surface refractivity	App. D
n	Number of pulses integrated	1.4
	Signal sample index	7.1
	Refractive index	App. D
n_e	Effective number of independent samples	5.1
n_t	Index of refraction in vicinity of target	App. D

Symbol	*Meaning*	*Section Number*
P	Probability	6.2
	Total atmospheric pressure	App. D
P_{av}	Average transmitter power	1.4
P_b	Beacon transmitter power	1.4
P_n	Probability of false alarm	3.4
P_r	Peak received power	1.4
P_t	Peak transmitter power	1.4
P_x	Probability distribution of x	5.1
p	Measure of correlator-to-signal matching	4.4
	Partial atmospheric pressure	App. D
p_n	False-alarm probability per range element	3.4
R	Range to target	1.4
R_c	Maximum clutter range	8.2
	Waveform correlation function	App. B
R_o	Reference range, for $S/N = 1$	8.2
R_1	Transition to short-range region	8.2
R_2	Transition to long-range region	8.2
r	Measure of filter-to-signal matching	3.3
	Radial aperture coordinate	App. A
r'	Normalized radial aperture coordinate	App. A
$\mathscr{R}$	Energy ratio $= 2E/N_o$	1.3
$\mathscr{R}_k$	Energy ratio at beam crossover	2.4
$\mathscr{R}_m$	Energy ratio on beam axis	1.4
$\mathscr{R}_o$	Energy ratio with uniform illumination	1.4
$\mathscr{R}_1$	Single-pulse energy ratio	1.4
S	Signal power	
S_a	Peak signal power	3.1
S_o	Signal power density	1.2
S_x	Peak filter output power	3.1
S'	Electrical length of frequency-scan network	8.3
$S'(\phi)$	Normalized cross section for two-point target	6.1
S/N	Signal-to-noise power ratio	1.4
$(S/N)_{av}$	Average signal-to-noise power ratio	1.4
$(S/N)_d$	Difference-channel S/N ratio	2.6
$(S/N)_f$	S/N ratio in Doppler filter	1.4
$(S/N)_m$	S/N ratio on beam axis	1.4
$(S/N)_o$	Output S/N ratio of processor	2.6
$(S/N)_v$	Video S/N ratio	2.6
$(S/N)_1$	Single-pulse S/N ratio	4.7
S/I	Signal-to-interference ratio	5.1
T	Temperature	

Symbol	*Meaning*	*Section Number*
T	Time between samples	7.1
T_i	Receiver input temperature	1.4
T_o	Reference temperature $= 290°K$	1.4
t	Time	
t_a	RC time constant	8.1
t_c	Correlation time	5.4
t_c'	Correlation time of video signal	6.2
t_d	Range delay time	1.2
t_d'	Pulse rise time	3.6
t_m	Point in time to which receiver is matched	4.2
t_n	Equivalent target span in time delay	6.2
t_{ns}	Equivalent target span after detection	6.2
t_o	Observation time	1.4
t_p	Interpulse period	2.6
t_s	Antenna scan period	8.3
	Data smoothing interval	App. C
U	$\sqrt{u^2 + v^2}$ = angular coordinate (circular aperture)	App. A
u	Antenna angular coordinate	App. A
u_i	Location of interference in u	5.2
u_s'	Normalized array scan angle	7.2
v	Antenna angular coordinate	App. A
	Voltage	3.1
v_a	Velocity of air mass	App. D
v_b	Beam velocity in atmosphere	App. D
v_r	Radial velocity	4.1
v_t	Target velocity	8.3
v_w	Wind velocity	5.4
$W(f)$	Power spectral density	
$W(x)$	Scatterer distribution in cross-range	6.2
W_a	Markoffian spectrum	8.1
W_c	Spectral density for sinusoidal error	8.1
W_n	Spectral density for white noise	8.1
W_N	Spectral density of refractivity variations	App. D
W_o	Maximum value of $W(f)$	6.2
$W'(f)$	Power spectrum of envelope	6.2
w	Aperture width	1.3
w_n	Effective (noise) aperture width	App. A
w_r	Effective (receiving) aperture width	2.1
x	Horizontal coordinate in aperture plane	2.1

Symbol	*Meaning*	*Section Number*
x'	Normalized aperture coordinate $= x/w$ (x is also used in Chap. 8 to denote the error in a generalized coordinate)	App. A
y	Coordinate normal to x in aperture plane	2.1
Z	Focal length of monopulse feed	2.4
z	Coordinate normal to aperture plane	2.1
	General coordinate, which can represent range, Doppler, or angle	5.1
z_i	Location of interference in z-coordinate	5.1
z_3	Half-power resolution width in z-coordinate	5.1
α	Rms time duration	App. B
	Direction cosine	App. A
α_a	Rms duration of signal power	4.2
α_{a1}	Rms duration of signal voltage	4.3
α_c	Rms duration of filter output voltage	4.2
$\dot{\alpha}_h$	Rms duration of squared weighting function	4.2
α_{h1}	Rms duration of weighting function	4.3
α_t	Angle between radar beam and target velocity	5.4
α_1	Rms duration of single pulse	1.3
β	Rms bandwidth	1.3
	Direction cosine	App. A
β_a	Rms width of signal power spectrum	3.1
β_{a1}	Rms width of signal voltage spectrum	3.2
β_h	Rms width of squared filter transfer function	3.1
β_{h1}	Rms width of filter transfer function	3.2
β_n	Servo noise bandwidth (single-sided)	1.4
β_x	Rms width of filter output spectrum	3.1
β_4	Fourth moment of filter transfer function	3.6
γ	Grazing angle	5.4
γ_{max}	Critical grazing angle	5.4
γ_r	Phase shift introduced by surface reflection	5.4
Δ	Edge illumination of aperture	2.4
	Beam pattern width to first null	2.4
	Monopulse difference-channel voltage	2.6
	A priori target selection interval	3.4
	Quantization interval	7.2
Δ_a	Lag error due to acceleration	8.3
ΔE_o	Elevation error caused by tropospheric refraction	App. D
Δf	Frequency error caused by multipath	5.4

Symbol	Meaning	Section Number
Δf	Frequency band over which measurements are taken	6.2
	Frequency-scan giving one beamwidth motion	6.3
	Bandwidth of sinusoidal error spectrum	8.1
Δ_f	Doppler frequency quantization interval	8.5
Δ_i	Difference-channel signal sample	8.3
ΔR	Error in range measurement	5.4
Δ_r	Range quantization interval	8.4
Δt	Time-delay bias caused by multipath	
Δv	Velocity error caused by refraction	App. D
$\Delta \alpha_t$	Error in direction of velocity vector	App. D
$\Delta \theta$	Angle error caused by monopulse network	8.4
$\Delta \phi$	Phase modulation caused by multipath	5.4
$\delta(t)$	Impulse or delta function	4.2
δ_f	Error in apparent Doppler	6.1
δ_r	Error in apparent range	6.1
δ_v	Error in apparent radial velocity	6.1
δ_θ	Error in apparent target angle	6.1
ϵ	Normalized monopulse error voltage	2.6
$\epsilon(t)$	Range error voltage waveform	3.5
$\bar{\epsilon}$	Discriminator output voltage	3.5
ϵ_f	Frequency error	4.2
ϵ_o	Monopulse bias shift	8.3
ϵ_t	Time error relative to point of maximum output	3.1
	Time error caused by sampling	7.1
ϵ_θ	Angle error caused by quantization	7.2
ϵ_2	Monopulse null shift with phase	8.3
η_a	Aperture efficiency	1.4
η_f	Filter matching efficiency	3.1
η_v	Volume reflectivity of clutter	5.3
η_x	x-axis illumination efficiency	2.1
η_y	y-axis illumination efficiency	2.1
Θ	Rms beamwidth	App. A
θ	Angular coordinate	2.1
θ_a	Half-power azimuth beamwidth	2.1
θ_{da}	Azimuth width of glistening surface	5.4
θ_{de}	Elevation width of glistening surface	5.4
θ_e	Half-power elevation beamwidth	2.1
θ_k	Crossover or squint angle of beam	2.1
θ_n	Effective (noise) beamwidth	App. A
θ_o	Beamwidth for uniform illumination	2.1

Symbol	*Meaning*	*Section Number*
θ_s	Array scan angle from broadside	7.2
	Width of tilted beam in V-beam system	8.3
θ_{u3}	Half-power width in u-coordinate	App. A
θ_{v3}	Half-power width in v-coordinate	App. A
θ_w	Angular resolution constant	App. A
θ_3	Half-power (one-way) beamwidth	2.1
λ	Wavelength	1.3
λ_N	Wavelength of refractivity fluctuation	App. D
ν	Doppler filter bandwidth	4.2
π	3.14159	
ρ	Surface reflection coefficient	5.4
	Normalized autocorrelation function	6.2
ρ'	Autocorrelation function of envelope	6.2
$\rho(i)$	Correlation between 0th and ith terms	6.2
ρ_d	Diffuse scattering factor	5.4
ρ_o	Reflection coefficient for smooth surface	5.4
ρ_s	Specular scattering factor	5.4
σ	Radar target cross section	1.4
$\bar{\sigma}$	Average cross section	6.2
σ_A	Rms azimuth error	5.4
σ_a	Standard deviation of Gaussian signal spectrum	3.3
	Rms error caused by sampling	7.1
	Rms error of Markoffian spectrum	8.1
σ_b	Rms of apparent bias error	8.1
σ_c	Rms of cyclic error	8.1
	Standard deviation of Gaussian weighting function	3.3
σ_e	Minimum error level (instrumental error)	8.2
σ_E	Rms elevation error	5.3
σ_f	Rms frequency error	1.3
	Rms frequency spread of power spectrum	5.3
σ_{fd}	Error in Doppler obtained from differentiated range data	4.6
σ_{fn}	Frequency error for n-pulse train	4.6
σ_{f1}	Single-pulse frequency error	4.4
σ_h	Standard deviation of Gaussian filter function	4.4
	Standard deviation in surface height	5.4
σ_i	Rms error caused by interference	5.2
σ_L	Rms error in linear measure	8.2
σ_n	Rms noise component of error	8.1

Symbol	Meaning	Section Number
σ_o	Standard deviation of Gaussian output pulse	3.3
	Rms of true bias error	8.1
σ_r	Rms range error	3.1
σ_{rm}	Rms range error caused by multipath	5.4
σ_{ro}	Apparent range fluctuation	App. D
σ_{rc}	Rms range fluctuation after correction	App. D
σ_s	Rms angle error caused by scintillation	6.3
σ_{sl}	Rms sidelobe level (voltage)	7.2
σ_t	Rms time-delay error	1.3
	Standard deviation of Gaussian waveform	3.3
σ_{t1}	Single-pulse time-delay error	3.2
	Rms error in radial velocity	4.6
σ_v	Rms velocity spread of scatterers	5.4
σ_{vA}	Velocity error caused by $\dot{A}$	App. D
σ_{vE}	Velocity error caused by $\dot{E}$	App. D
σ_x	Standard deviation of Gaussian illumination	2.4
	Standard deviation of filter output	3.3
	Rms error in x-coordinate	8.7
	Standard deviation of normal distribution	8.1
$\sigma_{\dot{x}}$	Rms error in derivative of x	App. C
$\sigma_{\dot{x}2}$	Rms error in two-point difference in x	App. C
σ_y	Rms error in y-coordinate	8.7
σ_z	Rms error in general coordinate z	5.1
σ_α	Rms surface slope	5.4
σ_θ	Rms error in angle measurement	1.3
$\sigma_{\theta q}$	Rms angle error caused by quantization	7.2
σ_ψ	Rms phase quantization error	7.2
σ^0	Surface reflectivity of clutter	5.3
σ_1	Rms error in off-axis angle	2.6
	Single-sample rms error	8.6
σ_2	Rms error caused by normalization	2.6
Σ	Monopulse sum-channel voltage	2.6
Σ_i	Sum-channel signal sample	4.5
Σ_o	Sum-channel voltage for matched filter	4.2
τ	Pulse width, signal duration	1.4
τ_a	Total signal duration	3.1
τ_c	Total duration of correlator output	4.2
τ_g	Range gate width	2.6
τ_h	Total duration of impulse response	3.1
τ_n	Equivalent (noise) width in time domain	App. B
τ_{na}	Noise width of signal	3.1
τ_{nh}	Noise width of weighting function	3.1

Symbol	*Meaning*	*Section Number*
τ_o	Half-power signal width out of matched filter	3.3
τ_x	Half-power signal width at filter output	3.1
τ_{3a}	Half-power width of signal	3.1
τ_{3h}	Half-power width of weighting function	3.1
τ_{3x}	Half-power width of filter output	3.1
ϕ	Angular coordinate measure around z-axis from x-axis	2.1
	Phase shift of multipath signal	5.4
	Pattern tilt in V-beam antenna	8.3
ϕ_1	Precomparator phase error	8.3
ϕ_2	Postcomparator phase error	8.3
ψ	Output voltage, response function	1.1
$\psi(t_d, f_d)$	Receiver response function	1.2
$\psi(u, v)$	Antenna response function (pattern)	1.2
$\psi_o(t_d, f_d)$	Matched-filter response function	1.2
ψ_d	Solid angle of glistening surface	5.4
ω	Scanning rate in search radar	1.4
ω_a	Target rotation rate	6.1

Appendix G

*Bibliography**

Barton, D. K., *Radar System Analysis* (Englewood Cliffs, N. J.: Prentice-Hall, Inc., 1964).

———, "Radar Equations for Jamming and Clutter," *Trans. IEEE*, AES-3, No. 6 (Nov. 1967) (Suppl.), pp. 340–55. [5.3]

Bean, B. R. and B. A. Cahoon, "The Use of Surface Weather Observations to Predict the Total Atmospheric Bending of Radio Rays at Small Elevation Angles," *Proc. IRE*, 45, No. 11 (Nov. 1957), pp. 1545–46. [App. D]

Bean, B. R., J. D. Horn, and A. M. Ozanich, Jr., "Climatic Charts and Data of the Radio Refractive Index of the United States and the World," Natl. Bureau of Standards Monograph No. 22 (Washington, D. C.: U. S. Govt. Printing Office, Nov. 25, 1960). [App. D]

Bean, B. R. and G. D. Thayer, "Models of the Radio Refractive Index," *Proc. IRE*, 47, No. 5 (May 1959), pp. 740–55. [App. D]

Beckmann, P. and A. Spizzichino, *The Scattering of Electromagnetic Waves from Rough Surfaces* (New York: Macmillan Co., 1963). [5.4]

Bernstein, R., "An Analysis of Angular Accuracy in Search Radar," *IRE Conv. Record* (1955), Part 5, pp. 61–78. [6.3]

Bowles, Kenneth L., "Lima Radio Observatory," Natl. Bureau of Standards Report 7201 (Suppl.) (April 30, 1961). [App. D]

Burdic, William S., *Radar Signal Analysis* (Englewood Cliffs, N. J.: Prentice-Hall, Inc., 1968). [1.2]

Cook, Charles E. and Marvin Bernfeld, *Radar Signals* (New York: Academic Press, 1967). [3.3]

Davenport, Wilbur B., Jr. and William L. Root, *An Introduction to the Theory of Random Signals and Noise* (New York: McGraw-Hill Book Company, 1958). [2.6]

Develet, Jean A., Jr., "Thermal-Noise Errors in Simultaneous-Lobing and Conical-

*Numbers in brackets indicate the handbook section in which the work is mentioned.

Scan Angle-Tracking Systems," *Trans. IRE*, SET-7, No. 2 (June 1961), pp. 42–51. [2.2, 2.4, 2.6]

Dunn, John H. and Dean D. Howard, "The Effects of Automatic Gain Control Performance on the Tracking Accuracy of Monopulse Radar Systems," *Proc. IRE*, 47, No. 3 (March 1959), pp. 430–35. [6.3]

George, Samuel F. and Arthur S. Zamanakos, "Multiple Target Resolution of Monopulse vs. Scanning Radars," *Proc. NEC*, 15 (1959), pp. 814–23. [2.4]

Hannan, Peter W., "Optimum Feeds for All Three Modes of a Monopulse Antenna," *Trans. IRE*, AP-9, No. 5 (Sept. 1961). [2.1, 2.2, 2.4]

Hansen, R. C., *Microwave Scanning Antennas, Vol. II* (New York: Academic Press, 1966). [7.1]

Hastings, A. E., J. E. Meade, and H. L. Gerwin, "Noise in Tracking Radars, Part II—Distribution Functions and Further Power Spectra," Naval Res. Lab. Report No. 3929 (Jan. 16, 1952). [6.1]

Howard, Dean D., "Radar Target Angular Scintillation in Tracking and Guidance Systems Based on Echo Signal Phase-Front Distortion," *Proc. NEC*, 15 (1959), pp. 840–49. [6.1]

———, "Analysis of the 29-Foot Monopulse Cassegrain Antenna of the AN/FPQ-6 and AN/TPQ-18 Precision Tracking Radars," U. S. Naval Res. Lab. Memo. 1776 (June 1967), DDC Document AD 816772. [1.2]

Howard, Dean D. and B. L. Lewis, "Tracking Radar External Range Noise Measurements and Analysis," Naval Res. Lab. Report 4602 (Aug. 31, 1955). [6.1]

Hynes, Robert, and R. E. Gardner, "Doppler Spectra of S-Band and X-Band Signals," *Trans. IEEE*, AES-3, No. 6 (Nov. 1967, Suppl.), pp. 356–65. [8.5]

Kirkpatrick, G. M., "Aperture Illuminations for Radar Angle-of-Arrival Measurements," *Trans. IRE*, PGAE-9 (Sept. 1953), pp. 20–27. [2.1, 2.2]

Lawson, James L. and George E. Uhlenbeck, *Threshold Signals* (New York: McGraw-Hill Book Company, 1950). [2.6]

Long, Robert W., "A New Approach to Sensing Angles and Angular Rates for Digital Systems," presented at the Northeast Electronics Research & Engineering Meeting, Boston, Mass., Nov. 4–6, 1965. [7.2]

Manasse, Roger, "Range and Velocity Accuracy from Radar Measurements," Lincoln Lab. Report 312–26 (Feb. 3, 1955), DDC Document AD 236236. [Pref., 3.4, 4.6]

———, "An Analysis of Angular Accuracies from Radar Measurements," Lincoln Lab. Group Report 32–24 (Dec. 6, 1955). [Pref.]

———, "Maximum Angular Accuracy of Tracking a Radio Star by Lobe Comparison," *Trans, IRE*, AP-3, No. 4 (Jan. 1960). [Pref., 2.1, 2.2, 2.6]

Meade, John E., "Target Considerations," Chap. 11 in *Guidance*, A. S. Locke, ed. (Princeton, N. J.: D. Van Nostrand Company, Inc., 1955). [6.1]

Millman, G. H., "Atmospheric Effects on VHF and UHF Propagation," *Proc. IRE*, 46, No. 8 (Aug. 1958), pp. 1492–1501. [App. D]

Muchmore, R. B. and A. D. Wheelon, "Line-of-Sight Propagation Phenomena," *Proc. IRE*, 43, No. 10 (Oct. 1955), pp. 1437–66. [App. D]

Nathanson, F. E. and J. P. Reilly, "Clutter Statistics Which Affect Radar Performance Analysis," *Trans. IEEE*, AES-3, No. 6 (Nov. 1967) (Suppl.), pp. 386–98. [5.3]

Nesline, F. William, "Polynomial Filtering of Signals," *Proc. 5th Natl. Conf. on Military Electronics*, IRE-PGMIL, Washington, D. C. (June 26–28, 1961). [App. C]

North, D. O., "An Analysis of the Factors Which Determine Signal/Noise Discrimination in Pulsed Carrier Systems," RCA Labs. Tech. Report PTR-6C (June 25, 1943). [1.1] [The preceding report has been reprinted in the *Proc. IEEE*, 51, No. 7 (July 1963), pp. 1015–27.]

Norton, K. A., "Effects of Tropospheric Refraction in Earth-Space Links," XIV Gen. Assembly of URSI, Tokyo, Japan (Sept. 1963); published in *Progress in Radio Science* 1960–1963, Vol. II, F. duCastel, ed. (New York: Elsevier Publishing Co., 1965), pp. 186–210.

Norton, K. A. and associated authors, "An Experimental Study of Phase Variations in Line-of-Sight Microwave Transmissions," Natl. Bureau of Standards Monograph No. 33 (Washington: U. S. Govt. Printing Office, Nov. 1, 1961). [App. D]

Pfister, W. and T. J. Keneshea, "Ionospheric Effects of Positioning of Vehicles at High Altitudes," Air Force Surveys in Geophysics No. 83 (March 1956), Cambridge Res. Center, DDC Document AD 98777. [App.D]

Rice, S. O., "Statistical Properties of a Sine Wave Plus Random Noise," *Bell System Tech. Journal*, 27, No. 1 (Jan. 1948), pp. 109–57. [6.1]

Rihaczek, A. W., "Radar Signal Design for Target Resolution," *Proc. IEEE*, 53, No. 2 (Feb. 1965), pp. 116–28. [1.2]

———, "Radar Accuracy of Chirp Signals," *Proc. IEEE*, 53, No. 4 (April 1965), pp. 412–13. [4.7]

Shannon, C. E., "Communication in the Presence of Noise," *Proc. IRE*, 37, No. 1 (Jan. 1949), pp. 10–21. [7.1]

Sharenson, S., "Angle Estimation Accuracy with a Monopulse Radar in the Search Mode," *Trans. IRE*, ANE-9, No. 3 (Sept. 1962), pp. 175–79. [2.6]

Silver, Samuel, *Microwave Antenna Theory and Design* (New York: McGraw-Hill Book Company, 1949). [2.4, App. A]

Skolnik, Merrill I., "Theoretical Accuracy of Pulse Measurements," *Trans. IRE*, ANE-7, No. 4 (Dec. 1960), pp. 123–29. [Pref., 2.1, 3.4]

———, *Introduction to Radar Systems* (New York: McGraw-Hill Book Company, 1962). [Pref., 2.1, 2.2, 2.4, 2.5]

Smith, P. Gene, "Atmospheric Distortion of Signals Originating from Space Sources," *Trans, IEEE*, AES-3, No. 2 (March 1967), pp. 207–16. [App. D]

Spencer, R.C., "Fourier Integral Methods of Pattern Analysis," Mass. Inst. of Technology Radiation Lab. Report 762–1 (Jan. 21, 1946). [2.3]

Swerling, Peter, "Probability of Detection for Fluctuating Targets," RAND Corp. Research Memo. RM-1217 (March 17, 1954). [6.2] [The preceding report has been reprinted in the *Trans. IRE*, IT-6, No. 2 (April 1960), pp. 269–308.]

———, "Maximum Angular Accuracy of a Pulsed Search Radar," *Proc. IRE*, 44, No. 9 (Sept. 1956), pp. 1146–55. [Pref., 2.6, 6.3]

Thompson, M. C., H. B. Janes, and W. B. Grant, "Study of Atmospheric Errors in Microwave Tracking Systems," Natl. Bureau of Standards Report 9139 (Dec. 28, 1965). [App. D]

Thompson, M.C., H. B. Janes, and R. W. Kirkpatrick, "An Analysis of Time Variations in Tropospheric Refractive Index and Apparent Radio Path Length," *Journal of Geophysics Research*, 65, No. 1 (Jan. 1960), pp. 193–201. [App. D]

Woodward, P. M., *Probability and Information Theory, with Applications to Radar* (New York: McGraw-Hill Book Company, 1953). [Pref., 1.3, 3.1, 3.2, 3.4, 4.3, 7.1]

Woodward, P. M. and I. L. Davies, "A Theory of Radar Information," *Phil. Mag.*, 41 (1950), pp. 1001–17. [Pref.]

Index